Mathematics and Geosciences: Global and Local Perspectives
Volume I

Edited by
Jesús I. Díaz
Rafael Orive
M. Luisa Osete
José Fernández

Previously published in *Pure and Applied Geophysics* (PAGEOPH), Volume 172, No. 1, 2015

 Birkhäuser

Editors

Jesús I. Díaz
Universidad Complutense de Madrid
Facultad de Ciencias Matemáticas
Instituto de Matemática
Interdisciplinar
Plaza de Ciencias 3
28040 Madrid
Spain

Rafael Orive
Instituto de Ciencias Matemáticas
CSIC-UAM-UCM-UC3M
Nicolás Cabrera 13-15
28049 Madrid
Spain

M. Luisa Osete
Universidad Complutense de Madrid
Departamento de Física de la Tierra I
Facultad de Ciencias Físicas
Plaza de las Ciencias 1
28040 Madrid
Spain

José Fernández
Institute of Geosciences
(IGEO) (CSIC, UCM)
Plaza de Ciencias 3
28040 Madrid
Spain

ISBN 978-3-0348-0929-0   ISBN 978-3-0348-0930-6 (eBook)
DOI 10.1007/978-3-0348-0930-6

Library of Congress Control Number: 2015945248

Springer Basel Heidelberg New York Dordrecht London
© Springer Basel 2015

Cover illustration: Based on Fig. 1 in "Wind Forecasting Based on the HARMONIE Model and Adaptive Finite Elements" by Albert Oliver et al.

Cover design: deblik, Berlin.

Printed on acid-free paper

Springer Basel AG is part of Springer Science+Business Media

www.springer.com

Contents

Pure Appl. Geophys. 172 (2015), 1–5
© 2014 Springer Basel
DOI 10.1007/s00024-014-0961-1

Introduction to Mathematics and Geosciences: Global and Local Perspectives, Volume I

RAFAEL ORIVE,[1,2] MARÍA L. OSETE,[3,4] JESÚS I. DÍAZ,[5,6] and JOSÉ FERNÁNDEZ[3]

Mathematics plays a fundamental role in all scientific fields and, of course, in the Geosciences. This has been especially well known since, for example, the beginning of geodesy in the Greek era (VANÍČEK and KRAKIWSKY 1992) or in the pioneering studies of planet Earth's interior and the description of the Earth's potential fields (geomagnetic and gravimetric). We can remember that Johann Carl Friedrich Gauss (1777–1855), who is recognized as one of the history's most influential mathematicians, had a remarkable influence in many fields of mathematics and science. He contributed significantly to the development of modern geophysics. In the field of geomagnetism, he developed the first method to obtain absolute geomagnetic measurements, worked out the mathematical theory for separating the inner and outer sources of Earth's magnetic field, and founded the "Magnetischer Verein" (Magnetic Union). Another pioneering natural topic was exploring Earth's interior using seismic waves for analyzing the Earth's inner structure and discovering underground resources. For instance, in 1936, the Danish mathematician Inge Lehman was the first to interpret P wave arrivals, which inexplicably appeared in the P wave shadow of the Earth's core as reflections at an inner core, and discovered the Earth's solid inner core by studying these anomalies. Now it is natural to use geophysical methods and remote sensing in exploring our soils to find water, oil, minerals, etc., and to use advanced mathematical tools for interpreting observation results.

Therefore, as described by CAMACHO et al. (2008a, b), mathematics is one of the branches of science, together with physics, chemistry, and Information Technology that studies and furthers knowledge of the Earth's structure and dynamics. This research is conducted in multidisciplinary studies that use and incorporate the most advanced methods of those sciences, in the framework of, or in close cooperation with, the different branches of Earth sciences such as geology, geophysics, and geodesy. The international dimension of this collaboration is evident (see, e.g., IUGS 2014; IUGG 2014).

This was the environment in which the Complutense International Seminar on "Earth Sciences and Mathematics" was organized and held in Madrid in 2006. The presentations at that meeting were published in two Pure and Applied Geophysics topical issues (CAMACHO et al. 2008c, d).

Since 2006, Earth sciences and, therefore, their "mathematical needs" have evolved, marked to a very significant extent by the consolidated use of artificial satellites for the observation of Earth, which plays a more important role everyday (e.g., ASI 2014; CSA 2014; DLR 2014; ESA 2014a; JAXA 2014; NASA 2014). Satellite observation is evolving from regional-scale coverage at medium or low spatial resolution, to regional-scale and global-scale coverage at medium and high spatial resolution and local-

[1] Instituto de Ciencias Matemáticas (ICMAT), CSIC-UAM-UCM-UC3 M, C. Nicolás Cabrera 13-15, 28049 Madrid, Spain. E-mail: rafael.orive@icmat.es

[2] Departamento de Matemáticas, Facultad de Ciencias, Universidad Autónoma de Madrid, 28049 Madrid, Spain.

[3] Institute of Geosciences (IGEO) (CSIC, UCM), Plaza de Ciencias 1&3, 28040 Madrid, Spain. E-mail: mlosete@fis.ucm.es; jft@mat.ucm.es

[4] Departamento de Física de la Tierra I, Facultad de Ciencias Físicas, Universidad Complutense de Madrid, Plaza de las Ciencias 1, 28040 Madrid, Spain.

[5] Instituto de Matemática Interdisciplinar, Facultad de Ciencias Matemáticas, Universidad Complutense de Madrid, Plaza de Ciencias 3, 28040 Madrid, Spain. E-mail: ji_diaz@mat.ucm.es

[6] Departamento de Matemática Aplicada, Facultad de Ciencias Matemáticas, Universidad Complutense de Madrid, Plaza de las Ciencias 3, 28040 Madrid, Spain.

scale coverage at ultra-high resolution. This evolution is bringing new needs of techniques/methodologies for data fusion (to obtain a unique, new data set by combining several sets from different satellites or from satellites and terrestrial observations) and for the joint interpretation of data of different sources (space and terrestrial) and kinds (displacements, gravity changes, seismicity,…).

Satellite synthetic aperture radar interferometry (InSAR) can serve as a typical example of this evolution, considering second-generation satellites (see, e.g., ASI 2014; DLR 2014; ESA 2014b; SANSOSTI *et al.* 2014). Data fusion is probably the only tool in this case to obtain a 3D displacement field with high spatial resolution (see, e.g., SAMSONOV *et al.* 2008) or to obtain a continuous time series of displacement data for a long time period (see, e.g., SAMSONOV and D'OREYE 2012; SAMSONOV *et al.* 2014). In addition, joint data interpretation is a powerful tool for studying Earth's active phenomena (see, e.g., CAMACHO *et al.* 2011).

Advances in geomagnetic field knowledge in recent times have come with the combined use of satellite data about magnetism and the world geomagnetic observatories network (INTERMAGNET) developed for the present International Geomagnetic Reference Field: IGRF11 (FINLAY 2010). A spherical harmonic expansion is truncated at degree 13 (195 coefficients). The Swarm mission, launched in November 2013, is providing unprecedented insight into the complex workings of Earth's magnetic field (ESA 2014c, d). Advanced models based on Swarm data describing each of the various sources of the measured field are being developed, and will lead to new insight into many natural processes, from those occurring deep inside the planet, to weather in space caused by solar activity. In turn, this information will yield a better understanding of why the magnetic field is weakening.

At present, mathematical models and methods are required for diverse studies such as global change models, geodynamo models, flows in porous media, wave propagation in continua medium, riskmap analysis in hazards, time series analysis, dynamics of interfaces, studies with multiples scales, nonlinear analysis in the Earth sciences, fractal analysis in geological structures, stochastic models for processing data, etc.

Simultaneously, new problems (such as fracking), or the continental or global data sharing promoted by different initiatives such as COPERNICUS (2014), the Global Earth Observation System of Systems (GEOSS 2014), the European Plate Observing System (EPOS 2014), or SUPERSITES (2014), are posing new needs.

During the year 2013, at the initiative of American and Canadian research mathematical institutes and societies, 2013 was declared a thematic year in Mathematics and the Planet Earth (MPE2013 2014), under the patronage of UNESCO. More than 120 mathematical organizations around the world took part in MPE2013 and gave the public and schools an exciting opportunity to discover and experience many aspects of mathematics, including its applications on life and Earth sciences, its relevance on societal and environmental issues, the impact of mathematics research on thematic topics specific to climate change, remote sensing, mathematical modelling to eradicate diseases as well as the global trends, and perspectives on mathematics for sustainable development.

The four MPE2013 subthemes were a planet to discover; a planet supporting life; a planet organized by humans; and a planet in danger. Thus, the mathematical family found natural topics where mathematics play a fundamental role in describing different events in Earth sciences, in particular, in the first and the last theme.

To mark the MPE2013, the Institute of Mathematical Sciences (ICMAT) and Institute of Geosciences (IGEO), under the patronage of the Spanish Council for Scientific Research (CSIC), Madrid Autonomous University (UAM), the Institute of Interdisciplinary Mathematics (IMI), Madrid Complutense University (UCM), and Technical University of Madrid (UPM) sought to highlight the aforementioned different aspects, the two-way street between Mathematics and Earth Sciences, by organizing the workshop "Mathematics and Geosciences: Global and Local Perspectives" in Madrid at ICMAT from 4 to 8 November 2013. The Organizing Committee's main idea, when it proposed such a title, was to emphase the combination of two different types of approaches: the global approaches that study natural phenomena arising from the whole planet (such as the climate, geomagnetism, and so on) and the almost infinite collection of local aspects, which are associated with natural, ecological, and even economic studies on more concrete problems. The following topics were addressed: climatology and paleoclimatology, oceanography, geomagnetic field, the Earth's rotation, remote sensing, natural hazards, structure and geodynamics, renewable energies, and social and environmental aspects. These topics were addressed from a global perspective on the Earth as well as from a local point of view of the Geosciences. Scientists from both areas, Mathematics and Earth Sciences, participated in this activity seeking to gain and learn from their work and proposing partnerships.

This is the first volume of the Topical Issue on "Mathematics and Geosciences: Global and Local Perspectives" and contains 13 papers, most of which were presented at the Congress. They address different topics such as the percolation theory to observe microseismicity in fracking, fractional differential analysis to model complex dynamics, climate change, earthquake statistics, earth rotation, geomagnetic field, fire models, Mars exploration, numerical methods in meteorology, environmental conservation, and probabilistic models in the study of soils and fluid mechanics.

The first paper is about hydraulic fracturing (fracking); technical processes used to extract oil and gas from tightly sealed shale reservoirs using high pressures and a low viscosity fluid. Norris et al., model the injection of fluid in the fracking process using invasion percolation, which is a 2D square lattice of random bonds to model the sealed natural fractures. This model exhibits burst dynamics (small earthquakes) associated with the microseismic activity generated by fracking injections and shows the role of anisotropic stress distributions.

Tejedor et al., apply a new probability density function, the negative binomial distribution, to study the recurrence of large earthquakes. The new distribution has been applied to the study of the Parkfield seismic series, finding that it performs very similarly to other distributions. In contrast to other statistical models, the new distribution has a plausible physical meaning, that is, it is based upon the idea of the elastic-rebound and Markov modelling.

The paper by Velasco et al., deals with an interesting topic about Mars exploration. The authors

present a new method of tomography-based signal analysis for the detection of events in the Martian atmosphere boundary layer, such as dust devils. The mathematical aspects are considered through fractional differential equations associated with diffusion processes and nonlocal problems. This area of work can provide important results of remarkable applications in the future.

In the paper by Fowler, the future evolution of atmospheric CO_2 concentration is assessed on a millennial timescale by using a simple box model of the ocean carbon cycle. The model is based on two equations, one for the atmospheric CO_2 concentration, and the other is for the carbonate concentration in the ocean. The model predicts a positive feedback in response to the long-term increase in atmospheric CO_2 as a result of the decrease in the carbonate concentration and the buffering effect. A further consequence of the repartitioning of ocean carbon is a dramatic rise in atmospheric carbon dioxide on a millennial time scale.

Ferrándiz et al., present a very interesting review of the crucial problem of the natural sciences, that is, to improve the theory of the Earth's rotation in order to reach the accuracy requirements that must be fulfilled in the near future. The accurate determination and prediction of the Earth's orientation and motion in Space has been needed in a broad variety of fields ever since the advent of the Space Age. Applications of geodetic measurements to the determination of sea level variations, mass movements in oceans, ice sheets, terrestrial water storages, displacement fields associated with earthquakes, etc., are demanding far higher levels of accuracy. The authors broach the problem and suggest new research directions aimed at creating adequate use of the Earth's rotation theory.

The paper by De Santis and Qamili is an overview of the complex characteristics of the Earth's magnetic field investigated from a holistic approach. Within the framework of geosystemics and by using information-theoretic tools such as Shannon entropy or other nonlinear tools borrowed from critical phenomena, they explore the possibility of an imminent change of the geomagnetic field dynamical regime.

The paper by Campuzano et al., analyzes the main error sources in the geomagnetic dipole moment computation from palaeomagnetic data (the influence of the nondipole terms in the average approach, the inhomogeneous distribution of the current palaeomagnetic database, and the averaging procedure used to obtain the evolution of the dipole moment), using synthetic data from a global model based on instrumental and satellite data, the 11th generation of the International Geomagnetic Reference Field.

Oliver et al., introduce a new numerical method to solve mathematical models for wind field forecasting, coupling the predictions of the HARMONIE mesoscale model as the input data for an adaptive finite element in a local mass-consistent wind model specifically suited for complex terrains. These wind models serve as tools for studying several atmosphere-related problems, such as the effect of wind on structures, pollutant, fire spreading, and wind farm location. At present, the use of wind power to produce electric power involves using these mathematical tools to design the wind farms. The final goal of the paper is to validate the model in several realistic applications on Gran Canaria Island, Spain, with some experimental data obtained by the AEMET (Spanish Meteorological Agency) at its meteorological stations.

The paper by Ferragut et al., presents a global physical model (heat and mass transfer) available for analysis and numerical simulation of forest fire propagation. They incorporate data assimilation techniques to the numerical solution in near real time, which improve the simulation results. From a numerical and computational point of view, a highly interesting issue is the successful implementation of Yosida regularizations to approximate the enthalpy, which consists of a nonlinear multivalued maximal monotone operator.

Martín et al., consider fragmentation schemes inspired in the theoretical results and conjectures of Kolmogorov, which are applied to produce particle size distributions of a different nature depending on fragmentation parameters. A 2-D computer simulation packing method is applied to the resulting distributions and the void fraction is evaluated. They study the relation between the void fraction and the characteristic parameters of the fragmentation process.

In the paper by Arregui and Vázquez, the authors propose new efficient numerical methods to solve some mathematical models related to the opportunity of starting an industrial project that provides some uncertain benefits but also involves some irreversible environmental effects. Of course, both the environmental and the industrial project's benefits, are uncertain and governed by stochastic processes. The authors consider a finite element discretization of the equivalent PDE problem posed on a suitable bounded domain and, to add instantaneous effects, propose a Lagrangian active set method (ALAS). Finally, the numerical method is validated through qualitative properties theoretically proven in the literature for different examples.

Muñoz-Ortega et al., show the utility of performing different geometric measures of soil pore space using 3-D images in order to characterize the soil structure. They characterize in quantitative terms several geometric factors of great interest in the study of biological and physical processes, such as porosity, pore surface area, connectivity of the pores, and pore size distribution. For the latter objective, they use mathematical morphology tools.

The paper by Berselli et al., studies the motion of a quasi-incompressible fluid with several dispersed particles. This model plays a fundamental role in the dynamics of both oceanic and atmospheric flows. They consider a reduced multiphase model with particular attention to the right evaluation of physical parameters that make the approximation effective. The resulting model is used in direct numerical simulations and also large eddy simulations of a dam-break (lock-exchange) problem, which is a well-known academic test case.

A second issue will be published later in 2015 with other reviewed papers presented at the workshop "Mathematics and Geosciences: Global and Local Perspectives".

We appreciate the great and generous work carried out by the many referees. They have worked in most of the cases, in the difficult intersection of two different fields such as Earth Sciences and Mathematics. The reviewers have been: G. Alguacil, J. Almendros, O. Arzel, L. D'Auria, S. Alvarez, G. Balasis, Y. Barkin, M. Bialecki, R. Carbonell, A. Castro, M. Crucifix, G. Diaz, J.I. Díaz, L. Dinis, J. Durany, P. García, R. Granero-Belinchón, R. Gross, A. Jimenez, J. von de Koppel, M. Korte, V. Kossobokov, M. Kravchenko, S. Lennartz-Sassinek, F. Luzón, J. Mandel, M. Montoya, J.J. Nieto, J.H. Ortega, J.F. Padial, A. Pazoto, E. Perfect, M. Sahimi, T. Sengupta, C. Trenchea, D.L. Turcotte, F. Valero, and C. Wang.

Acknowledgments

We would like to take this opportunity to thank the Instituto de Ciencias Matemáticas and the Institute of Geosciences for their financial and technical support in organizing the Mathematical and Geosciences Congress, and especially E. Fuentes, E. Frechilla, and A. Chacón from ICMAT. We are also grateful to CEI Campus Moncloa (UCM-UPM) and CEI UAM + C-SIC for their support. This workshop has been also partially supported with funds from research projects AYA2010-17448, MTM2011-26696, SEV-2011-0087, CGL2011-24790 of MINECO, MTM2011-26119 of the DGISPI (Spain), the UCM Research Group MOMAT (Ref. 910480), and the ITN FIRST of the Seventh Framework Program of the European Community's (grant agreement number 238702). The editors also thank Dr. Renata Dmowska and Ganesh Priyanka for the help, suggestions, and support received during the editing of this topical issue. Finally, we wish to thank all the authors of this volume for their contributions.

References

ASI, Italian Space Agency (2014). http://www.asi.it/en.

Camacho, A.G., Díez, J.I., Fernández, J. (2008a). *Introduction: Linking Earth Sciences and Mathematics*. Pure and Applied Geophysics, *165*, no. 6, 997–1001. doi:10.1007/s00024-008-0343-7.

Camacho, A.G., Díez, J.I., Fernández, J. (2008b). *Introduction to Earth Sciences and Mathematics, Volume II*. Pure and Applied Geophysics, *165*, no. 8, 1459–1463. doi:10.1007/s00024-008-0396-7.

Camacho, A.G., Díez, J.I., Fernández, J. (Editors) (2008c). Pure and Applied Geophysics Topical Issue "Earth Sciences and Mathematics. Volume I." 165, no. 6.

Camacho, A.G., Díez, J.I., Fernández, J. (Editors) (2008d). Pure and Applied Geophysics Topical Issue "Earth Sciences and Mathematics. Volume II." 165, no. 8, 1459–1706.

Camacho, A. G., P. J. González, J. Fernández, and G. Berrino (2011). *Simultaneous inversion of surface deformation and gravity changes by means of extended bodies with a free geometry: Application to deforming calderas*, J. Geophys. Res., *116*, B10401, doi:10.1029/2010JB008165.

Copernicus (2014). http://www.copernicus.en/main/overview/.

CSA, Canadian Space Agency (2014). http://www.asc-csa.gc.ca/eng/.

DLR, National Aeronautics and Space Research Centre of the Federal Republic of Germany (2014). http://www.dlr.de/dlr/en/desktopdefault.aspx/tabid-10376.

ESA, European Space Agency-Observing the Earth (2014a). http://www.esa.int/Our_Activities/Observing_the_Earth/.

ESA, European Space Agency, Sentinel-1 (2014b). http://www.esa.int/Our_Activities/Observing_the_Earth/Copernicus/Sentinel-1.

ESA, European Space Agency-Observing the Earth (2014c). http://www.esa.int/Our_Activities/Observing_the_Earth/Swarm/Swarm_reveals_Earth_s_changing_magnetism.

ESA, European Space Agency (2014d). ESA, 3rd Swarm Scientific meeting, June 2014, Denmark. http://congrexprojects.com/2014-events/Swarm/home.

EPOS, European Plate Observing System (2014). http://www.epos-eu.org/.

Finlay, C. C. (2010). *International Geomagnetic Reference Field: the eleventh generation*. Geophys. J. Int. *183*, 1216–1230.

GEOSS, Global Earth Observation System of Systems (2014). http://www.epa.gov/geoss/.

IUGG, International Union of Geodesy and Geophysics (2014). http://www.iugg.org/.

IUGS, International Union of Geological Sciences (2014). http://www.iugs.org/.

JAXA, Japan Aerospace Exploration Agency (2014). http://global.jaxa.jp/.

MPE2013, Mathematics of Planet Earth (2014). http://mpe2013.org/.

NASA, National Aeronautics and Space Administration (2014). http://www.nasa.gov.

Samsonov, S., and N. d'Oreye (2012), *Multidimensional time series analysis of ground deformation from multiple InSAR data sets applied to Virunga Volcanic Province*, Geophys. J. Int., *191*(3), 1095–1108, doi:10.1111/j.1365-246X.2012.05669.x.

Samsonov, S.V., Tiampo, K.F.; Rundle, J.B. (2008). *Application of DInSAR-GPS optimization for derivation of three-dimensional surface motion of the southern California region along the San Andreas fault*. Computers and Geosciences, *34*, 503–514, doi:10.1016/j.cageo.2007.05.013.

Samsonov, S.V., Tiampo, K.F., Camacho, A.G., Fernández, J., González, P.J. (2014). *Spatiotemporal analysis and interpretation of 1993–2013 ground deformation at Campi Flegrei, Italy, observed by advanced DInSAR*. Geophysical Research Letters, 41, pp. 6101–6108, doi:10.1002/2014GL060595.

Sansosti, E., Berardino, P., Bonano, M., Calò, F., Castaldo, R., Casu, F., Manunta, M. Manzo, M., Pepe, A., Pepe, S., Solaro, G., Tizzani, P., Zeni, G., Lanari, R. (2014). *How second generation SAR systems are impacting the analysis of ground deformation*. International Journal of Applied Earth Observation and Geoinformation, *28*, pp. 1–11, doi:10.1016/j.jag.2013.10.007.

SUPERSITES (2014). http://supersites.earthobservations.org/.

Vaníček, P. and Krakiwsky, E. (1992). Geodesy, the concepts. Second Edition. Elsevier, Amsterdam, The Netherlands, ISBN: 0-444-87777-0, 697 pp.

(Received October 17, 2014, accepted October 18, 2014, Published online November 29, 2014)

Pure Appl. Geophys. 172 (2015), 7–21
© 2014 Springer Basel
DOI 10.1007/s00024-014-0921-9

Anisotropy in Fracking: A Percolation Model for Observed Microseismicity

J. Quinn Norris,[1] Donald L. Turcotte,[2] and John B. Rundle[1,2,3]

Abstract—Hydraulic fracturing (fracking), using high pressures and a low viscosity fluid, allow the extraction of large quantiles of oil and gas from very low permeability shale formations. The initial production of oil and gas at depth leads to high pressures and an extensive distribution of natural fractures which reduce the pressures. With time these fractures heal, sealing the remaining oil and gas in place. High volume fracking opens the healed fractures allowing the oil and gas to flow to horizontal production wells. We model the injection process using invasion percolation. We use a 2D square lattice of bonds to model the sealed natural fractures. The bonds are assigned random strengths and the fluid, injected at a point, opens the weakest bond adjacent to the growing cluster of opened bonds. Our model exhibits burst dynamics in which the clusters extend rapidly into regions with weak bonds. We associate these bursts with the microseismic activity generated by fracking injections. A principal object of this paper is to study the role of anisotropic stress distributions. Bonds in the *y*-direction are assigned higher random strengths than bonds in the *x*-direction. We illustrate the spatial distribution of clusters and the spatial distribution of bursts (small earthquakes) for several degrees of anisotropy. The results are compared with observed distributions of microseismicity in a fracking injection. Both our bursts and the observed microseismicity satisfy Gutenberg–Richter frequency-size statistics.

Key words: Fracking, percolation, anisotropy, microseismicity.

1. Introduction to Fracking

Fracking, the common term for hydraulic fracturing, dates back to the late 1940s with the first commercial applications in 1949. The original process was a secondary recovery method designed to enhance production in reservoirs where primary recovery had decreased to the point where production was no longer economical. By injecting a high viscosity fluid at high pressures into the reservoir rock, one or two large fractures were created that extended from the borehole. Also injected were large quantities of sand which "propped" the generated fractures open and allowed oil and gas to flow through the fractures to the borehole. We refer to this process as low-volume or traditional fracking. Traditional fracking is applied to conventional reservoirs with high permeability, typically sandstone reservoirs. Because of the high permeability, the oil and gas can readily migrate to the generated fractures and flow to the borehole where it can be extracted.

Two developments in the 1980s allowed fracking to extract oil and gas from tight shale reservoirs where the natural formation permeability is too low for economic extraction using traditional methods. The first development was horizontal drilling. Because many formations are relatively thin and lie nearly horizontally, a vertical well can only access a limited volume of reservoir rock. By turning the well bore horizontal, using directional drilling, a single well can access a much larger volume. This reduces the cost of well drilling making the extraction more economical. The second development was "slickwater." In many reservoirs the low natural matrix permeability prevents the flow of oil and gas to a well. In these reservoirs, significant flow can only occur along fractures. By injecting a low viscosity fluid at high pressure, a distributed network of fractures is generated. These fractures increase the permeability in the rock surrounding the borehole allowing oil and gas to flow to the borehole. The combination of horizontal drilling with "slickwater" has changed the nature of fracking to a method that can be used to extract oil and gas from tight shales.

[1] Department of Physics, University of California, One Shields Ave., Davis, CA 95616, USA. E-mail: jqnorris@ucdavis.edu

[2] Department of Earth and Planetary Sciences, University of California, One Shields Ave., Davis, CA 95616, USA.

[3] Santa Fe Institute, Santa Fe, NM 87501, USA.

Reprinted from the journal

Shales are important source rocks for oil and gas (TOURTELOT 1979; ARTHUR SAGEMAN and SAGEMAN 1994). It is estimated that a large fraction of gas and oil has been formed in black shales during anoxic periods (ULMISHEK and KLEMME 1990; KLEMME and ULMISHEK 1991; TRABUCHO-ALEXANDRE et al. 2012). As oil and gas develop in a shale, they generate pressures sufficient to fracture the rock (OLSON et al. 2009). Typical shales have extensive fracture networks and joint sets. If a shale is relatively old, there is a greater chance that the natural fractures have been sealed by the deposition of silica or carbonates (GALE et al. 2007). Fracking seems to be effective only in tight reservoirs where the natural fractures have been sealed (KING 2012); examples are the Barnett Shale in Texas and the Bakken Shale in North Dakota. Large quantities of natural gas are now being extracted from the Barnett Shale, and large quantities of oil are being extracted from the Bakken Shale. Fracking appears to be ineffective in increasing the production from shales where the natural fractures are open; examples are the Antrim Shale in Michigan and the Monterey Shale in California. In both cases, production of oil and gas continues to decrease. Fracture permeability of shales allow the migration of oil and gas into overlying strata which typically have a higher permeability. In the overlying strata, the oil and gas flow into structural or stratigraphic traps. These traps are the traditional reservoirs from which the majority of oil and gas recovery has occurred.

In order to understand the process and to optimize recovery, the fracking injections are often monitored using sensitive seismometers (WARPINSKI 2013). In addition to recovery boreholes, one or two monitoring boreholes are often drilled. Sensitive seismometer arrays are placed along their lengths. The recorded microseismic data is used to determine the locations and magnitudes of the microseismic events that occur during a fracking treatment. This information can be used in real time to control the pressure, rate, and composition of the injected fluid. Because of the great depths, analysis of microseismic data is one of the primary methods used to understand the fracking process, yet only 3 % of the fracking treatments performed in 2009 were microseismically monitored (ZOBACK et al. 2010). An example of microseismic data recorded during a four-stage frack are shown in

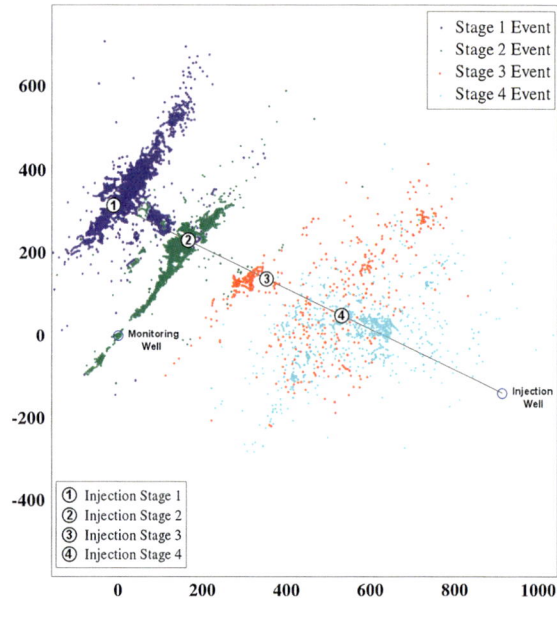

Figure 1
Map of the epicenters of microseismicity associated with four fracks of the Barnett Shale (MAXWELL 2011). The *colors* correspond to the four injections and the axes are distances in meters from the monitoring well

Fig. 1. Anisotropy plays a large role in fracking, with stress anisotropy being common. In a reservoir, the principle stresses are often not equal. Fractures grow perpendicular to the minimum principle stress and tend to be confined to the horizontal plane because the maximum principal stress in generally vertical. The distribution of microseismicity in Fig. 1 is clearly anisotropic.

As fracking has spread to more populated areas, such as the Marcellus Shale in Pennsylvania, there has been an increase in public concern over the safety of the process. The most common concerns are the potential to generate large earthquakes and the potential to contaminate drinking water. The largest recorded earthquake from a fracking treatment had a magnitude around three (ELLSWORTH 2013), much too small to cause any significant damage. However, the waste water generated during a fracking treatment is often re-injected into deep saline aquifers. There is increasing evidence that this re-injection causes larger earthquakes (KERANEN et al. 2013). Additionally, because the microseismicity follows a power-law (Gutenberg–Richter) frequency-magnitude distribution (MAXWELL 2011), it is impossible to rule out the possibility of large events. It has been observed that drinking wells near fracking wells contain elevated levels of methane (OSBORN *et al.* 2011). This contamination is not likely to be the result of the fracture network extending from the borehole at depth to the near-surface freshwater aquifer. Shale layers are typically 3 to 5 km deep, whereas freshwater aquifers are on the order of 100 m deep. Fractures running several kilometers would have seismic energy releases much greater that those observed during fracking treatments. These large fractures would also be undesirable from an engineering perspective. If the fracture network extends beyond the shale layer into the overlying strata, which often has a much lower permeability, significant leak-off can occur which reduces the effectiveness of the frack. The observed contamination is likely due to poor quality or damaged cement well casings.

Despite the lack of understanding of the processes and increased concern over the safety of fracking, there is very little publicly available research on fracking. This paper is part of an attempt to increase the availability of fracking research that can be used to understand the processes and risks associated with fracking.

2. Introduction to Percolation

Percolation theory has been used to study everything from conductivity (SEAGER and PIKE 1974) to economics (CONT and BOUCHAUD 2000). Within geophysics, percolation theory has been used to study rock transport properties (GUEGUEN and DIENES 1989), earthquakes (OTSUKA 1971; SAHIMI *et al.* (1993), and oil production KING *et al.* (1999). At its core, percolation is the study of connectivity. The classical random percolation problem is as follows: given a random lattice of sites or bonds, what fraction of those bonds must be occupied for a cluster (a group of connected sites) to span from one side of a lattice to the other. This cluster is called the spanning cluster. The minimum occupation probability for which a spanning cluster exists on an infinite lattice (percolation threshold) has been shown to be a classical critical point. Near the percolation threshold, the statistics of the clusters are governed by power-laws

analogous to the critical point of the liquid-gas phase transition. For a more complete introduction to percolation theory see STAUFFER and AHARONY (1994).

There have been many variants of this initial model (SAHIMI 1994). The variant closest to our model is called invasion percolation (WILKINSON and WILLEMSEN 1983) and has been used to study water flooding for oil recovery. Water flooding is a secondary recovery method where water is injected into a reservoir to drive oil and gas to the production well. In practice, several injection wells are drilled along one edge of an oil field and several production wells are drilled on the opposite edge. Water is injected from the injection wells and drives oil or gas to the production wells. In their model, the sites in a lattice are assigned random numbers on the interval [0, 1). The sites along one edge are added to the perimeter of a growing cluster. The site on the perimeter with the smallest random number is invaded, and all sites adjacent to the invaded site are added to the perimeter. At each time step the site on the perimeter with the smallest random number is added to the single connected cluster. A later study involved injection from a single site with the cluster growing radially (WILKINSON and BARSONY 1984). There are two major variants of the model, trapping and non-trapping, depending on whether the defending fluid is incompressible (trapping) or compressible (non-trapping) (KNACKSTEDT *et al.* 2002). Non-trapping invasion percolation has been shown to belong to the same universality class as random percolation (KNACKSTEDT *et al.* 2002). One of the properties of invasion percolation is that it displays self-organized criticality, i.e the dynamics take the system to a critical state. For a review of invasion percolation see EBRAHIMI (2010).

Despite its relative simplicity, there are many aspects of percolation theory which are still unknown or not well studied. One of those aspects, addressed in this paper, is the role of anisotropy. Anisotropy is commonly introduced by occupying bonds in the horizontal direction with one probability p_h and the bonds in the vertical direction with another probability p_v. The critical line for 2D bond percolation on a square lattice was determined by SYKES and ESSAM (1963). Since that time, renormalization approaches have been used to explore anisotropic percolation both on and away from criticality (IKEDA 1979; LOBB *et al.* 1981; KIM and LEE 1992). The primary experimental work has been done in the field of material conductivity (SMITH and LOBB 1979; MENDELSON and KARIORIS 1980; BALBERG *et al.* 1983). Initially, experimental results suggested that the introduction of anisotropy would cause changes to previously universal critical exponents (BALBERG 1987). More recent work suggests that anisotropic percolation should share universal isotropic exponents (HAN *et al.* 1991; CELZARD and MARÉCHÉ 2003). The applicability of these results may be limited as our model provides a method for exploring the cross-over from one- to two-dimensional percolation. This cross-over requires a change in the critical exponents from their values in 2D. Similar cross-overs in percolation models have been studied previously (CHAME *et al.* 1984; SOTTA and LONG 2003).

HERRMANN *et al.* (1993) studied anisotropic fracture propagation using a lattice of elastic beams subject to tensional failure. Fractal branching structures were obtained, but the density of fractures were much lower than in percolation models

3. Model

Our model is an extension of the radial invasion model first studied by WILKINSON and BARSONY (1984). The isotropic version of our model has been given previously (NORRIS *et al.* 2014). The reservoir rock is assumed to have a network of natural fractures that have been sealed by deposition. We assume a point injection of a low viscosity fluid that breaks the seals as the fluid flows from the point of injection. We neglect the viscous pressure drop during flow and assume the fluid breaks the weakest seals as it flows through the matrix of preexisting fractures. The sealed fractures are represented by a lattice of bonds. Each bond is assumed to have a effective strength (*s*) which we represent with a random number. For simplicity we use a 2D square lattice of bonds.

Our justification for the applicability of the 2D model is our interest in layered sedimentary deposits that have remained nearly horizontal. In many cases, the target reservoir strata (the black shale) is relatively thin (say 100 m) and the horizontal well is drilled within this strata. We hypothesize that fractures are confined to this target strata. The anisotropy we model is due to the anisotropy of the stress field in the layer. We assume the least principal stress is in the *y*-direction and the intermediate principal stress is in the *x*-direction. Thus, the induced fractures will tend to propagate in the *x*-direction. Fractures tend to be oriented perpendicular to the direction of the least principal stress, the *y*-direction, and for this reason, horizontal wells are drilled in this direction.

In order to model the preferential fracture orientation due to the existing stress field in the rock, we assign random numbers (effective strengths) to bonds oriented in the *x*-direction on the interval [0, 1) and bonds oriented in the *y*-direction on the interval [0, a) with $a > 1$. When $a = 1$ the model is isotropic. When $a = \infty$ the model becomes one-dimensional, with propagation only in the *x*-direction. Additionally, the tuning of *a* provides a simple way of exploring the cross-over from 2D ($a = 1$) to 1D ($a = 0$) percolation.

The variable *a* is a measure of the anisotropy and gives the relative likelihood of a fracture to propagate in a given direction. Thus, in a simulation with $a = 2$, the fracture network is twice as likely to grow in the *x*-direction than the *y*-direction. We relate this choice of anisotropy to the stresses in the rock.

$$\sigma_{xx} = a\sigma_{yy} \tag{1}$$

We can then interpret *a* as being the ratio of the two principal horizontal stresses in the rock.

Fluid is injected from a single site and the fracture network can grow in one of four directions as illustrated in Fig. 2a. These bonds are assigned anisotropic effective strengths as explained previously. The weakest bond (smallest *s*) fails and the fluid-filled fracture network grows in that direction as illustrated in Fig. 2b. The bond fails and fluid flows into the opened crack due to the pressure difference between the injected fluid and the surrounding rock. The new nearest-neighbor bonds are assigned effective strengths and the process repeats. At each time step, the weakest bond on the perimeter of the growing fracture network fails, the fluid-filled fracture network grows in that direction, and new perimeter bonds are given effective strengths. Although the

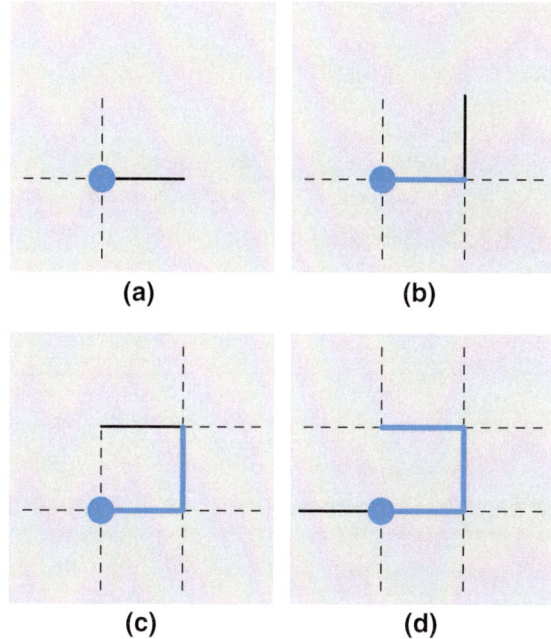

(a)   **(b)**

(c)   **(d)**

Figure 2

Illustration of our model. **a** Fluid is injected at the site shown. The four bonds to adjacent sites are also shown; the weakest bond (smallest s) is a *solid line*. **b** Three bonds to adjacent sites are added; the weakest of the six available bonds is a *solid line*. **c** Step b is repeated. **d** Step b is repeated, but the internal bond is removed

effective bond strength is not assigned until the bond joins the perimeter, the bond strength does not change once assigned (quenched disorder). If at anytime, the two ends of a bond belong to the growing fracture network as shown in Fig. 2c, the bond is removed from the simulation as shown in Fig. 2d. Because the differences in fluid pressure within the fluid-filled fracture network are much smaller than the difference between the pressure in the surrounding rock, these bonds are much less likely to open and can be removed from the simulation. This bond removal step leads to a non-intersecting (loopless) fracture network. In our simulations we do not include an external boundary, and the fracture network can grow indefinitely. Typically, we grow a cluster until a specified number of bonds have been added to the cluster. This can be thought of as limiting the volume of fluid injected during the fracking treatment. We refer to the number of bonds in a cluster as the mass (M) of the cluster.

In our and other invasion percolation models, growth occurs in bursts, the failure of a relatively

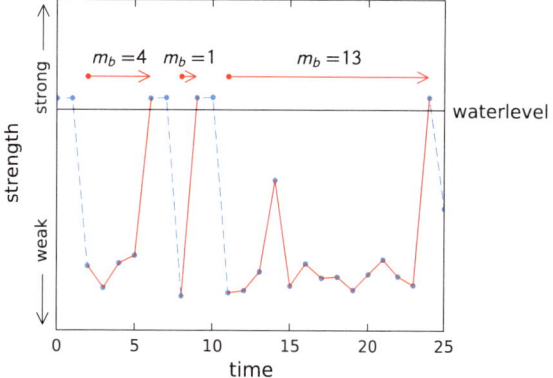

Figure 3

Illustration of our definition of a burst. A typical sequence of 25 opened bond strengths is given. A burst begins when an opened bond strength is below the water-level and ends when the strength is above the water-level. Three bursts with masses $m_b = 4$, 1, and 13 are illustrated

strong bond is followed by the failure of a series of relatively weak bonds. We previously (NORRIS *et al.* 2014) introduced a definition of a burst involving a water-level. A water-level is chosen. A burst begins when the strength of a failed bond falls below the chosen water-level. The burst continues until a failed bond's strength is greater than the chosen water-level. We refer to the number of bonds in a burst as the mass (m_b) of the burst. This definition is illustrated in Fig. 3. By choosing a water-level just below the strength of the strongest failed bond in the fracture network, we obtain a power-law distribution of bursts (NORRIS *et al.* 2014). In our and other percolation models, the strength of the strongest failed bond lies just below the percolation threshold of the lattice. This makes sense because in the absence of external constraints (stopping growth at a certain mass) the fracture network would grow to infinite size, percolating the infinite lattice. The power-law distribution of bursts lets us interpret bursts as the observed microseismic event generated during fracking treatments.

4. Results

We have performed simulations for several different values of anisotropy (a). These simulations take seconds on a desktop computer for even the largest runs, making our model ideal for exploring parameter space. Our simulations are currently

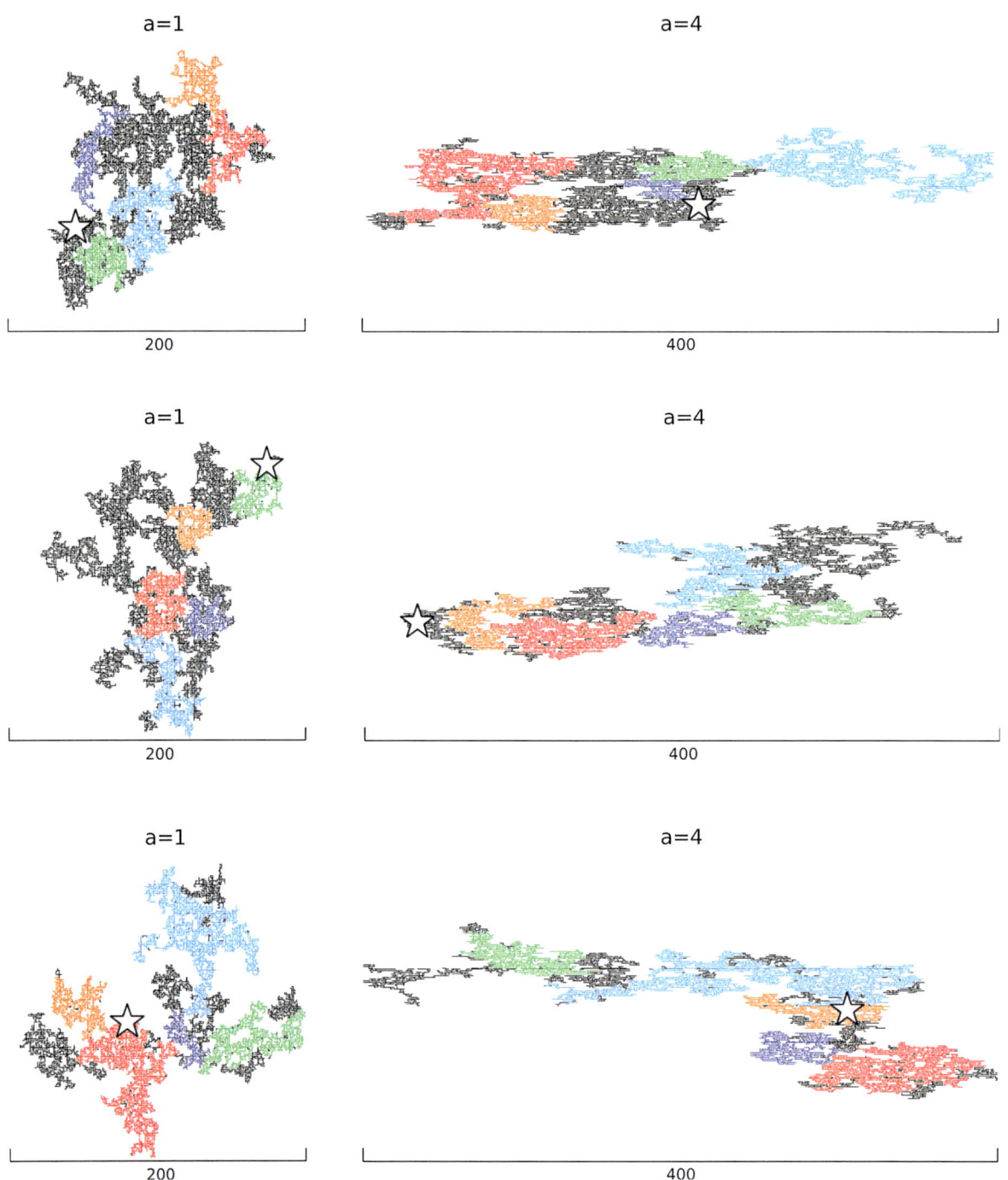

Figure 4
Three examples of clusters with burst structures for $a = 1$ and for $a = 4$. In each realization $M = 10,000$. The four largest bursts are shown in *color*. Smaller bursts and non-burst bonds are shown in *black*. The injection site for each cluster is shown as a *star*

memory limited due to the large number of perimeter bonds that must be stored. One of the goals of this paper is to understand how the simulated microseismic events (bursts) are related to the underlying fracture network and the role anisotropy plays in the structure of both. We will first examine the statistics of grown clusters (simulated fracture networks) and then examine the statistics of bursts.

4.1. Cluster Statistics

In our model, the clusters represent the connected fracture network generated during a fracking treatment. It is important to understand the properties of this network to minimize risk and optimize production. For this paper, we are primarily interested in determining how anisotropy affects cluster properties.

4.1.1 Images of Clusters

To get an idea of the general geometry of and variations between cluster realizations, we have generated three, relatively small ($M = 10,000$) clusters for two different anisotropies ($a = 1$ and $a = 4$). These clusters are shown in Fig. 4. The five largest bursts in each cluster have been colored, while the black bonds are smaller bursts and non-bursting bonds.

The clusters become more elongated with increasing anisotropy. This is expected as large anisotropies lead to weaker bonds in the x-direction. These weaker bonds provide the most likely paths for growth. As with other percolation models, there are many regions within a cluster that are completely surrounded by the cluster. The bonds on the boundaries of these cutoff regions are relatively strong with strengths greater than approximately $s = 0.5$ and prevent further expansion of the cluster into the cutoff region. Bonds within the cutoff regions which are not on the boundaries can have any value of s, $0 < s < 1$. These cutoff regions are most easily observed in simulations where $a = 1$. This is similar to the prevention of loops in self-avoiding random walks.

Cluster growth is often asymmetrical about the point of injection despite a homogeneous distribution of strengths. In a few cases, growth occurs in nearly in a single direction. This shows that while the distribution may be symmetrical, a single realization may not exhibit that symmetry. It also shows that even small degrees of heterogeneity in an otherwise homogeneous material can result in large-scale inhomogeneous structures.

4.1.2 Occupation Probability

One property of interest is the distribution of bond strengths. In the isotropic case, we found that the frequency density ($f = \frac{dN}{ds}$) of bond strengths in the

cluster shows a sharp cutoff near the critical point of the lattice (NORRIS et al. 2014). We have generated a single cluster of mass $M = 10^7$ for six different anisotropies $a = 1, 2, 4, 8, 16, 100$ and calculated the frequency density of bond strengths as shown in Fig. 5.

As in the isotropic case, the distributions of bond strengths for all the anisotropies considered show a sharp cutoff. The cutoff moves closer to 1 as anisotropy increases. Additionally, as anisotropy increases, the distribution becomes more step-like. The critical line for 2D anisotropic bond percolation has been given previously from graph theory (SYKES and ESSAM 1963) and renormalization (AROVAS et al. 1983; CHAVES et al. 1979). The critical line in terms of p_x and p_y, the occupation probability for bonds oriented in the x and y directions, is

$$p_x + p_y = 1 \tag{2}$$

In the isotropic case this equation gives $p_c = p_x = p_y = 0.5$. In our case, we do not have two different occupation probabilities, but a largest strength which in the isotropic case was equal to the critical occupation probability. To compare the largest strengths in the anisotropic case we rewrite Eq. 2 in terms of our anisotropic parameter a

$$p + \frac{p}{a} = 1 \Rightarrow p = \frac{a}{a+1} \tag{3}$$

To determine whether the sharp cutoffs observed in Fig. 5 are near the critical value given in Eq. 2, we look for the largest strength in a cluster of mass $M = 10^7$. Because small clusters often contain strengths above the cutoff, we determine the largest bond strength after an initial transient of 10,000 bonds. The largest bond strengths as a function of inverse anisotropy are shown in Fig. 6 along with the critical curve predicted by Eq. 3. We find excellent agreement between the cutoffs and the critical values predicted by Eq. 3, indicating that the largest bond strength is near the critical occupation probability for the lattice, even with the introduction of anisotropy.

4.1.3 Fractal Dimension

One common measure used to distinguish between clusters of different types is the fractal dimension. To our knowledge, the fractal dimension of anisotropic percolation clusters has not been measured previously.

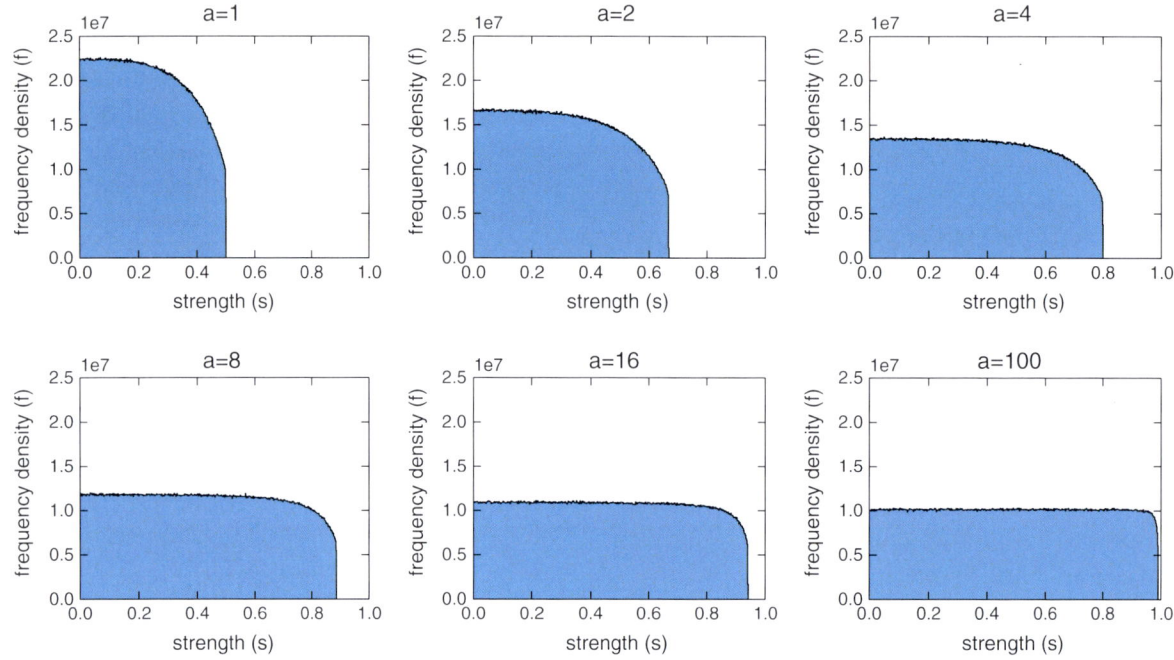

Figure 5
Frequency densities of open bond strengths $f(s)$ are given as a function of bond strength s for several values of a

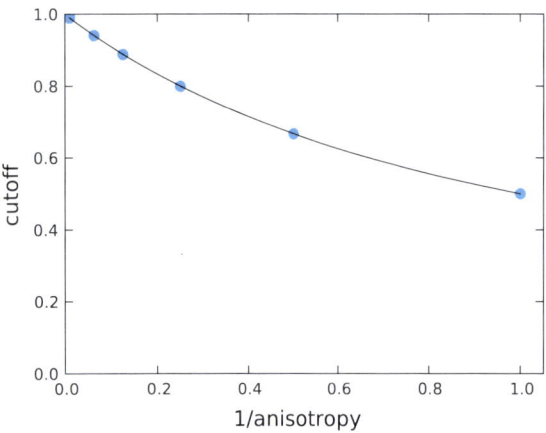

Figure 6
Bond strength cutoff as a function of 1/anisotropy ($\frac{1}{a}$). *Solid circles* are from the data given in Fig. 5 and the *solid line* is the critical point prediction from Eq. 3

We follow the convention presented by Bunde and Havlin (2012) and define the fractal dimension (D) as the scaling exponent between the mass of the cluster M and the radius from its center r

$$M(r) \sim r^D \qquad (4)$$

Because we are interested in how much the reservoir is connected to the borehole, we measure the distance r from the injection site (the borehole). In general, the borehole is not the location of the center of the mass of the cluster so different results may be obtained if distances are measured from the center of mass. To obtain good statistics, we generate 1,000 clusters of mass $M = 10^7$ for each value of anisotropy. For each cluster we center circles of varying radii $r_1, r_2, ..., r_i$ on the injection site. For each circle we determine the mass of the cluster (number of bonds) contained within each circle $M_1, M_2, ..., M_i$. We then take the logarithm of the radii and mass data and do a least-squares fit of aggregate log–log data to

$$\log M = D \log r + C \qquad (5)$$

The average cluster mass as a function of radius along with the fit for several different anisotropies are shown on a log–log plot in Fig. 7.

For large values of r, the cluster mass flattens out as entire clusters are contained within a circle of radius r, and the cluster begins to look more point-like. For small values of r, the discrete nature of the lattice causes variations in the masses. Because of these two

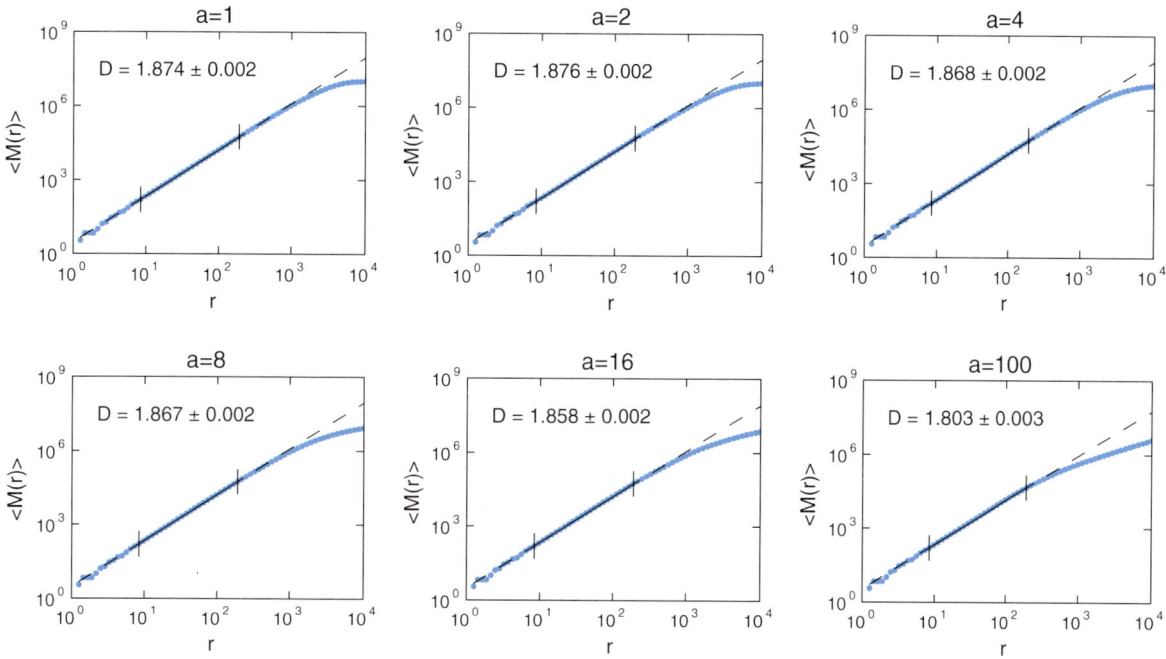

Figure 7

Dependence of the number of opened bonds M contained within a circle of radius r centered on the injection site on the radius r for several values of the anisotropy parameter a. The best fits of Eq. 5 to the data over the region between the vertical lines gives the fractal dimension D

factors, we look for a linear region free from both types of variation. The region used for the fit is shown in Fig. 7. Because the region used is arbitrary, the uncertainty given is the uncertainty in the fit.

We see that the fractal dimension is weakly dependent on the anisotropy parameter. The fractal dimension only differs by 0.07 between $a = 1$ and $a = 100$. Initially, we thought that we could explore the cross-over to one-dimensional percolation; however, we see that anisotropies orders of magnitudes greater than those observed in reservoirs are required to significantly alter the fractal dimension.

4.2. Burst Statistics

Having quantified the anisotropic clusters generated by our model, we now turn out attention to bursts. It is important to understand how the properties of the bursts are related to the properties of the underlying cluster. In fracking, this translates into understanding how the properties of the microseismic data are related to the underlying reactivated fracture network. Our definition of a burst requires

the specification of a water-level just below the largest bond strength in the cluster. In this paper, we have shown that the largest bond strength in the cluster is near the critical occupation probability for the lattice. How close the largest bond strength is to the critical occupation probability depends on the size of the cluster. We have found that as clusters grow larger, the largest bond strength becomes asymptotically close to the critical occupation probability.

For the six clusters shown in Fig. 4 we have determined the locations and sizes of the bursts using cutoffs of 0.47 and 0.77 for $a = 1$ and $a = 4$, respectively. We have plotted these bursts in Fig. 8. The burst markers are colored and scaled according to the size m_b of the burst.

4.2.1 Images of Bursts

We see that the bursts and clusters cover roughly the same area and have a similar outline. This indicates that bursts are useful in determining the extent of cluster growth. This suggests that

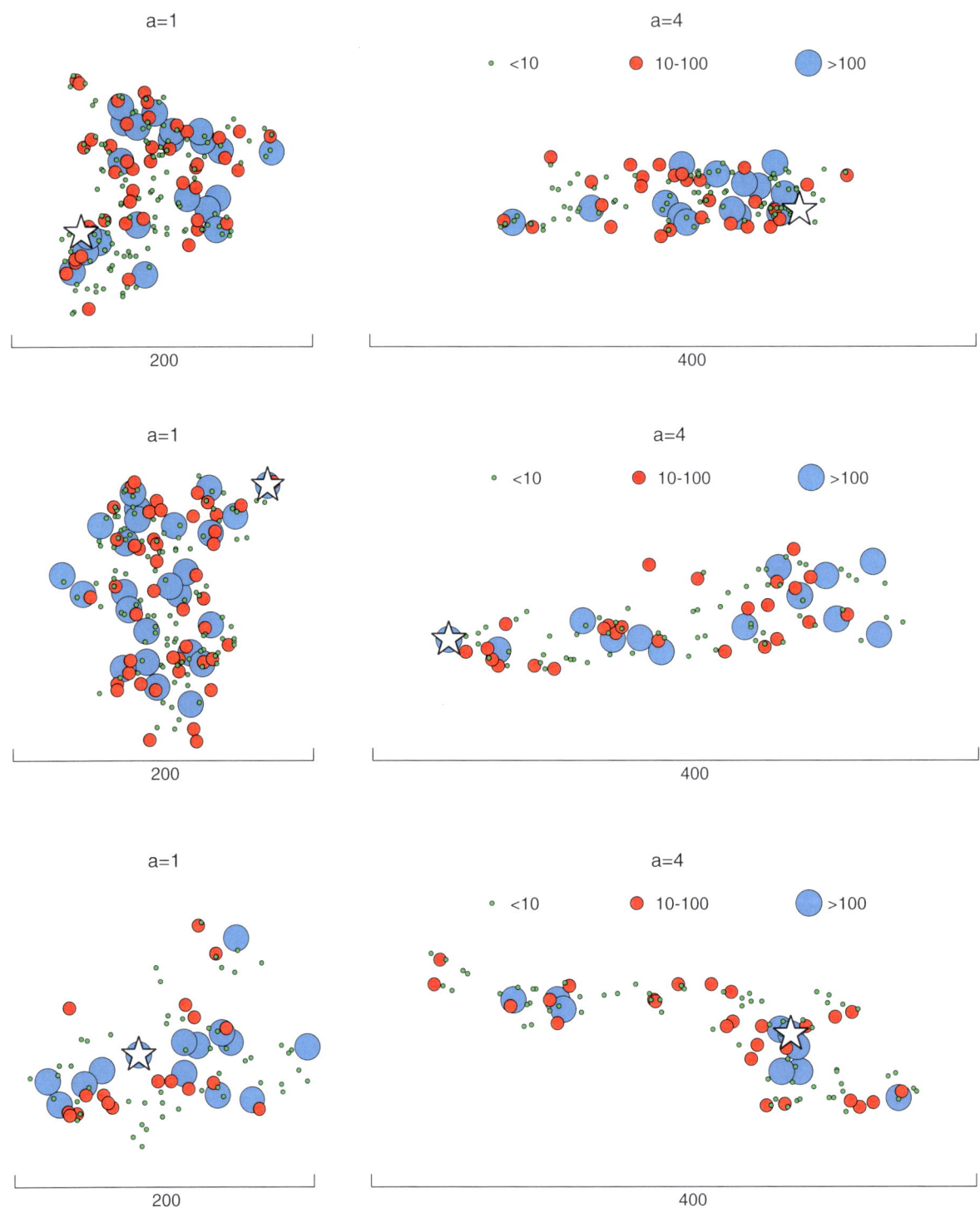

Figure 8

Epicenter locations of the bursts in the six realizations given in Fig. 4. *Different size circles* correspond to different ranges of burst masses m_b

bursts could be used in real time to monitor the location and direction of cluster growth. We see that the bursts of different sizes seem to be more or less evenly distributed over the cluster, rather than larger bursts bunching in one location and small bursts bunching in another. The bursts shown are just a small sample and are meant to give a qualitative illustration of the spatial distribution of bursts.

To provide a more quantitative description of bursts, we have determined the bursts for the same set of 1,000 clusters of mass $M = 10^7$ used to calculate the fractal dimension shown in Fig. 7. Because these are relatively large clusters, a water-level very close to the critical occupation probability must be chosen. The critical occupation probabilities and chosen water-levels for the six anisotropies considered are given in Table 1.

Using these values, we determine the non-cumulative burst frequency-size distribution for each realization. For each anisotropy considered, we aggregate the data. For larger bursts, the aggregate data are sparse with zero or one bursts of a given mass. In this case, it is standard treatment to bin the data (MALAMUD et al. 2004). For each anisotropy considered, we bin the data and do a linear least-squares fit of the log–log data to the power-law distribution

$$N_b \sim m_b^{-\tau} \qquad (6)$$

For larger anisotropies there is a rollover for small bursts. This region becomes significant for $a = 16$ and $a = 100$ and has been excluded from the fit. In Fig. 9, we give the binned data and fit for each value of anisotropy considered.

4.2.2 Frequency-Size Distributions

For each anisotropy value considered, we find excellent agreement with Eq. 6. The exponent takes values near 1.46, suggesting that the slope is

Table 1

For the anisotropy values a we consider, we give the critical probabilities from Eq. 3 and our water-level values

Anisotropy	p_c	Water-level
$a = 1$	0.50000	0.49950
$a = 2$	0.66667	0.66610
$a = 4$	0.80000	0.79950
$a = 8$	0.88889	0.88830
$a = 16$	0.94118	0.94060
$a = 100$	0.99010	0.98995

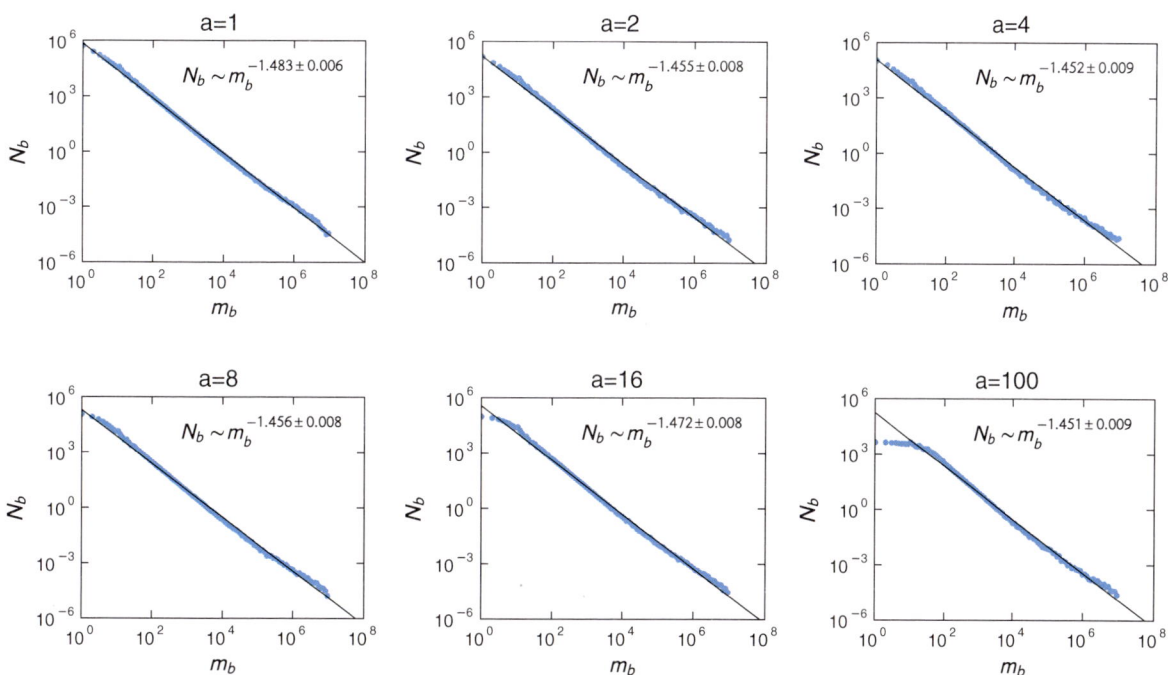

Figure 9

Dependence of the number of bursts N_b with mass m_b on m_b. These data are aggregated from 1,000 runs with $M = 10^7$ for each run. The best fit correlation with Eq. 6 is shown for each value of a

unchanged by the introduction of anisotropy. The rollover observed for $a = 16$ and $a = 100$ appears to be increasing with anisotropy. As the anisotropy becomes very large, this rollover may dominate the distribution.

Because this is a non-cumulative distribution, the b value for bursts generated by our model is $b = \tau - 1$ and $b \approx .46$ for all anisotropies considered. This is significantly lower than the $b = 2$ typically reported for microseismic data MAXWELL (2011). More recent work by TAFTI et al. (2013) on injection into geothermal reservoirs has found $b \approx 1.3$. We note that while the b value of our model is significantly lower than that observed during fracking treatments, our simulation is only in two-dimensions and moving to three dimensions might significantly change the b value.

4.2.3 Burst Fractal Dimension

To obtain a comparison between clusters and bursts, we calculate the fractal dimension for the bursts generated using our model. To obtain the fractal dimension we use the correlation function (HIRATA et al. 1987)

$$C(r) = \frac{2}{N(N-1)} N_r(R < r), \qquad (7)$$

where N is the total number of pairs of events and $N_r(R < r)$ is the number of pairs of events whose separation is less than r. If the burst distribution is fractal, the correlation should follow a power-law

$$C(r) \sim r^D. \qquad (8)$$

The fractal dimension D is sometimes called the correlation dimension and has been used previously to compare the fractal dimension of earthquakes to the fractal dimension of percolation clusters TAFTI et al. (2013). Using the same realizations as before, we calculate the correlation defined in Eq. 7 for the burst hypocenters. For each realization we fit the linear region to Eq. 8. If the realization does not have a large enough linear region or has two different linear realizations, it is discarded. For the six aniso-tropies considered, we have calculated the mean and standard deviation of the fractal dimensions of the remaining realizations. These are given in Table 2.

Table 2

The mean and standard deviations of the burst fractal dimensions D are given for the anisotropy values a that we consider

Anisotropy	Burst fractal dimension (D)
$a = 1$	1.84 ± 0.11
$a = 2$	1.82 ± 0.15
$a = 4$	1.81 ± 0.17
$a = 8$	1.79 ± 0.16
$a = 16$	1.77 ± 0.17
$a = 100$	1.49 ± 0.25

In all cases, the average fractal dimension of the bursts is less than the fractal dimension of the clusters. We also find that the difference increases with increasing anisotropy. However, for relatively small anisotropies, the variation in fractal dimension is small and might not be significant in practical applications.

4.2.4 Burst Anisotropy

We also want to determine how the bursts are related to the underlying anisotropy in the bond strength distribution. As a simple measure of burst anisotropy, we calculated the standard deviations s_x and s_y in the x and y locations of the burst hypocenters relative to the injection point. This measure gives the aspect-ratio of the spatial distribution of bursts. Using the same 1,000 realizations of $M = 10^7$, we aggregated the burst hypocenters and computed the means and standard deviations. In performing the averages, we found that the results did not change significantly if the bursts were weighted by their size. For simplicity, the results presented here are not weighted by burst size. Additionally, we calculated the anisotropy using a common method introduced by FAMILY et al. (1985). This method utilizes the ratio of the eigenvalues (λ_1 and λ_2) of the gyration tensor and gives an anisotropy in the range 0 (completely anisotropic) to 1 (fully isotropic). In order to compare this measure of anisotropy to our results, we take the square root of the inverse of the anisotropy defined by FAMILY et al. (1985).

For the simulations discussed above, we have determined the mean values of the ratio of standard deviations $\frac{s_x}{s_y}$ and the mean values of the square root of the ratio of eigenvalues $\sqrt{\frac{\lambda_2}{\lambda_1}}$. These results are given in Table 3 for several values of the anisotropy

Table 3

The mean values of the ratio of the standard deviations $\frac{s_x}{s_y}$ and the mean values of the square root of the ratio of eigenvalues $\sqrt{\frac{\lambda_2}{\lambda_1}}$ are given for several values of the anisotropy parameter a

Anisotropy	$\frac{s_x}{s_y}$	$\sqrt{\frac{\lambda_2}{\lambda_1}}$
$a = 1$	0.987	0.980
$a = 2$	1.843	1.851
$a = 4$	3.526	3.527
$a = 8$	6.589	6.589
$a = 16$	12.73	12.74
$a = 100$	78.29	78.94

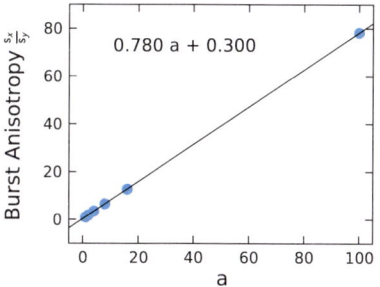

Figure 10

Dependence of the burst anisotropy $\frac{s_x}{s_y}$ on the bond strength anisotropy a. A linear correlation of the data is also shown

parameter a. The two methods give similar values which are somewhat less than the anisotropy of the strength distribution a. In Fig. 10 we give the dependence of $\frac{s_x}{s_y}$ on a. We find a good fit to a linear dependence.

5. *Discussion*

High-volume fracking allows the extraction of oil and gas from tightly sealed shale reservoirs. During the initial formations of the reservoirs, natural hydraulic fractures are generated by the high pressure associated with gas and oil generation. With time, these natural hydraulic fractures are sealed by chemical deposition. The injection of a high pressure, low viscosity fluid penetrates the formation reopening the sealed fractures. This allows the migration of oil and gas to the horizontal injection/production wells.

In order to model the injection process, we used invasion percolation. We assume a 2D square lattice of bonds. These bonds represent the preexisting array of natural fractures. Each bond is assigned a random strength and the weakest bond breaks at each time step. This represents the migration of the injected fluid through the sealed network of natural fractures. This migration occurs in bursts as the fluid enters a region of weak bonds. We associate these bursts with the microseismicity that occurs during fracking injections. A primary focus of this paper is on the role of anisotropic strengths on injection patterns.

The examples of microseismicity associated with four fracking injections into the Barnett Shale, illustrated in Fig. 1, clearly show strong anisotropy. It is of interest to compare this microseismicity with the modeled microseismicity given in Fig. 8. The state 1 injection in Fig. 1 is similar to the $a = 4$ injections illustrated in Fig. 8. Both the modeled and the observed microseismicity exhibit Gutenberg–Richter frequency magnitude statistics; however, the b values differ.

Our model certainly involves a number of serious approximations. Our model uses a two-dimensional square grid. Actual fluid injections are clearly three dimensional, but seismic observations indicate that the flow tends to be confined to a relatively narrow horizontal layer. The preexisting natural fractures that the injection reopens tend to have spacings in the range 0.1–1 m, but are only approximated by a square grid.

Our model neglects the pressure drops associated with the fluid flow. This is probably a good approximation between "bursts" (microseismic events), but significant pressure drops may occur during a "burst." Our model also assumes that the assigned bonds strengths are uncorrelated in space, some spacial correlations may be expected in actual reservoirs.

Despite the assumptions, the geometries of the invading cluster and associated modeled microseismicity are certainly qualitatively similar to the patterns of injection indicated by observed microseismicity. As discussed in our introduction, high-volume fracking is successful only if the natural fractures are largely sealed. Unsealed fractures allow the inject fluid to flow through them without producing the distributed damage required for production. An interesting future extension of this model would include some open fractures prior to injection to quantify the problems associated with fluid leakage through these fractures.

Acknowledgments

The research of JQN and JBR has been supported by a grant from the US Department of Energy to the University of California, Davis #DE-FG02-04ER15568.

REFERENCES

AROVAS D, BHATT RN, SHAPIRO B (1983) *Anisotropic bond percolation in two dimensions.* Phys Rev B *28*(3):1433–1437, doi:10.1103/PhysRevB.28.1433.

ARTHUR MA, SAGEMAN BB (1994) *Marine black shales: Depositional mechanisms and environments of ancient deposits.* Annual Review of Earth and Planetary Sciences *22*(1):499–551, doi:10. 1146/annurev.ea.22.050194.002435.

BALBERG I (1987) *Tunneling and nonuniversal conductivity in composite materials.* Phys Rev Lett *59*(12):1305–1308, doi:10. 1103/PhysRevLett.59.1305.

BALBERG I, BINENBAUM N, BOZOWSKI S (1983) *Anisotropic percolation in carbon black-polyvinylchloride composites.* Solid State Communications *47*(12):989–992.

BUNDE A, HAVLIN S (2012) Fractals and Disordered Systems, 2nd Ed. Springer London.

CELZARD A, MARÉCHÉ JF (2003) *Non-universal conductivity critical exponents in anisotropic percolating media: a new interpretation.* Physica A: Statistical Mechanics and its Applications *317*(34):305–312.

CHAME A, DE QUEIROZ SLA, DOS SANTOS RR (1984) *Dimensional crossover in directed percolation.* Journal of Physics A: Mathematical and General *17*(12):L657.

CHAVES CM, OLIVEIRA PM, DE QUEIROZ SLA, RIERA R (1979) *Remarks on the percolation problem in anisotropic systems.* Progress of Theoretical Physics *62*(6):1550–1555, doi:10.1143/ PTP.62.1550.

CONT R, BOUCHAUD JP (2000) *Herd behaviour and aggregate fluctuations in financial markets.* Macroeconomic Dynamics *4*(02):170–196.

EBRAHIMI F (2010) *Invasion percolation: A computational algorithm for complex phenomena.* Computing in Science and Engineering *12*(2):84–93, doi:10.1109/MCSE.2010.42.

ELLSWORTH WL (2013) *Injection-induced earthquakes.* Science *341*(6142), doi:10.1126/science.1225942.

FAMILY F, VICSEK T, MEAKIN P (1985) *Are Random Fractal Clusters Isotropic?* Phys Rev Lett *55*(7):641–644, doi:10.1103/ PhysRevLett.55.641.

GALE JFW, REED RM, HOLDER J (2007) *Natural fractures in the Barnett Shale and their importance for hydraulic fracture treatments.* AAPG Bulletin *91*(4):603–622, doi:10.1306/11010606061.

GUEGUEN Y, DIENES J (1989) *Transport properties of rocks from statistics and percolation.* Mathematical Geology *21*(1):1–13.

HAN KH, LEE JO, LEE SI (1991) *Confirmation of the universal conductivity critical exponent in a two-dimensional anisotropic system.* Phys Rev B *44*(13):6791–6793, doi:10.1103/PhysRevB. 44.6791.

HERRMANN HJ, SAHIMI M, TZSCHICHHOLZ F (1993) *Examples of fractals in soil mechanics.* Fractals *01*(04):795–805, doi:10.1142/ S0218348X93000824

HIRATA T, SATOH T, ITO K (1987) *Fractal structure of spatial distribution of microfracturing in rock.* Geophysical Journal of the Royal Astronomical Society *90*(2):369–374, doi:10.1111/j.1365-246X.1987.tb00732.x.

IKEDA H (1979) *Percolation in anisotropic systems: Real-space renormalization group.* Progress of Theoretical Physics *61*(3):842–849, doi:10.1143/PTP.61.842.

KERANEN KM, SAVAGE HM, ABERS GA, COCHRAN ES (2013) *Potentially induced earthquakes in Oklahoma, USA: Links between wastewater injection and the 2011 Mw 5.7 earthquake sequence.* Geology *41*(6):699–702, doi:10.1130/G34045.1.

KIM CS, LEE MH (1992) *Monte Carlo renormalization group studies of anisotropic bond percolation.* International Journal of Modern Physics B *06*(09):1505–1515, doi:10.1142/ S0217979292000700.

KING G (2012) Hydraulic fracturing 101: What every representative, environmentalist, fegulator, feporter, investor, university researcher, neighbor and engineer should know about estimating frac risk and improving frac performance in unconventional gas and oil wells. In: SPE Hydraulic Fracturing Technology Conference, 6–8 February, The Woodlands, Texas, USA.

KING PR, JR JS, BULDYREV SV, DOKHOLYAN N, LEE Y, HAVLIN S, STANLEY H (1999) *Predicting oil recovery using percolation.* Physica A: Statistical Mechanics and its Applications *266*(14):107–114.

KLEMME HD, ULMISHEK GF (1991) *Effective petroleum source rocks of the world; stratigraphic distribution and controlling depositional factors.* AAPG Bulletin *75*(12):1809–1851.

KNACKSTEDT MA, SAHIMI M, SHEPPARD AP (2002) *Nonuniversality of invasion percolation in two-dimensional systems.* Physical Review E *65*(3):35,101, doi:10.1103/PhysRevE.65.035101.

LOBB CJ, FRANK DJ, TINKHAM M (1981) *Percolative conduction in anisotropic media: A renormalization-group approach.* Phys Rev B *23*(5):2262–2268, doi:10.1103/PhysRevB.23.2262.

MALAMUD BD, TURCOTTE DL, GUZZETTI F, REICHENBACH P (2004) *Landslide inventories and their statistical properties.* Earth Surface Processes and Landforms *29*(6):687–711, doi:10.1002/ esp.1064.

MAXWELL S (2011) *Microseismic hydraulic fracture imaging: The path toward optimizing shale gas production.* The Leading Edge *30*(3):340–346, doi:10.1190/1.3567266.

MENDELSON KS, KARIORIS FG (1980) *Percolation in two-dimensional, macroscopically anisotropic systems.* Journal of Physics C: Solid State Physics *13*(33):6197.

NORRIS JQ, TURCOTTE DL, RUNDLE JB (2014) *Loopless nontrapping invasion-percolation model for fracking.* Phys Rev E *89*(2):22,119, doi:10.1103/PhysRevE.89.022119.

OLSON JE, LAUBACH SE, LANDER RH (2009) *Natural fracture characterization in tight gas sandstones: Integrating mechanics and diagenesis.* AAPG Bulletin *93*(11):1535–1549, doi:10.1306/ 08110909100.

OSBORN SG, VENGOSH A, WARNER NR, JACKSON RB (2011) Methane contamination of drinking water accompanying gas-well drilling and hydraulic fracturing. Proceedings of the National Academy of Sciences *108*(20):8172–6, doi:10.1073/pnas.1100682108.

OTSUKA M (1971) *A simulation of earthquake occurrence part 1. A mechanical model.* Zisin (Journal of the Seismological Society of Japan 2nd ser) *24*(1):13–25.

SAHIMI M (1994) Applications of percolation theory. Taylor & Francis, London; Bristol, PA.

SAHIMI M, ROBERTSON MC, SAMMIS CG (1993) *Fractal distribution of earthquake hypocenters and its relation to fault patterns and percolation*. Phys Rev Lett *70*(14):2186–2189, doi:10.1103/PhysRevLett.70.2186.

SEAGER CH, PIKE GE (1974) *Percolation and conductivity: A computer study. II*. Phys Rev B *10*(4):1435–1446, doi:10.1103/PhysRevB.10.1435.

SMITH LN, LOBB CJ (1979) *Percolation in two-dimensional conductor-insulator networks with controllable anisotropy*. Phys Rev B *20*(9):3653–3658, doi:10.1103/PhysRevB.20.3653.

SOTTA P, LONG D (2003) *The crossover from 2D to 3D percolation: Theory and numerical simulations*. The European Physical Journal E *11*(4):375–388, doi:10.1140/epje/i2002-10161-6.

STAUFFER D, AHARONY A (1994) Introduction to Percolation Theory, 2nd Ed. Taylor & Francis Group.

SYKES MF, ESSAM JW (1963) *Some exact critical percolation probabilities for bond and site problems in two dimensions*. Phys Rev Lett *10*(1):3–4, doi:10.1103/PhysRevLett.10.3.

TAFTI TA, SAHIMI M, AMINZADEH F, SAMMIS CG (2013) *Use of microseismicity for determining the structure of the fracture network of large-scale porous media*. Phys Rev E *87*(3):32,152, doi:10.1103/PhysRevE.87.032152.

TOURTELOT HA (1979) *Black shale; its deposition and diagenesis*. Clays and Clay Minerals *27*(5):313–321.

TRABUCHO-ALEXANDRE J, HAY WW, DE BOER PL (2012) *Phanerozoic environments of black shale deposition and the Wilson Cycle*. Solid Earth *3*(1):29–42, doi:10.5194/se-3-29-2012.

ULMISHEK GF, KLEMME HD (1990) Depositional controls, distribution, and effectiveness of world's petroleum source rocks. Tech. rep., USGS Bulletin 1931.

WARPINSKI NR (2013) Understanding Hydraulic Fracture Growth, Effectiveness, and Safety Through Microseismic Monitoring. In: Effective and Sustainable Hydraulic Fracturing.

WILKINSON D, BARSONY M (1984) *Monte Carlo study of invasion percolation clusters in two and three dimensions*. Journal of Physics A: Mathematical and General *17*(3):L129–L135, doi:10.1088/0305-4470/17/3/007.

WILKINSON D, WILLEMSEN JF (1983) *Invasion percolation: a new form of percolation theory*. Journal of Physics A-Mathematical and General *16*(14):3365–3376, doi:10.1088/0305-4470/16/14/028.

ZOBACK M, KITASEI S, COPITHORNE B (2010) Addressing the Environmental Risks from Shale Gas Development. Tech. rep., Worldwatch Institute: Natural Gas and Sustainable Energy Initiative.

(Received April 10, 2014, revised August 1, 2014, accepted August 6, 2014, Published online August 24, 2014)

Pure Appl. Geophys. 172 (2015), 23–31
© 2014 Springer Basel
DOI 10.1007/s00024-014-0871-2

The Negative Binomial Distribution as a Renewal Model for the Recurrence of Large Earthquakes

ALEJANDRO TEJEDOR,[1] JAVIER B. GÓMEZ,[2] and AMALIO F. PACHECO[3]

Abstract—The negative binomial distribution is presented as the waiting time distribution of a cyclic Markov model. This cycle simulates the seismic cycle in a fault. As an example, this model, which can describe recurrences with aperiodicities between 0 and 0.5, is used to fit the Parkfield, California earthquake series in the San Andreas Fault. The performance of the model in the forecasting is expressed in terms of error diagrams and compared with other recurrence models from literature.

Key words: Negative binomial distribution, renewal process, seismic cycle, earthquake forecasting.

1. Introduction

The *elastic-rebound model* is the canonical "macroscopic" theory of great earthquakes (REID 1910; SCHOLZ 2002). It states that a great earthquake will occur where large elastic strains have accumulated in the crust. The earthquake itself will relieve most of the strain, which will then accumulate slowly again by a steady input of tectonic stress until the elastic strain becomes sufficiently large for another earthquake to ensue. The duration of this *earthquake cycle* (the time between two consecutive large earthquakes) is the ratio of the strain released during an earthquake to the rate of input of tectonic strain by plate motion.

Because the Earth's crust is heterogeneous and faults are not isolated from each other, the earthquake cycle of a specific fault is not periodic. So, although the elastic-rebound model is in essence a deterministic theory, its application to a heterogeneous and interacting crust implies its translation into a probabilistic framework.

The variability of the duration of a cycle (either real earthquakes on a fault or synthetic earthquakes in a model) can be appropriately defined in the context of a probability density function (pdf) by means of the coefficient of variation, α, the ratio of the standard deviation σ to the mean μ of the pdf:

$$\alpha = \frac{\sigma}{\mu}. \tag{1}$$

In the seismological literature the coefficient of variation is also known as the *aperiodicity*, a very descriptive name when applied to the duration of the earthquake cycle: $\alpha = 0$ gives perfectly periodic cycles, $0 < \alpha < 1$ quasiperiodic cycles, and $\alpha > 1$ clustering of events. The case $\alpha = 1$ is particularly important because the exponential distribution has this property, and the exponential distribution is the pdf of an earthquake cycle where large earthquakes occur in time following a Poisson distribution (i.e., they are random in time). In actual seismic faults, the aperiodicity of the earthquake series is always <1 (SYKES and MENKE 2006; ELLSWORTH *et al.* 1999; ABAIMOV *et al.* 2007).

RIKITAKE (1974) was the first to formally introduce a probabilistic description of the occurrence times of specific earthquakes. He treated earthquake recurrence as a *renewal process*, in which the times between successive events (in this case, the large earthquakes in a specific fault) are assumed to be independent and independently distributed random variables.

---

[1] Saint Anthony Falls Laboratory, Department of Civil Engineering, University of Minnesota, 2 Third Avenue SE, Minneapolis, MN 55455, USA.

[2] Department of Earth Sciences, University of Zaragoza, 50009 Zaragoza, Spain. E-mail: jgomez@unizar.es

[3] Department of Theoretical Physics, University of Zaragoza, 50009 Zaragoza, Spain.

Reprinted from the journal

Since then, several authors have proposed probabilistic versions of the elastic-rebound model in the shape of a plethora of probability distribution functions (pdfs) for the duration of the earthquake cycle: exponential (UTSU 1984; SORNETTE and KNOPOFF 1997; MATTHEWS et al. 2002), Weibull (UTSU 1984; SORNETTE and KNOPOFF 1997; MATTHEWS et al. 2002; FERRÁES 2003; GÓMEZ and PACHECO 2004; YAKOVLEV et al. 2006; ABAIMOV et al. 2007, 2008; GOLTZ et al. 2009), log-normal (UTSU 1984; SORNETTE and KNOPOFF 1997; MATTHEWS et al. 2002; GÓMEZ and PACHECO 2004; FERRÁES 2005; ABAIMOV et al. 2007, 2008), gamma (UTSU 1984; MATTHEWS et al. 2002; GÓMEZ and PACHECO 2004; FERRÁES 2005), power-law (SORNETTE and KNOPOFF 1997), Brownian passage time (MATTHEWS et al. 2002; WGCEP 2003; MICHAEL 2005; YAKOVLEV et al. 2006; ABAIMOV et al. 2007; ZÖLLER et al. 2008), among others. However, due to the scarcity of registered large earthquakes in a specific fault (usually 4–10 earthquakes), the statistics upon which the selection of a specific pdf is based are poor. This means that different pdfs can fit the empirical distribution function.

Most of the probability distributions have been used solely for their statistical properties, with no relationships with the physics of the underlying process (elastic rebound theory). However, a subset of them has a physical rationale and from this point of view can be considered as better motivated. One example is the Brownian passage time distribution (BPT; MATTHEWS et al. 2002), where the seismic cycle in a fault is modeled by the time evolution of the so-called Brownian relaxation oscillator.

Also, the majority of the probability distributions used in the context of earthquake recurrence are continuous. However, in the last 10 years, several *discrete* probability distributions that are the outcome of cellular automata models have been proposed (VÁZQUEZ-PRADA et al. 2002; GONZÁLEZ et al. 2005; TEJEDOR et al. 2009). These discrete, cellular automata-based probability distributions share with the BPT distribution their physical motivation, as the models behind these discrete probability distributions try to reproduce in a few cellular automata rules the physics of a seismic fault under the elastic rebound assumption.

The aim of this paper is to present a discrete probability distribution, the negative binomial distribution (NBD) for the recurrence of large earthquakes. The study of one-way Markov cycles was presented in TEJEDOR et al. (2012), together with two of its limits, the so-called box model and the NBD. Here we focus on the NBD for its particular importance: the NBD seems to be the unique distribution that derives from the dynamics of a cellular automaton and simultaneously appears in general textbooks in probability and statistics. In Sect. 2, the NBD and its first moments are introduced. Section 3 recalls that the NBD is a special case of a waiting time distribution for a one-way Markov cycle, as deduced in TEJEDOR et al. (2012); Sect. 4 then uses this distribution as a renewal model for large earthquakes, using the earthquake series of the Parkfield segment of the San Andreas Fault as an example. The quality of the fit of the NBD to the empirical distribution function of the Parkfield series is compared to other renewal models used in the literature. Section 5 assesses the forecasting capabilities of the NBD by means of a reference prediction strategy and error diagrams. Finally, in Sect. 6, the most important conclusions drawn from the paper are stated. The computation of the asymptotic limit of the hazard rate for the NBD is detailed in the Appendix.

2. The Negative Binomial Distribution

As there are some different modalities of defining the NBD, we will now specify the form used in this paper.

A *negative binomial experiment* is a statistical experiment that has the following properties: The experiment consists of n repeated trials. Each trial can result in just two outcomes, a success or a failure. The probability of success, denoted by $1 - a$ ($a < 1$), is the same on every trial. In consequence, the probability of failure is a. The trials are independent. And the experiment continues until N successes are observed. N is specified in advance.

The *negative binomial random variable* is the number n of repeated trials to produce N successes in a negative binomial experiment. The probability distribution of the negative binomial random variable is called an NBD. Its form is:

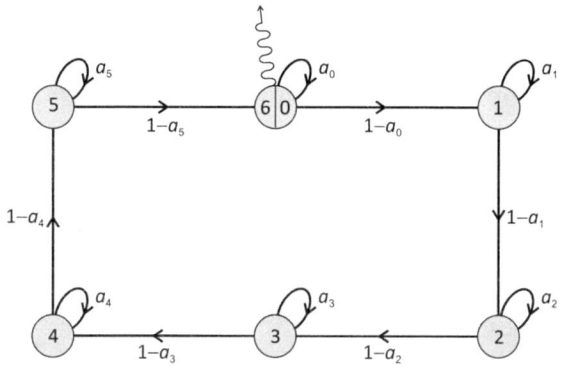

Figure 1
Scheme of a one-way Markov cycle with $N = 6$. The probability of staying in state i is a_i and the probability of jumping from state i to state $i + 1$ is $(1 - a_i)$. Jumping from one state to the next means that the fault has accumulated more strain energy. The *wavy line* between states *6* and *0* indicates that at the end of the cycle, all the stored energy is released

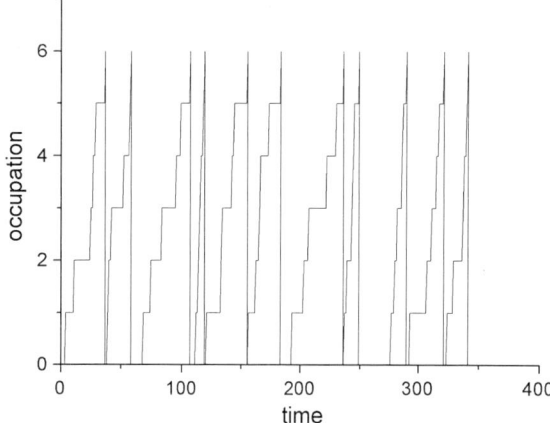

Figure 2
Occupation (number of occupied sites) of an $N = 6$ system as a function of time for 11 consecutive cycles. Note the repetitive pattern but the lack of perfect periodicity

$$P_{N,a}(n) = (1 - a)^N a^{n-N} \binom{n - 1}{N - 1} \qquad (2)$$

The mean, variance, and coefficient of variation—or aperiodicity—of this distribution are:

$$\mu = \frac{N}{1 - a}, \qquad (3)$$

$$\sigma^2 = \frac{Na}{(1 - a)^2}, \qquad (4)$$

and

$$\alpha \equiv \frac{\sigma}{\mu} = \sqrt{\frac{a}{N}}, \qquad (5)$$

respectively.

3. The NBD as the Waiting Time Distribution in an Specific Markov Cycle Model

Let us suppose a Markov chain with N sites forming a closed loop that has gone over clockwise (see Fig. 1 for illustration). The N sites are ordered by the index i, $i = 0, 1, \ldots, N - 1, N$. As a genuine cellular automaton, time increases in discrete steps. At the beginning of each cycle, our system occupies the first position, $i = 0$. In the first time step, it makes a trial to pass to site $i = 1$. The probability of success is $(1 - a_0)$ and that of failure is a_0. Typically, after some trials, the system will occupy site 1. Now all is

identical to the first case, except that the probability of passing from site 1 to site 2 is $(1 - a_1)$. Then the turn of sites is $2, 3, \ldots, N - 1$.

When site N is occupied, the cycle ends. The system automatically passes to site 0 and a new cycle starts. Figure 2 shows an example of this process of slow filling and abrupt emptying for 11 consecutive cycles for a system with $N = 6$ and $(1 - a_i) = 1/N = 1/6$, for all states i.

The traveling in successive discrete steps around the cycle can be interpreted as a process of gradual increase of strain in fault, and thus this Markov cycle represents the seismic cycle in a fault. Site 0 represents the state with no strain and site N represents the state of maximum strain that is automatically released to pass to site 0. This sudden release of strain simulates the occurrence of a characteristic earthquake in the fault. Thus, in this model, a decrease in the strain, such as could take place in a random walk type model, is forbidden. This model is illustrated in Fig. 1 and materialized in the following Markov matrix:

$$[M] = \begin{pmatrix} a_0 & 1 - a_0 & 0 & 0 & 0 & 0 \\ 0 & a_1 & 1 - a_1 & 0 & 0 & 0 \\ 0 & 0 & a_2 & 1 - a_2 & 0 & 0 \\ 0 & 0 & 0 & a_3 & 1 - a_3 & 0 \\ 0 & 0 & 0 & 0 & a_4 & 1 - a_4 \\ 1 - a_5 & 0 & 0 & 0 & 0 & a_5 \end{pmatrix}$$

$$(6)$$

Note that the number of parameters in this discrete model is $N + 1$: the length of the cycle, N, plus the value of the N parameters a_i. Using standard techniques of Markov chains (Tejedor et al. 2012), one can obtain, in a closed form, the distribution function of the cycle lengths in this model:

$$P_N(n) = \prod_{i=0}^{N-1} (1 - a_i) \sum_{i=0}^{N-1} \left[\frac{a_i^{n-1}}{\prod_{j(\neq i)=0}^{N-1} (a_i - a_j)} \right], \quad (7)$$
$$n = N, N+1, \ldots, \infty$$

It is clear that until time step $n = N$, the probability of completing a cycle is null. In seismology this is called a stress-shadow.

A property of this general model is that no matter what the value of its parameters are, the aperiodicity is lower than 1.

When the N parameters a_i are equal,

$$a = a_1 = a_2 = \cdots = a_N \quad (8)$$

Equation (7) becomes Eq. (2). That is, if Eq. (8) is fulfilled, an NBD is the waiting time distribution of the Markov cycle.

After this hypothesis, the pdf has only two parameters, N and a. This bi-parametric freedom can be used for fitting purposes, including, of course, the seismic cycles. In this paper, however, we will step forward with an additional simplification by relating them in the form:

$$1 - a = \frac{1}{N} \quad (9)$$

After this new hypothesis, there is only one free parameter and each cycle of the model can be intuitively associated with the *ordered* filling of a box with N positions. The new simplified NBD is

$$P_N(n) = \left(\frac{1}{N}\right)^N \left(\frac{N-1}{N}\right)^{n-N} \binom{n-1}{N-1}, \quad (10)$$
$$n = N, N+1, \ldots \infty$$

In the next section, we will see that $N = 6$ is the most appropriate size of the model to fit the recurrence of earthquakes in the Parkfield, California section of the San Andreas Fault. For this case, the pdf in Eq. (10) is simply:

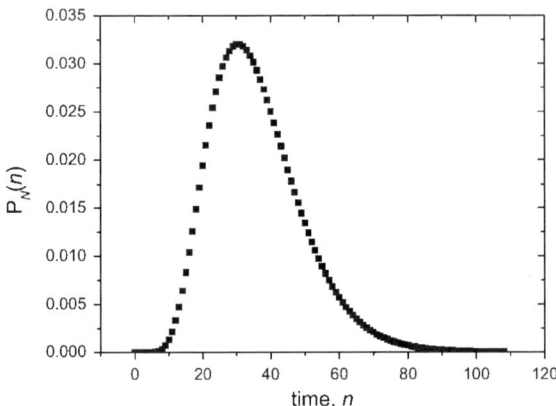

Figure 3
Probability density function of the NBD for the case $N = 6$, $(1 - a) = 1/6$

$$P_6(n) = \left(\frac{1}{6}\right)^6 \left(\frac{5}{6}\right)^{n-6} \binom{n-1}{5}, \quad n = 6, 7, \ldots \infty$$
$$(11)$$

The values of its mean ad aperiodicity are:

$$\mu_6 = 36 \quad \text{and} \quad \alpha_6 = 0.373 \quad (12)$$

Figure 3 plots the NBD written in Eq. (11). To mark the discrete nature of the probability distribution, only points for integer time steps have been drawn, with no line connecting them.

4. Applications of the NBD in Seismicity and Earthquake Forecasting: the Parkfield Series

Including the latest event, the Parkfield series (Bakun and Lindh 1985; Bakun 1988; Michael and Jones 1998) consists of seven $M_w \approx 6$ mainshocks, which occurred on 9 January 1857; 2 February 1881; 3 March 1901; 10 March 1922; 8 June 1934; 28 June 1966 and 28 September 2004. In consequence, the duration (in years) of the six observed inter-event times are: 24.07, 20.08, 21.02, 12.25, 32.05 and 38.25. The mean value μ_{Pk}, the sample standard deviation σ_{Pk} (the square root of the bias-corrected sample variance), and the aperiodicity α_{Pk} of this six-data series are:

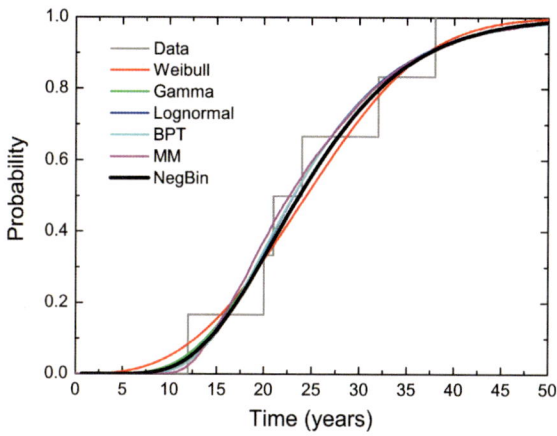

Figure 4

Fit of the NBD model (*black continuous line*) to the Parkfield series (*gray step-like line*) and comparison with other statistical models used in the literature

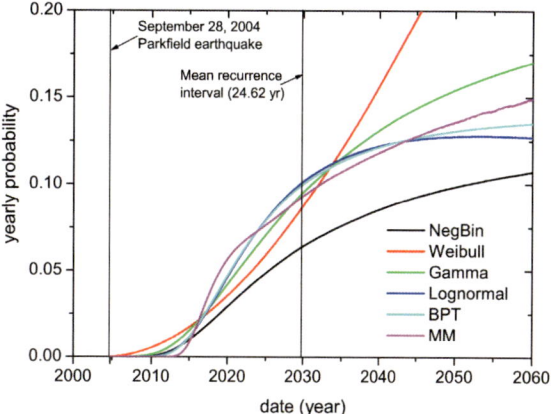

Figure 5

Yearly conditional probability for the Parkfield series as predicted by the negative binomial model compared to other statistical models used in the literature

$$\mu_{\text{Pk}} = 24.62 \text{ years}; \quad \sigma_{\text{Pk}} = 9.25 \text{ years};$$
$$\alpha_{\text{Pk}} = 0.3759 \tag{13}$$

Now, we will proceed to fit these data using the simplified NBD written in Eq. (11). Its aperiodicity is given by

$$\alpha_{\text{NBD}} = \sqrt{\frac{N-1}{N^2}} \tag{14}$$

As we want a distribution with the same aperiodicity (and mean) as the Parkfield series, taking α_{Pk} from Eq. (13) and substituting it in Eq. (14), we have $N = 5.8$. But because N is a discrete quantity, we use the nearest integer, $N = 6$.

However, for fitting the data, it is necessary to assign a definite number of years to the non-dimensional time step of the model. This second parameter will be called τ. From Eqs. (3) and (9), we have that for the NBD, $\mu = N^2 = 36$ time steps. This mean cycle length (in non-dimensional time steps) should be equal to the mean recurrence time of the Parkfield series, $\mu_{\text{Pk}} = 24.62$ years, so that $\tau = 0.68$ years per time step of the model. In Fig. 4, we have plotted the empirical distribution function of the Parkfield series (gray step-like line) and the fit to the cumulative NBD with $N = 6$ (black continuous line), together with five other (cumulative) distribution functions used as renewal models in the literature: Weibull, gamma, log-normal, BPT and minimalist model (MM; VAZQUEZ-PRADA *et al.* 2002). It is quite obvious

from the figure that the performance of all six models is good and very similar, including the NBD. Indeed, the residuals for the NBD evaluated at the midpoints of the horizontal segments of the empirical distribution function are the lowest of the six tested models.

The NBD (and any of the other models shown in Fig. 4) can be used to estimate the time-dependent probability of having an earthquake as a function of the time elapsed since the last earthquake in the series (28 September 2004). This estimation can be carried out with the hazard rate function,

$$h_{N,a}(n) = \frac{P_{N,a}(n)}{\sum_{i=n}^{\infty} P_{N,a}(i)} \tag{15}$$

For discrete distributions like the NBD, the hazard rate is the probability for an earthquake to occur at time step n on the condition that it has not occurred until time step $n - 1$. However, in the seismological literature, it is customary to express the likelihood of a future earthquake using the *yearly conditional probability* of earthquake occurrence, $P(n|\Delta t = 1 \text{ year})$, instead of the hazard rate. This function gives the probability of having an earthquake during the next year, provided it has not occurred before:

$$P_{N,a}(n|\Delta t = 1 \text{ year}) = \frac{S_{N,a}(n + \Delta t) - S_{N,a}(n)}{1 - S_{N,a}(n - 1)} \tag{16}$$

where $S_{N,a}(n) = \sum_{n'=N}^{n} P_{N,a}(n')$ is the cumulative distribution function. The yearly conditional probability function for the Parkfield series is illustrated in

Fig. 5. Again, as in Fig. 4, the NBD and five other models are compared. The present yearly probability of earthquake occurrence is 0.004, i.e., there is a 0.4 % probability of having an earthquake in the following 12 months. Obviously this probability is low because the earthquake cycle is in its early stages. When the cycle is at its average duration, 24.62 years, the yearly probability of earthquake occurrence will be 6 %.

Both the hazard rate and the yearly conditional probability functions for the NBD reach a constant value for large times. Inserting Eq. (2) into Eq. (15), one obtains that, for long times,

$$\lim_{n \to \infty} h_{N,a}(n) = 1 - a \qquad (17)$$

The derivation of this equation can be found in the Appendix. If Eq. (9) is used instead (i.e., the one-parameter simplification of the NBD), the asymptotic limit of the hazard function is equal to $1/N$.

5. Error Diagrams for the Parkfield Example

A hint of the predictability of the large relaxations in this type of model is given by the *aperiodicity* of their time series. The aperiodicity, as stated in Sect. 1, is a quantitative measure of the lack of regularity of a time series. As the aperiodicity of this model is always <1, the occurrence of the large events is a quasi-periodic phenomenon. A robust way to assess the predictability of a time series is by trying to forecast its events by declaring alarms at particular times.

The aim is to declare alarms before all the events in order not to miss any, but to declare them just before each event in order to minimize the total alarm time. Many strategies can be devised to declare the alarms, but there is a *reference strategy* to which all others can be compared (NEWMAN and TURCOTTE 1992; VÁZQUEZ-PRADA et al. 2002; KEILIS-BOROK and SOLOVIEV 2003). This strategy consists of waiting a fixed time after each event (waiting time w), then setting the alarm, and maintaining it until the occurrence of the next event (Fig. 6). If the following event in the time series occurs before the alarm is raised, it is counted as a prediction error; if the following event in the time series occurs after the alarm is raised, it is counted as a prediction of success and the alarm is then cancelled.

The events that are to be predicted (large earthquakes) are the vertical red bars numbered correlatively. An alarm (vertical black lines with rounded top) is set a fixed time interval after each event (waiting time) and the prediction is labeled error (E) or success (S), depending on whether the alarm was off or on when the event occurred, respectively. The fraction of errors is the number of events not predicted (one in the example, the second event) divided by the total number of events (five events), i.e., $f_e = 0.2$; and the fraction of alarm time is the total alarm time (blue sections of the time line: 29 time units) divided by the total duration of the time series (86 time units), i.e., $f_a = 0.34$ in the example shown in the figure.

The fraction of errors f_e (number of missed events divided by the total number of events) and the fraction of alarm time f_a (total alarm time divided by the total duration of the time series) can be computed as a function of the above mentioned waiting time w, and the purpose is to find the optimum waiting time. This optimum waiting time depends on the relative importance that failing to predict an event has compared to keeping the alarm on. An objective function, called loss function, L, that incorporates this trade-off in each particular case can be defined. Here, we will use the simplest of them, $L = f_e + f_a$, where failure to predict an event and a longer alarm time are equally penalized.

Thus, the aim is to find the waiting time $w = w^*$ that minimizes $L(w)$. This minimum value is denoted by $L^* \equiv L(w^*)$. The best way to graphically display this is by means of an error diagram, where the fraction of alarm time f_a runs along the horizontal axis and the fraction of errors f_e runs along the vertical axis. Error diagrams were introduced in earthquake forecasting by MOLCHAN (1997), who contributed to the optimization of the earthquake prediction strategies with rigorous mathematical analysis.

A good strategy of forecasting must produce both small f_e and f_a, because both the prediction failures and the alarms are costly. A random guessing strategy (randomly turning the alarm on and off) will yield $L = 1$, a result which can be easily understood. The

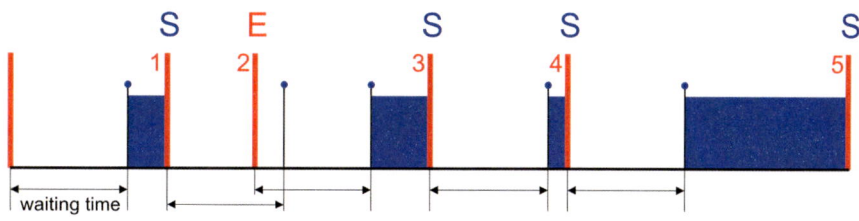

· Figure 6
Reference strategy for the assessment of the predictability of a time series. *Red bars* are the earthquakes to be forecasted (five in the example). An *S* (success) above a *red bar* means that the earthquake has been successfully predicted, whereas an *E* (error) means that the earthquake has not been predicted. The *blue strips* stand for the time with the alarm on before each earthquake

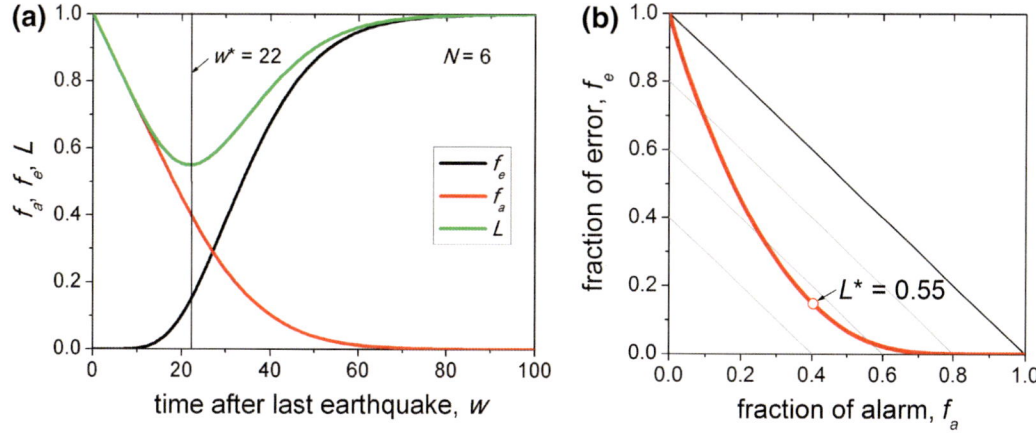

Figure 7
a Fraction of error f_e, fraction of alarm, f_a and loss function L as a function of the time after the last earthquake for a NBD model with $N = 6$.
b *Error diagram* for the prediction strategy shown in **a**. The minimum value of the loss function is $L^* = 0.55$ for $w^* = 22$

alarm will be on, randomly, during a certain fraction of time, f_a. Thus, there will be a probability equal to f_a for it being on when an earthquake eventually occurs (and a probability of $1 - f_a$ for it being off). The result is that $f_e = 1 - f_a$. As a trivial special case, if the alarm is always on ($f_a = 1$), then all the earthquakes are "forecasted" ($f_e = 0$). Conversely, all the earthquakes are failures to predict if the alarm is always off. The random guessing strategy is considered as a baseline, so a forecasting procedure makes sense only if it gives $f_a + f_e < 1$.

Both functions, f_a and f_e, together with the loss function $L = f_e + f_a$ are plotted in Fig. 7a for the case $N = 6$, while Fig. 7b plots the error diagram for the same data. For each value of N, $L(w)$ has a minimum at a specific value of w, $w^*(N)$. As can be seen in Fig. 7, $w^*(6) = 22$, for which

$$f_a(w^*) = 0.403, \quad f_e(w^*) = 0.147, \quad L(w^*) = 0.550 \tag{18}$$

For the Parkfield sequence, w^* corresponds to

$$\tau w* = 15.0 \, \text{years}$$

If the distribution derived from the NBD model correctly describes the recurrence of large earthquakes at Parkfield, an alarm connected 15 years after the last earthquake (beginning of the cycle) and disconnected just after the occurrence of each shock would yield the results given in Eq. (18). Note that this time is approximately equal to the difference between the mean and the standard deviation. This is reasonable because $w^* = 15$ years would capture most of the probability curve, as can be seen in Fig. 3.

6. Conclusions

We have introduced the NBD as a renewal model to describe the recurrence of large earthquakes in faults.

As a test ground of application, we have used the Parkfield series. The yearly conditional probability and other functions as predicted by the NBD are compared to other statistical models used in the literature, and a simple forecasting strategy has been evaluated using error diagrams.

Our results show that the NBD is competitive against other models, but general conclusions cannot be drawn because of the smallness of the sample.

The NBD seems to be the unique discrete distribution coming from a cellular automaton whose properties can be found in textbooks of probability and statistics.

In this paper, we have reduced one parameter of the distribution by relating the probability of advancing in the Markov process to the total number of steps in the cyclic chain. With this simplification, this model can be intuitively understood as the progressive ordered filling of a finite box.

Appendix: Asymptotic Behavior of the Hazard Rate Function

Recall that the N-step Markov-cycle distribution, Eq. (7), collapses to a NBD when all transition probabilities are equal, $a = a_1 = a_2 = \cdots = a_N$:

$$
\begin{aligned}
P_{N,a}(n) &= (1-a)^N a^{n-N} \binom{n-1}{N-1} \\
&= \left(\frac{1-a}{a}\right)^N a^n \frac{(n-1)\ldots(n-N+1)}{(N-1)!}.
\end{aligned}
\tag{19}
$$

Using the definition of hazard rate for a discrete distribution, Eq. (15) we can write

$$
\begin{aligned}
h_{N,a}(n) &= \frac{P_{N,a}(n)}{\sum_{i=n}^{\infty} P_{N,a}(i)} \\
&= \frac{a^n(n-1)\ldots(n-N+1)}{\sum_{i=n}^{\infty} a^i(i-1)\ldots(i-N+1)} \\
&= \frac{1}{\sum_{i=1}^{\infty} a^{i-n}\frac{i-1}{n-1}\cdots\frac{i-N+1}{n-N+1}}.
\end{aligned}
\tag{20}
$$

To proceed further, we make the following change of variable:

$$
i - n = m.
\tag{21}
$$

With this change of variable, the hazard rate of the general, two-parameter NBD, Eq. (20), can be written as

$$
h_{N,a}^{-1} = \sum_{m=0}^{\infty} a^m \left(1 + \frac{m}{n-1}\right)\cdots\left(1 + \frac{m}{n-N+1}\right).
\tag{22}
$$

In the long-time limit, i.e., when n tends to infinity, we have

$$
\begin{aligned}
\lim_{n\to\infty} h_{N,a}^{-1} &= \sum_{m=0}^{\infty} a^m (1 \times 1 \times 1 \times \cdots \times 1) \\
&= \sum_{m=0}^{\infty} a^m = \frac{1}{1-a}.
\end{aligned}
\tag{23}
$$

So, in the general, two-parameter NBD the asymptotic limit of the hazard rate is:

$$
\lim_{n\to\infty} h_{N,a} = 1 - a.
\tag{24}
$$

REFERENCES

ABAIMOV, S.G., TURCOTTE, D.L. and RUNDLE, J.B. (2007), *Recurrence-time and frequency-slip statistics of slip events on the creeping section of the San Andreas fault in central California,* Geophys. J. Int. *170,* 1289–1299.

ABAIMOV, S.G., TURCOTTE, D.L., SHCHERBAKOV, R, RUNDLE, J.B. YAKOVLEV, G., GOLTZ, C., and NEWMAN, W.I. (2008), *Earthquakes: Recurrence and Interoccurrence Times,* Pure Appl. Geophys. *165,* 777–795.

BAKUN, W.H. (1988), *History of significant earthquakes in the Parkfield area,* Earthq. Volcano. *20,* 45–51.

BAKUN, W.H., and LINDH, A.G. (1985), *The Parkfield, California, earthquake prediction experiment,* Science *229,* 619–624.

ELLSWORTH, W.L., MATTHEWS, M.V., NADEAU, R.M., NISHENKO, S.P., REASENBERG, P.A., SIMPSON, R.W. (1999), A physically-based earthquake recurrence model for estimation of long-term earthquake earthquake probabilities. United States Geological Survey Open-File Report 99, 552pp.

FERRÁES, S. (2003), *The conditional probability of earthquake occurrence and the next large earthquake in Tokyo, Japan,* J. Seismol. *7,* 145–153.

FERRÁES, S. (2005), *A probabilistic prediction of the next strong earthquake in the Acapulco-San Marcos segment, Mexico,* Geofísica Internacional *44*(4), 347–353.

GOLTZ, C., TURCOTTE, D.L., ABAIMOV, S.G., NADEAU, R.M., UCHIDA, N., and MATSUZAWA, T. (2009), *Rescaled earthquake recurrence time statistics: application to microrepeaters*, Geophys. J. Int. *176*, 256–264.

GÓMEZ, J.B. and PACHECO, A.F. (2004), *The Minimalist Model of characteristic earthquakes as a useful tool for description of the recurrence of large earthquakes*, Bull. Seismol. Soc. Am. *94*, 1960–1967.

GONZÁLEZ, Á., GÓMEZ, J.B. and PACHECO, A.F. (2005), *The occupation of a box as a toy model for the seismic cycle of a fault*, Am. J. Phys. *73*, 946–952.

KEILIS-BORK D. V. and SOLOVIEV A. (2003), Nonlinear Dynamics of the Lithosphere and Earthquake Prediction, Springer Verlag, Berlin.

MATTHEWS, M.V., ELLSWORTH, W.L. and REASENBERG, P.A. (2002), *A Brownian model for recurrent earthquakes*, Bull. Seismol. Soc. Am. *92*, 2233–2250.

MICHAEL, A.J. and JONES, L.M. (1998), *Seismicity alert probabilities at Parkfield, California, revisited*, Bull. Seismol. Soc. Am. *88*, 117–130.

MICHAEL, A.J. (2005), *Viscoelasticity, postseismic slip, fault interactions, and the recurrence of large earthquakes*, Bull. Seismol. Soc. Am. *95*, 1594–1603.

MOLCHAN, G.M. (1997), *Earthquake prediction as a decision-making problem*, Pure Appl. Geophys. *149*(1), 233–247.

NEWMAN W. I. and TURCOTTE D.L. (1992), *A simple model for the earthquake cycle combining self-organized complexity with critical point behavior*, Nonlinear Process. Geophys. *9*, 453–61.

REID, H.F. (1910), The mechanics of the earthquake, In: The California Earthquake of April 18, 1906, Report of the State Earthquake Investigation Commission, Carnegie Institution, Washington, DC, Vol. 2, pp. 1–192.

RIKITAKE, T. (1974), *Probability of earthquake occurrence as estimated from crustal strain*, Tectonophysics *23*(3), 299–312.

SCHOLZ, C.H. (2002), The Mechanics of Earthquakes and Faulting, Cambridge University Press.

SORNETTE, D. and KNOPOFF, L. (1997). *The paradox of the expected time until the next earthquake*. Bull. Seismol. Soc. Am. *87*, 789–798.

SYKES, L.R., and MENKE, W. (2006), *Repeat Times of Large Earthquakes: Implications for Earthquake Mechanics and Long-Term Prediction*. Bull. Seismol. Soc. Am. *96*(5), 1569–1596.

TEJEDOR, A., GÓMEZ, J.B., and PACHECO, A.F. (2009), *Earthquake size-frequency statistics in a forest-fire model of individual faults*, Physical Review E *79*, 046102.

TEJEDOR, A., GÓMEZ, J.B., and PACHECO, A.F. (2012), *One-way Markov process approach to repeat times of large earthquakes in faults*, J. Stat. Phys. *149*(5), 951–963.

UTSU, T. (1984), *Estimation of parameters for recurrence models of earthquakes*, Bull. Earthq. Res. Inst. Univ. Tokyo *59*, 53–66.

VÁZQUEZ-PRADA, M., GONZÁLEZ, Á., GÓMEZ, J.B. and PACHECO, A.F. (2002), *A minimalist model of characteristic earthquakes*. Nonlinear. Process. Geophys. *9*, 513–519.

WORKING GROUP ON CALIFORNIA EARTHQUAKE PROBABILITIES (2003), Earthquake Probabilities in the San Francisco Bay Region: 2002–2031, United States Geological Survey Open-File Report 03-214, 234 p.

YAKOVLEV, G., TURCOTTE, D.L., RUNDLE, J.B., and RUNDLE, P.B. (2006), *Simulation-Based Distributions of Earthquake Recurrence Times on the San Andreas Fault System*, Bull. Seismol. Soc. Am. *96*(6), 1995–2007.

ZÖLLER, G., HAINZL, S., and HOLSCHNEIDER, M. (2008), *Recurrent Large Earthquakes in a Fault Region: What Can Be Inferred from Small and Intermediate Events?*, Bull. Seismol. Soc. Am. *98*, 2641–2651.

(Received January 14, 2014, revised May 19, 2014, accepted May 29, 2014, Published online June 26, 2014)

Reprinted from the journal

Pure Appl. Geophys. 172 (2015), 33–47
© 2014 Springer Basel
DOI 10.1007/s00024-014-0870-3

Pure and Applied Geophysics

Mathematics and Mars Exploration

M. P. Velasco,[1] D. Usero,[2] S. Jiménez,[3] C. Aguirre,[4] and L. Vázquez[5]

Abstract—In this study we consider modelization associated with study of solar radiation at the surface of Mars and the Martian atmosphere. In particular, we present elements concerning retrieval of the solar irradiance spectrum on the surface of Mars from data collected by arrays of photodiodes, such as those onboard the "Curiosity" MSL-rover and other missions currently under design. By using these techniques we are able to provide an approximate description of the expected measures. In this work we have also developed a new method of tomography-based signal analysis for detection of events in the Martian atmosphere boundary layer, such as dust devils. In general, this method enables detection of events that occur briefly in time and are localized in space. This tomographic method allows us to identify the presence of more dust devils than detected previously using the same data. Finally we show new scenarios of modelization through fractional differential equations associated with diffusion processes and nonlocal problems. Such approaches could be used to model complex Martian dynamics.

Key words: Mars, solar irradiance, retrieval, tomographic analysis, dust devils, fractional calculus, eractional differential equations, nonlinear differential equations, nonlocal models, solitary-wave solutions, numerical methods.

1. Introduction

The history of humanity is associated with technological evolution. Probably one of the next great steps will be a manned mission to Mars and to establish there a human colony. To achieve the above an international collaboration (MEIGA 2014) of scientists and engineers collaborating on the same ground is necessary.

The study of space defines an environment that is interdisciplinary, transdisciplinary, very international and with a natural link to the industry (Dehant *et al.* 2012; Preston and Dartnell 2014). In this contribution, we present some of the research activity carried out at the Martian Group Studies at the Universidad Complutense de Madrid. As an illustration of the mathematical problems that arise in the scientific definition of the instruments for the Mars exploration as well as in the modeling of the observed phenomena, we have the nonlocal problems either in/or space associated to the retrieval issues. This is the case for retrieval of ground solar radiation from information given by photodiodes (Sect. 2 for the ultraviolet radiation). In Sect. 3 we analyze some of the basic questions regarding Martian data mining. We have huge amounts of data (Viking, Pathfinder, Phoenix, MER and MSL Missions) readily available but spread over many files and not uniquely sampled. Some of the integral equations can be formulated in the framework of fractional calculus by using fractional derivatives. Fractional calculus provides a suitable instrument for analyzing possible interpolating dynamics between the properties and dynamics of the integer derivatives. A relevant reference case is the possible interpolations between the classical diffusion and wave equations through the fractional derivative in time. In this context, we present in Sect. 4 some remarks about fractional diffusion equations, while in Sect. 5 we show some examples of numerical simulation of simple fractional differential equations.

[1] Área de Matemáticas, Estadística e Investigación Operativa, Centro Universitario de la Defensa - IUMA, Zaragoza, Spain. E-mail: velascom@unizar.es

[2] Departamento de Matemática Aplicada, Facultad de Química, Universidad Complutense de Madrid, Madrid, Spain. E-mail: umdavid@mat.ucm.es

[3] Departamento de Matemática Aplicada, E.T.S.I. Telecomunicación, Universidad Politécnica de Madrid, Madrid, Spain. E-mail: s.jimenez@upm.es

[4] Departamento de Ingeniería Informática, Escuela Politécnica Superior, Universidad Autónoma de Madrid, Madrid, Spain. E-mail: carlos.aguirre@uam.es

[5] Departamento de Matemática Aplicada, Facultad de Informática, Universidad Complutense de Madrid, Madrid, Spain. E-mail: lvazquez@fdi.ucm.es

2. Estimation of the Irradiance Intensity on the Surface

2.1. Solar Irradiance Sensors

Solar radiation that reaches the surface of mars is not affected by a global dipole field or by a thick atmospheric layer as it is on earth. Although the intensity is lower due to a greater distance from the sun, the harmful effects of ultraviolet (UV) radiation to life as we know it can be a serious hindrance to human habitability in the future. Also, it is important to determine if earth-like life could be found on or near the surface under the first layers of the Martian soil. Radiation also affects the atmosphere and has a direct influence on the Martian climate.

For all these considerations, probes that have been sent to mars or are under consideration to be sent in the near future carry solar irradiance sensors (SIS). Due to payload restrictions, both on mass and dimension, those sensors are composed of an array of photodiodes that measure radiation intensity in different bands.

Primary information provided by the SIS is the intensity over a certain range of wavelengths reaching the device. Choosing, for instance, an UV or an Infrared (IR) band, the SIS can provide comparative information about evolution of the intensity for different measuring periods.

Also, information over different bands can provide a global picture of the irradiance spectrum for the whole region covered by the measures. In this section we present some models that can be used to achieve this.

Depending on the photodiode design some bands are quite selective around a specific wavelength while others span over larger wavelength values. For instance, the Rover Environmental Monitoring Station (REMS) suite of instruments (REMS 2014; Gómez-Elvira et al. 2012; Portal of the Mars Science Laboratory 2014), on board the Rover "Curiosity" presently on mars, carries an SIS composed of 6 photodiodes that cover different UV ranges (Table 1).

Not all the photodiode ranges are disjointed and in some cases they may have intersecting or overlapping ranges. This is the case for REMS where photodiodes 2 to 6 have intersecting ranges that partially cover the spectral range of photodiode 1.

The METNET probes (Arruego et al. 2010; METNET 2014) carry a solar irradiance sensor (METSIS) with an array of eleven different photodiodes. They cover the UV range, similar to REMS, but also the visible and the near infrared part of the spectrum with some specific bands devoted to obtaining information on dust related effects. The design is such that some are duplicated on the different faces of the instrument in the shape of a rectangular parallelepiped (Table 2).

A simplified version of METSIS, SISDREAMS (Esposito et al. 2013) that has the shape of a truncated tetrahedron, is designed to have three different photodiodes, one on the top face, the other two being duplicated on the three lateral faces of the instrument (Table 3).

Table 1

SIS for REMS

No.	Range (nm)
1	Global UV: 200–370
2	UV-A: 320–370
3	UV-B: 280–320
4	UV-C: 220–280
5	Ozone absorption: 230–290
6	Complementary range: 300–350

Table 2

METSIS

No.	Range (nm)
1	Total luminosity reference: 230–1200
2	UV-A: 315–400
3	UV-B: 280–315
4	Hartley band: 200–310
5	Huggins band: 300–345
6	Dust optical depth: 440
7	Dust optical depth: 600
8	IR: 700–1100
9	VIS: 400–700
10	UV MRO: 245–290
11	H_2O: 930–950

Table 3

SISDREAMS

No.	Range (nm)
1	UV-A: 315–400
2	IR: 700–1100
3	Total luminosity reference: 220–1100

2.2. Mathematical Model

Besides providing specific information on the irradiance received by each photodiode over its corresponding measuring range, the collection of data at a given time can be used to estimate the dependence of the irradiance on the wavelength, through an approximated function $I(\lambda)$.

We can consider that ideally a given photodiode measures all wavelength contributions from $\lambda_{\min}$ to $\lambda_{\max}$ and gives as output a measure M_{ideal} that can be expressed as an integral:

$$M_{\text{ideal}} = \int_{\lambda_{\min}}^{\lambda_{\max}} I(\lambda)\, d\lambda, \qquad (1)$$

where $I(\lambda)$ is the intensity corresponding to wavelength λ. The term "ideally" indicates that the actual situations are far more complex. But we will consider this as a first approximation and describe below how to take into account some realistic features of the actual problem.

In general, we can assume that we have ideal measures provided by n photodiodes M_j, with $j = 1, \ldots, n$, with corresponding wavelength ranges given by $[\alpha_j, \beta_j]$:

$$M_j = \int_{\alpha_j}^{\beta_j} I(\lambda)\, d\lambda, \quad j = 1, \ldots, n. \qquad (2)$$

As we have seen in the previous tables the ranges are not necessarily disjointed and some even overlap each other or are entirely cover by others.

We may describe our problem as an interpolation issue, on data corresponding to a function $F(\lambda)$ that is a primitive version of our goal function $I(\lambda)$. We will suppose subsequently that such a function exists. The idea is to construct F and by derivation, to obtain I.

Starting from the data we have

$$M_j = F(\beta_j) - F(\alpha_j), \quad j = 1, \ldots, n. \qquad (3)$$

We then construct a polynomial approximation. We can use in general some other basis of functions provided our approximation belongs to some suitable space in which we perform a Fourier series: for instance, in (JIMÉNEZ and VÁZQUEZ 2014) a general formulation is described.

We represent F as a polynomial of degree n, where the constant term is arbitrary; for instance, we choose the constant term to be zero:

$$F(\lambda) = \sum_{k=1}^{n} a_k \lambda^k. \qquad (4)$$

Once the derivation is preformed we have

$$I(\lambda) = \sum_{k=1}^{n} k a_k \lambda^{k-1} = \sum_{\ell=0}^{n-1} (\ell - 1) a_{\ell-1} \lambda^{\ell}, \qquad (5)$$

a polynomial of degree $n - 1$.

The coefficients a_k are the solutions of a set of simultaneous linear equations given by (3):

$$\forall j = 1, \ldots, n, \quad M_j = \sum_{k=1}^{n} a_k \left[\alpha_j{}^k - \beta_j{}^k \right]. \qquad (6)$$

This approach was used for REMS in VÁZQUEZ *et al.* (2007). This is similar to a van der Monde matrix and the corresponding interpolation matrix can have a large condition number. Other approaches to minimizing the effect this may have on errors can be considered (see JIMÉNEZ and VÁZQUEZ 2014).

As a first approximation the spectrum of radiation emitted by the Sun can be estimated as that of a black body at a temperature about 5,755 K. We have applied this model to the configurations corresponding to the photodiodes of both METSIS and SISDREAMS and reconstructed the expected irradiance spectrum of the black body radiation. We represent the results for METSIS in Fig. 1.

The case for SISDREAMS is different: the three measures must fit a parabola, which is too crude of a representation for the general shape of the black body radiation. The idea is to include an extra measure that would correspond to a "virtual sensor." That measure is an additional variable and allows to fit a third order polynomial provided an extra condition or equation is also added. We

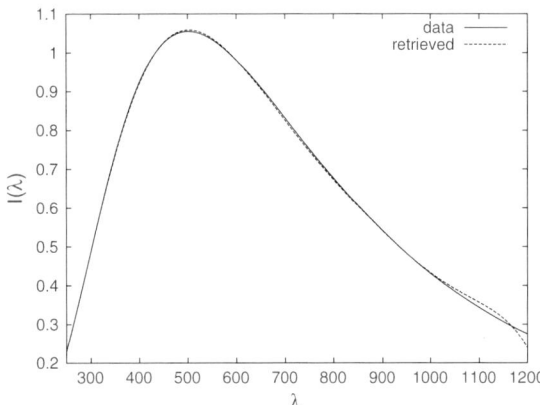

Figure 1
Approximation of the spectrum of the black body using METSIS photodiodes

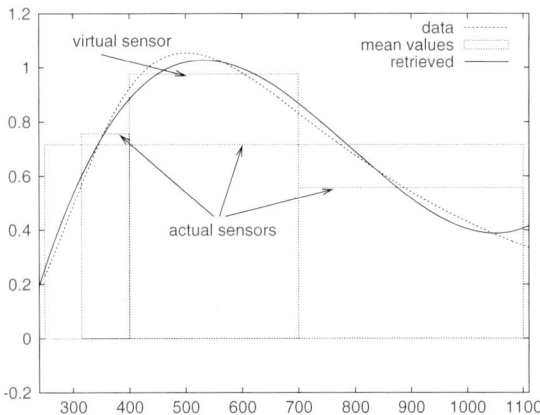

Figure 2
Approximation of the spectrum of the black body using SIS-DREAMS photodiodes and a virtual one

$$M = \int_{\lambda\min}^{\lambda\max} I(\lambda) f(\lambda)\, d\lambda, \qquad (7)$$

instead of (1).

The values of the filter function $f(\lambda)$ can be tabulated by experimental measures on the laboratory with sufficient precision. In this sense the presence of f would not suppose a significant difference in the approach, but some other factors have to be considered: dust will be deposited erratically on the surfaces of the photodiodes (sometimes the deposition increases, but at other times winds will clear the device) and also the response of the photodiodes can be altered due to exposure to the radiation itself.

To adjust for these changing conditions we consider an iterative process for approximating the actual values. We start by instead of (7) considering the values M_j^* given by:

$$M_j^*(0) = \int_{\alpha_j}^{\beta_j} I(\lambda) f_j(\lambda) R_j(\lambda)\, d\lambda, \quad j = 1,\ldots,n, \quad (8)$$

where $R_{j(\lambda)}$ is some positive function that represents the effect of the alterations of the filter function f_j acting on the photodiode j in such a way that efficiency of the filtered photodiodes can only decay: this supposes that the values of R_j belong to $[0,1]$. To estimate the global effect functions R_j over each photodiode range we define the average $\bar{R}_j$ as

$$\bar{R}_j(0) = \frac{M_j^*}{M_j}, \quad j = 1,\ldots,n. \qquad (9)$$

We now suppose that the actual measures given by the photodiodes correspond to M_j^* and we build the corresponding intensity, say, $I^0(\lambda)$. We then substitute in (7) and we obtain from (9) a first approximation $\bar{R}_j^0$. We can then rectify I^0 and define $I^1 = I^0/\bar{R}_j^0$. This process can then be iterated with further values I^p, $p = 2,\ldots$, until convergence. It was successfully applied to the REMS photodiode configuration in ZORZANO et al. (2009).

A second point that should, realistically, be taken into account is that the emission from the sun is really not that of a black body and has some irregularities. Also, the radiation is altered as it travels through the Martian atmosphere. Radiative transfer techniques give

have used as the new condition to minimize the integral of the third degree polynomial we obtain. We represent the results in Fig. 2.

2.3. More Realistic Settings

As a first correction, we may consider that real photodiodes do not provide the same response to all wavelengths inside their range. Also, this response depends on the incidence angle. We may represent this effect as if a filtering function was applied, in such a way that some wavelengths have more impact on the measure. In this sense, a more realistic representation would be:

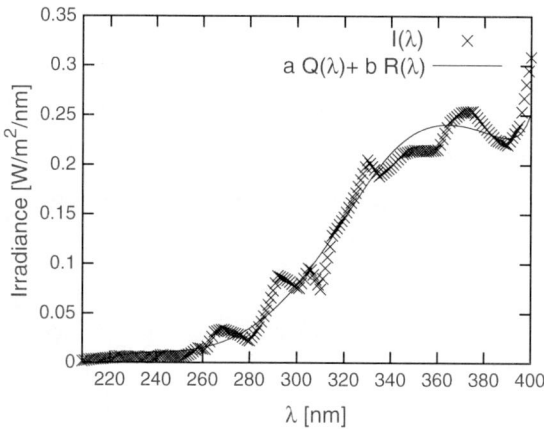

Figure 3
Approximation in the UV region minimizing the norm of a linear
combination of polynomials and a rational function

information on the different possible scenarios and it is possible to adapt the interpolation techniques to attempt matching the measures of some of these cases.

For instance, in VÁZQUEZ *et al.* (2007) a combination of approximations by rational functions and polynomials was used to reproduce the expected curve in the UV region from 220 to 400 nm. Using a linear combination of the polynomial $Q(\lambda)$ given by the right hand-side of (5) and the rational function approximation $R(\lambda)$ and choosing the coefficients in a way to minimize the total integral a better qualitative fit was achieved. The result can be seen in Fig. 3.

3. Detection of Dust-Devils in the Mars Atmosphere by Means of Tomographic Signal Analysis

Dust devils are convective vortices generated by surface heating, thereby generating convective plumes of rising air with internal pressure variation. Some of these vortices obtain horizontal wind speeds large enough for dust particles to be lifted off the surface and into the vortex and thus becoming dust devils. Dust devils are presented as sudden drops for a short range of time, of about 3 % in the value of the measured pressure data.

There have been several attempts at detection of dust devils on the martian atmosphere based mainly on checking several ad-hoc conditions in the measured data (SMITH *et al.* 2008) or by means of FPGAs (DE LUCAS 2012).

For this study the pressure data used are those obtained directly by NASA's Phoenix spacecraft, operational May 25, 2008 through May 2010. The objective is to find pressure drops more or less pronounced but well-localized in time and with a very short duration. Special attention is devoted to non-commutative tomography that provides very robust and strictly positive probability densities in the presence of noise and also provides filtering, resulting in separate signal components.

3.1. Tomography and Adapted Tomography for Signal Analysis

Recently a new kind of bilinear transforms, called tomograms, have been proposed. Tomograms are strictly positive probability densities and provide a full characterization of a signal. Tomograms are obtained by projecting a given signal $f(t)$ over the eigenvector set of a linear operator $B(\alpha)$. One of the most typical tomograms is the time-frequency tomogram (a particular case of the Radon transform). Let's consider the operator

$$B(\mu, \vartheta) = \mu t + \vartheta \omega = \mu t + \vartheta i \frac{\partial}{\partial x} \qquad (10)$$

where t is the time operator and ω is the frequency operator. Using the previously-defined operator the time–frequency operator can be obtained (MANKO *et al.* 2001); an explicit expression for the tomogram may be found at MANKO and VILELA-MENDES (1999). Even when time-frequency tomograms have proven to be a powerful tool for filtering or signal component detection, there are situations where the the signal components we are interested in are not well represented by their frequency spectrum, as is the case with a dust devil inside the Mars atmospheric pressure data. In order to detect appearances of dust devils, represented as brief and sudden drops in atmospheric pressure level, we have developed a tomographic technique to fully characterize such behavior.

3.1.1 Adapted Data Tomography

Consider a matrix $U \in M_{k \times N}$ where each row of U is a typical signal that contains the component we are interested in sampled at intervals δt. This set of typical signals can be obtained from data or can be artificially generated. Now construct the square matrix $A = U^T U$. The diagonalization U of provides k non-zero eigenvalues $\alpha_1, \ldots, \alpha_k$ and its corresponding orthogonal N-dimensional eigenvectors $\{\Phi_1, \ldots, \Phi_k\}$ with $\Phi_j \in R^N$. The linear operator S constructed from the set of typical signals is

$$S = \sum_{i=1}^{k} \alpha_i \Phi_i \Phi_i^T \quad \text{where} \quad S \in M_{N \times N} \quad (11)$$

For the tomogram consider now the linear operator

$$B(\mu, \vartheta) = \mu t + \vartheta S = \mu I \delta t + \vartheta \sum_{i=1}^{k} \alpha_i \Phi_i \Phi_i^T \quad (12)$$

where I is the identity matrix. The tomogram adapted to the operator pair (t, S) is obtained from the projection of the signal on the eigenvectors of $B(\mu, \vartheta)$.

3.2. Dust Devils Detection by Means of Adapted Data Tomography

For detection of dust devils we have generated a set of 278 signals that resemble the shape that a dust devil produces on real data, that is, a sudden drop of about 3 % from the baseline and different time extents that range from 60 to 80 time units. The upper right panel of Fig. 1 presents several of these typical signals. Some of the signals have been shifted up or down for representation purposes. Then a tomogram is constructed for 20 different values of θ at intervals $\Delta\theta = \pi/40$.

A contour plot of the first 999 coefficients of the tomogram is shown in the lower left panel of Fig. 4. Coefficient $n = 1,000$ corresponds to the biggest eigenvalue (and its corresponding eigenvector). This eigenvector contains most of the signal energy and is several orders of magnitude bigger than any other coefficient; so, again, for representation purposes, the value of the coefficient has not been plotted at the tomogram. Again, by direct inspection, we observe that, beside the coefficient $n = 1,000$, the strongest component is concentrated close to the value $n = 400$. The lower right panel in Fig. 4 shows projection on the eigenvectors 340–450 and 1,000 at $\theta = 19\pi/40$. One sees that the pressure drop

produced by the dust devil is very well reconstructed and separated from any other components present in the signal such as noise or lower pressure variations.

4. New Scenarios of Fractional Modeling

Fractional calculus (SAMKO et al. 1993) offers a very suggestive and stimulating scenario where we have convergence of deep and fundamental mathematical questions, development of appropriate numerical algorithms, as well as application to modelizations in different frameworks. So, fractional calculus has many applications in different areas, as it is cited in MAGIN (2006):

"The purpose of this book is to explore the behavior of biological systems from the perspective of fractional calculus. Fractional calculus, integration and differentiation of an arbitrary or fractional order, provides new tools that expand the descriptive power of calculus beyond the familiar integer-order concepts of rates of change and area under a curve."

"Fractional calculus adds new functional relationships and new functions to the familiar family of exponentials and sinusoids that arise in the realm of ordinary linear differential equations."

Fractals and fractional calculus create intermediate-order parameters: dimensions, integration and derivatives of arbitrary order. This has been studied in the literature broadly (ROCCO and WEST 1999; WEST et al. 2003), and it has enabled better modeling in different applications.

From a mathematical point of view modelization of the long-range dependence and systems with memory are associated with integrodifferential equations in a broad sense. On the other hand, in many cases such integrodifferential equations can be understood as fractional differential equations and they can be studied in the fractional calculus framework.

For example, we can consider different contexts of classical physics where equations are supported by similar laws:

- Hooke's law: $F(t) = kx(t)$
- Newton's fluid law: $F(t) = k\frac{dx}{dt}(t)$
- Newton's second law: $F(t) = k\frac{d^2x}{dt^2}(t)$
- Other possible fractional context: $F(t) = k\frac{d^\gamma x}{dt^\gamma}(t)$

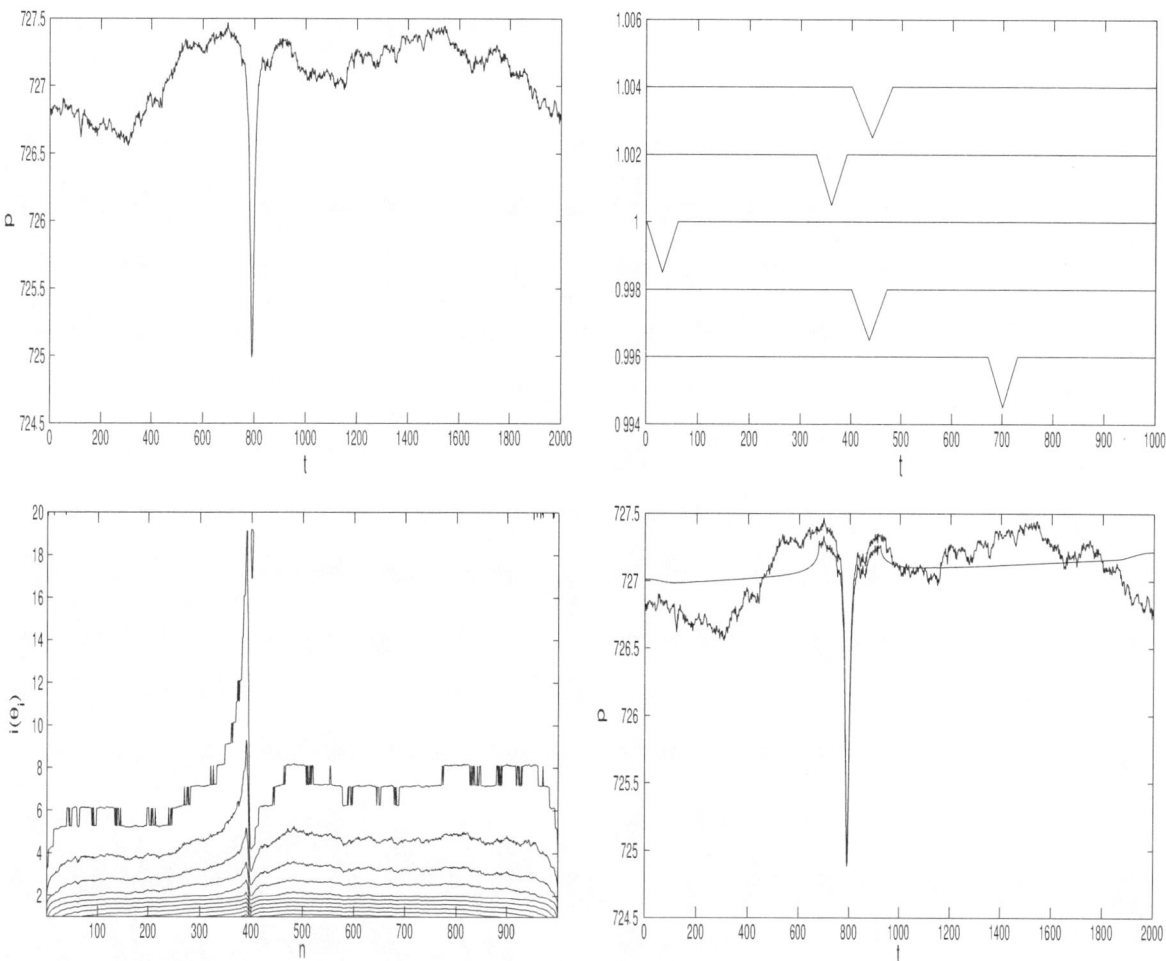

Figure 4

Signal, typical signals, the tomogram (coefs 1-999) and projection on the eigenvectors 340–450 and 1,000 at $\theta = 19\pi/40$

Table 4

Diffusion processes

Law	Darcy $\vec{q} = -K\,\overrightarrow{Grad}\,h$	Fourier $\vec{Q} = -\kappa\,\overrightarrow{Grad}\,T$	Fick $\vec{f} = -D\,\overrightarrow{Grad}\,C$	Ohm $\vec{j} = -\sigma\,\overrightarrow{Grad}\,V$
Flux	Subterranean Water: q	Heat: Q	Solute: f	Charge: j
Potential	Hydrostatic Charge: h	Temperature: T	Concentration: C	Voltage: V
Medium's property	Hydraulic conductivity: K	Thermal conductivity: κ	Diffusion coefficient: D	Electrical conductivity: σ

Also, time-fractional derivatives have been introduced to interpolate the diffusion and wave equations (EL-SAYED 1996) and to model anomalous diffusion (PIERANTOZZI and VÁZQUEZ 2005, 2006; VÁZQUEZ 2003; VÁZQUEZ and VILELA-MENDES 2003). In this sense there exist many contexts where the diffusion process appears to be associated with the same basic equation

39

$$\frac{\partial u}{\partial t} = \frac{\partial^2 u}{\partial x^2}, \qquad (13)$$

as it is shown in Table 4.

The above equation can be generalized through fractional operators and this allows one to obtain a natural interpolation between equations:

$$\text{Diffusion equation (parabolic)} : \frac{\partial u}{\partial t} = \frac{\partial^2 u}{\partial x^2} \qquad (14)$$

$$\text{Interpolation} : \frac{\partial^\alpha u}{\partial t^\alpha} = \frac{\partial^2 u}{\partial x^2} \qquad (15)$$

$$\text{Waves equation (hiperbolic)} : \frac{\partial^2 u}{\partial t^2} = \frac{\partial^2 u}{\partial x^2} \qquad (16)$$

Associated with this, other fractional context is the use of dirac fractional equations, following this scheme:

$$A\frac{\partial \psi}{\partial t} + B\frac{\partial \psi}{\partial x} = 0 \quad\xrightarrow{\begin{array}{c} A\frac{\partial^\alpha \psi}{\partial t^\alpha} + B\frac{\partial \psi}{\partial x} = 0 \\ \psi = \begin{pmatrix} \varphi \\ \xi \end{pmatrix} \end{array}}\quad A\frac{\partial^{1/2} \psi}{\partial t^{1/2}} + B\frac{\partial \psi}{\partial x} = 0$$

$$A^2 = I$$

$$B^2 = I$$

$$\{A, B\} = 0$$

$$\frac{\partial^2 u}{\partial t^2} - \frac{\partial^2 u}{\partial x^2} = 0 \quad\xrightarrow{\begin{array}{c} \gamma = 2\alpha \\ \frac{\partial^\gamma u}{\partial t^\gamma} - \frac{\partial^2 u}{\partial x^2} = 0 \end{array}}\quad \frac{\partial u}{\partial t} - \frac{\partial^2 u}{\partial x^2} = 0$$

Then, the equation

$$A\frac{\partial^{1/2} \psi}{\partial t^{1/2}} + B\frac{\partial \psi}{\partial x} = 0 \qquad (17)$$

can be explained as the description of two coupled diffusion processes or one diffusion process with two internal degrees of freedom. In this equation both components φ and ξ satisfy the standard diffusion equation and they are referred to as *difunors*, analogous to the *spinors* of quantum mechanic. This provides another form to study interpolation between the hyperbolic operator of the waves equation and the

parabolic operator of classical diffusion. By following the representation of Pauli's Algebra for A and B we have a system of coupled or no-coupled equations

$$A_1 = \begin{pmatrix} 0 & 1 \\ 1 & 0 \end{pmatrix} \quad B_1 = \begin{pmatrix} 0 & 1 \\ -1 & 0 \end{pmatrix}$$

$$\implies \begin{cases} \partial_t^\alpha \varphi = \varphi \\ \partial_t^\alpha \xi = -\xi \end{cases} \qquad (18)$$

$$A_2 = \begin{pmatrix} 1 & 0 \\ 0 & -1 \end{pmatrix} \quad B_2 = \begin{pmatrix} 0 & 1 \\ -1 & 0 \end{pmatrix}$$

$$\implies \begin{cases} \partial_t^\alpha \varphi = -\varphi \\ \partial_t^\alpha \xi = -\xi \end{cases} \qquad (19)$$

$$A\frac{\partial^\alpha \psi}{\partial t^\alpha} + B\frac{\partial \psi}{\partial x} = 0 \xrightarrow{\ \gamma = 2\alpha\ } \frac{\partial^\gamma u}{\partial t^\gamma} - \frac{\partial^2 u}{\partial x^2} = 0$$

From study of time inversion $(t \rightarrow -t)$ we have:

- For $\alpha = 1$ dirac and waves equations are invariant by time inversion.
- For $\alpha = 1/2$ the classical diffusion equation and its square root are not invariant by time inversion.
- Interpolation for: $0 < \alpha < 1$. The invariance by time inversion is satisfied by

  – Dirac fractional equation: $\alpha = \frac{1}{3}, \frac{1}{5}, \frac{1}{7}, ..., \frac{3}{5}, \frac{3}{7}, \frac{3}{9}, ..., \frac{5}{7}, \frac{5}{9}, \frac{5}{11}, ...$
  – Diffusion fractional equation: $\alpha = \frac{1}{3}, \frac{2}{3}, \frac{1}{5}, \frac{2}{5}, \frac{3}{5}, \frac{4}{5}, \frac{1}{7}, \frac{2}{7}, ..., \frac{6}{7}, \frac{1}{9}, ...$

From the study of the space-time inversion ($x \to -x$, $t \to -t$) we observe that both equations are invariant by space inversion and for the interpolation $0 < \alpha < 1$ the invariance by space-time inversion is satisfied for the same values of α in both equations:

$$\alpha = \frac{1}{3}, \frac{2}{3}, \frac{1}{5}, \frac{2}{5}, \frac{3}{5}, \frac{4}{5}, \frac{1}{7}, \frac{2}{7}, \dots, \frac{6}{7}, \frac{1}{9}, \dots$$

Dirac fractional equation is not invariant by time translation because of the non-local character of the time fractional derivative.

Other fractional differential equations are obtained by considering the cube root of the waves and diffusion equations:

$$\text{Waves equation} : P\partial_t^{2/3}\varphi + Q\partial_x^{2/3}\varphi = 0 \qquad (20)$$

$$\text{Diffusion Equation} : P\partial_t^{1/3}\varphi + Q\partial_x^{2/3}\varphi = 0 \qquad (21)$$

where

$$P^3 = I \qquad Q^3 = -I \qquad PPQ + PQP + QPP = 0$$
$$QQP + QPQ + PQQ = 0$$
$$(22)$$

A possible development is in terms of the 3×3 matrix associated with Sylvester algebra:

$$P = \begin{pmatrix} 0 & 0 & 1 \\ \omega^2 & 0 & 0 \\ 0 & \omega & 0 \end{pmatrix} \qquad Q = \Omega \begin{pmatrix} 0 & 0 & 1 \\ \omega & 0 & 0 \\ 0 & \omega^2 & 0 \end{pmatrix} \qquad (23)$$

where ω is a cube root of unity and Ω is a cube root of negative unity. In this case φ has three components.

As an example of related mathematical problems we can consider the general Cauchy problem in the space $LF = L(R^+) \times F(R)$ of functions whose Laplace and Fourier transforms exist.

$$^C D_t^\alpha u(t,x) - \lambda^L D_x^\beta u(t,x) = 0, \quad t > 0, \qquad (24)$$
$$x \in \mathbb{R}, \quad 0 < \alpha \le 1, \quad \beta > 0$$

$$\lim_{x \to \pm\infty} u(t,x) = 0, \quad u(0+,x) = g(x) \qquad (25)$$

where $^C D_t^\alpha$ is the Caputo fractional partial derivative that is defined as

$$D_t^\alpha u(t,x) = {}^C D_t^\alpha u(t,x) = \frac{1}{\Gamma(1-\alpha)} \int_0^t \frac{u_\tau(\tau,x)}{(t-\tau)^\alpha} d\tau \qquad (26)$$

and where D_x^β is the Liouville fractional partial derivative

$$D_x^\beta u(t,x) = {}^L D_x^\beta u(t,x) = \frac{1}{\Gamma(m-\beta)} \frac{\partial^m}{\partial x^m}$$
$$\int_{-\infty}^x \frac{u(t,z)}{(x-z)^{\beta-m+1}} dz$$
$$(27)$$

with $m = [\beta]$.

The solution of the Cauchy problem is:

$$u(t,x) = \frac{1}{2\pi} \int_{-\infty}^\infty G(k) E_\alpha(\lambda(-ik)^\beta t^\alpha) e^{-ikx} dk \qquad (28)$$

where $G(k)$ is the Fourier transform of $g(x)$ and E_α is the Mittag-Leffler function on the complex plane, defined as

$$E_\alpha(z) = \sum_{j=0}^\infty \frac{z^j}{\Gamma(\alpha j + 1)} \qquad (29)$$

For example, for $\beta = 1$ and $g(x) = e^{-\mu|x|}$, $\mu > 0$:

$$u(t,x) = e^{-\mu|x|} E_\alpha(-\mu\lambda t^\alpha) \qquad (30)$$

and the moments of the fundamental solution ($g(x) = \delta(x)$, $G(k) = 1$) for the case $\beta = 1$ are obtained as

$$\langle x^n \rangle = \int_{-\infty}^\infty x^n u(t,x) dx = (-\lambda t^\alpha)^n \frac{\Gamma(n+1)}{\Gamma(\alpha n + 1)}, \qquad (31)$$
$$n = 0, 1, 2, \dots$$

This last relation leads to one think that possible application of this kind of fractional equations could be modeling of movement and absorption properties of dust on the Martian atmosphere. So a part of incident energy on the atmosphere is scattered by this dust and it is observed that the coefficient of molecular scattering τ is a function of the wave-length of the radiation (ANGSTROM 1929; CÓRDOBA-JABONERO and VÁZQUEZ 2003), in the form:

$$\tau = \frac{\beta}{\lambda^\alpha} \qquad (32)$$

where α and β are characteristic parameters of the Martian dust. In this sense it is interesting to analyze what fractional differential equations could be related to this Angstrom exponent.

5. Numerical Integration of Nonlocal and Fractional Equations

The time-fractional derivatives, first introduced in the constitutive equations of linear viscoelasticity (CAPUTO and MAINARDI 1971), have been considered in linear mechanical problems where they lead to fractional oscillation equations (GORENFLO et al. 1997). The effect of a linear term with a fractional derivative is similar to a damping force, as is also the case in Hamilton's nonlinear equations (SEREDINSKA and HANYGA 2000).

In this section we present two different non-local problems that are described in terms of fractional derivatives. However, both problems are different from each other. The first one uses the Caputo fractional time derivative defined below. This derivative is non-local since its value depends also on the previous values. The second problem uses the Riesz fractional operator. This operator is defined in the Fourier space and thus uses all the values of the system at present time.

5.1. Fractional Hamilton–Jacobi models

We consider the Hamiltonian for a one-dimensional system with unit mass defined by $H = \frac{1}{2}p^2 + V(x)$ so that the equations of motions are

$$\dot{x} = \frac{\partial H}{\partial p} = p, \qquad \dot{p} = -\frac{\partial H}{\partial x} = -V'(x) \qquad (33)$$

This is equivalent to the second order equation $\ddot{x} + V'(x) = 0$. We generalize here the first order system by replacing the first order time derivative with the Caputo fractional derivative ${}^{C}D_t^{\alpha}$ where $0 < \alpha \leq 1$ namely

$$\begin{cases} {}^{C}D_t^{\alpha}x = p, \\ {}^{C}D_t^{\alpha}p = -V'(x). \end{cases} \qquad (34)$$

We recall here that the Riemann-Liouville time-fractional integral of order α is defined by

$$ {}^{RL}I_t^{\alpha}f(t) = \frac{1}{\Gamma(\alpha)} \int_0^t (t - \tau)^{\alpha - 1}f(\tau)d\tau \qquad (35)$$

where $J_0 = I$ is the identity, $J_1 = J$ is the ordinary integral and $J_{\alpha}J_{\beta} = J_{\alpha+\beta}$. The derivatives of order α are defined by

$$ {}^{RL}D_t^{\alpha} = D {}^{RL}I_t^{1-\alpha}, \qquad {}^{C}D_t^{\alpha} = {}^{RL}I_t^{1-\alpha}D,$$

$$(\,{}^{RL}D_t^{\alpha} - {}^{C}D_t^{\alpha})f(t) = \frac{f(0)\,t^{-\alpha}}{\Gamma(1 - \alpha)} \qquad (36)$$

where ${}^{RL}D_t^{\alpha}$ is the Riemann–Liouville fractional derivative of order α $(0 < \alpha \leq 1)$ and ${}^{C}D_t^{\alpha}$ is the Caputo fractional derivative of order α.

We notice that $D{}^{RL}I_t^{\alpha}$ is the left-inverse of ${}^{RL}I_t^{1-\alpha}$. Then the fractional differential equation ${}^{C}D_t^{\alpha}x = f(t)$ becomes $Dx = D{}^{RL}I_t^{\alpha}f(t)$ which is equivalent to $x(t) = x(0) + {}^{RL}I_t^{\alpha}f(t)$. This is simply a fractional generalization of the integral form of the differential equation.

$$\begin{aligned} Dx = f(t) &\longleftrightarrow x(t) = x(0) + {}^{RL}I_t^{1}f(t) \\ {}^{C}D_t^{\alpha}x = f(t) &\longleftrightarrow x(t) = x(0) + {}^{RL}I_t^{\alpha}f(t) \end{aligned} \qquad (37)$$

As a consequence our system (34) with an initial condition $x(0) = x_0$, $p(0) = p_0$ is equivalent to

$$\begin{cases} x(t) = x(0) + {}^{RL}I_t^{\alpha}p(t) \\ p(t) = p(0) - {}^{RL}I_t^{\alpha}V'(x(t)) \end{cases} \qquad (38)$$

For numeric solution of system (38) we developed a map (see TURCHETTI et al. 2002)

$$\begin{cases} p_n = p_0 - \dfrac{(\Delta t)^{\alpha}}{\Gamma(\alpha + 1)} \sum_{k=0}^{n-1} V'(x_k)[(n - k)^{\alpha} - (n - k - 1)^{\alpha}] \\ x_n = x_0 + \dfrac{(\Delta t)^{\alpha}}{\Gamma(\alpha + 1)} \sum_{k=0}^{n-1} p_{k+1}[(n - k)^{\alpha} - (n - k - 1)^{\alpha}] \end{cases}$$

$$(39)$$

When $\alpha = 1$ this is equivalent to a second order symplectic integrator $p_n = p_{n-1} - \Delta t\, V'(x_{n-1})$, $x_n = x_{n-1} + \Delta t\, p_n$. For $\alpha = 1$ it provides an orbit (x_n, p_n) approaching the exact orbit at $t = n\Delta t$ when $\Delta t \to 0$. If in the second equation p_{k+1} is replaced by p_k one recovers the second order Euler scheme which is not symplectic as $\Delta t \to 0$. As a consequence of the memory kernel in ${}^{RL}I_t^{\alpha}$ the mapping is infinitely dimensional since the orbit at step n depends on all the previous states up to the initial one. The computational complexity of the orbit up to (x_n, p_n) is of order n^2 whereas it is of order n for $\alpha = 1$.

This map has been tested by taking $\alpha = 1$ with standard models and using different potentials whose solutions are known. In particular, for the harmonic oscillator and initial conditions $x_0 = 1$, $p_0 = 0$ and $\Delta t = 0.01$, the solution has been compared with

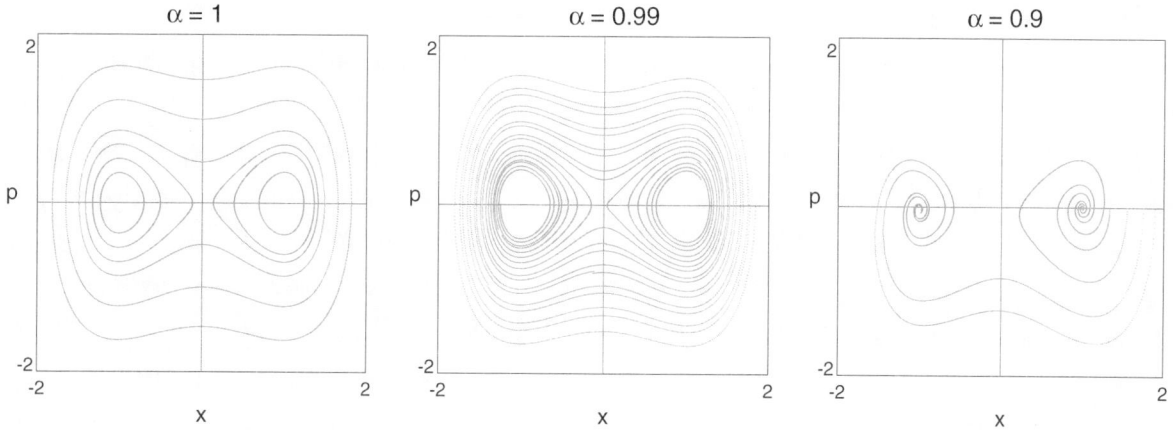

Figure 5
Phase portrait for the nonlinear oscillator with double well potential $V = \frac{1}{4}x^4 - \frac{1}{2}x^2$ (corresponding to the time interval $0 \le t \le 20$)

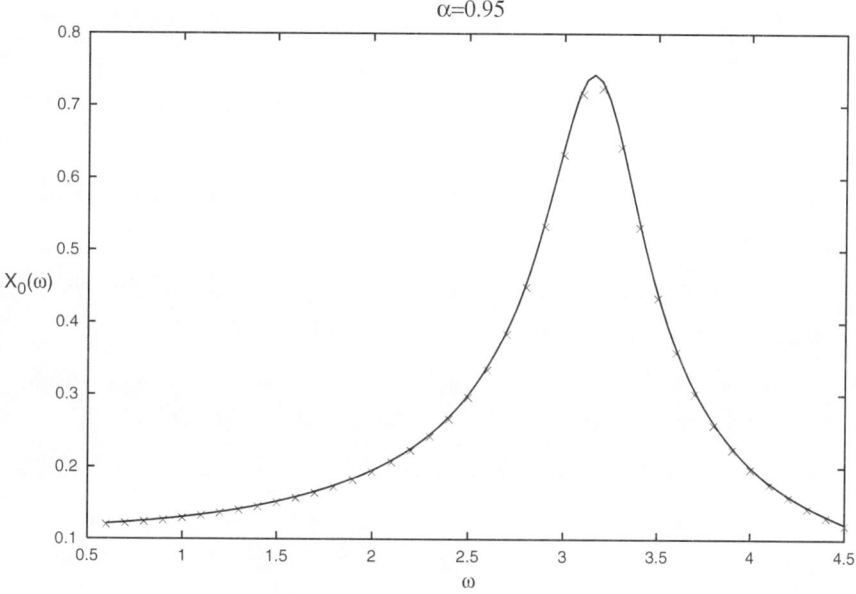

Figure 6
Plot of the amplitude of the limit cycle of $x(t)$ for different forcing frequencies ω versus the theoretical amplitude (41) with $\alpha = 0.95$

$x(t) = \cos(t)$, $p(t) = \sin(t)$ with a rate of error smaller than 0.5 % after 10,000 steps.

The special form of map (39) makes it good for obtaining numerical solutions to fractional generalized Hamiltonian problems with a potential $V(x)$. In USERO (2014) it was used to simulate standard academic cases like free particle motion ($V = 0$) and a uniformly accelerated particle ($V = kx$). In TURCHETTI *et al.* (2002) with the simple oscillator $V = \frac{1}{2}\omega^2 x^2$ the double well

potential $V = \frac{1}{4}x^4 - \frac{1}{2}x^2$ (see Fig. 5) and the pendulum $V = \cos(x)$. These former cases are significant since they are non-linear and an explicit solution cannot be found.

The Riemann–Liouville time fractional problem could be integrated in this way by using relation (36), substituting the Riemann–Liouville derivative with Caputo's and applying a similar mapping.

The same map can be generalized for studying non-homogeneous systems. In USERO and VÁZQUEZ

(2003) the same scheme is used with an external force to simulate a forced-damped oscillator

$$\begin{cases} {}^{C}D_t^{\alpha}x = p, \\ {}^{C}D_t^{\alpha}p = -\omega_0^2 x + f(t). \end{cases} \qquad (40)$$

being $f(t)$ an external force. Scheme (39) must be modified in order to include this external force $f(t)$. This change is not a problem and the same structure can be used and simply introduce an extra force term $f_n = f(n\Delta t)$ in the first equation of (39).

An harmonic forcing $(f(t) = A_0 \cos(\omega t))$ induces the system to evolve to a limit cycle as also occurs in the classic $(\alpha = 1)$ case with the forced-damped oscillator. Varying the forcing frequency ω a resonance motion is reproduced with amplitude

$$A_{res} = \frac{A_0}{2\omega_0} \left| \frac{1}{(i\omega)^{\alpha} - i\omega} \right| \qquad (41)$$

(see Fig. 6).

5.2. Nonlinear Nonlocal Equations

In the present section we present preliminary numerical estimations of the singularities of solutions of equation

$$\mathcal{L}u - u + u^p = 0. \qquad (42)$$

where $u = u(x)$ is a function of a single variable x and the integral operator $\mathcal{L}$ can be expressed in Fourier space as

$$\widehat{\mathcal{L}\sqcap}(\lambda) = D|\lambda|^{\alpha}\hat{u}(\lambda), \qquad (43)$$

being $1 \leq \alpha \leq 2$. This definition of the operator $\mathcal{L}$ is called the Riesz fractional operator and it is one of the possible generalizations of the derivative operator (ALFIMOV et al. 2000).

Equation (43) is used in study of traveling wave solutions of many different physical systems (ALFIMOV et al. 2000; VÁZQUEZ 2005). In LI and BONA (1996), BONA et al. (1976) and BONA (1997) the existence of periodic and localized solutions for $\alpha \geq 1$ is shown and the asymptotic behavior and the analyticity of such localized solutions is studied. The operator $k(\lambda)$ defined by its Fourier transform as

$$k(\lambda) = (1 + |\lambda|^{\alpha})^{-1} \qquad (44)$$

determines the asymptotic behavior of the solution. For $\alpha = 2$ we have the nonlinear wave equation with exponential asymptotic behavior. For $\alpha = 1$ we have an equation resulting from studying traveling waves of the generalized Benjamin–Ono equation which shows algebraic asymptotic behavior.

For $\alpha = 1$ Eq. (42) becomes

$$\mathcal{H}u_x u + u^p = 0. \qquad (45)$$

where the subscript x denotes derivation with respect to x and $\mathcal{H}$ is a nonlocal operator called the Hilbert transform and is defined as

$$\mathcal{H}u(x) = \frac{1}{\pi} p.v. \int\limits_{-\infty}^{+\infty} \frac{u(x')}{x' - x} dx'. \qquad (46)$$

Taking into account that Fourier transforms are $\widehat{\partial_x u}(\lambda) = -i\lambda\hat{u}(\lambda)$ for a derivative operator and $\widehat{\mathcal{H}u}(\lambda) = -i\,\text{sign}(\lambda)\hat{u}(\lambda)$ for the Hilbert transform, it is possible to combine both in order to obtain

$$\widehat{\mathcal{H}u_x}(\lambda) = -|\lambda|\hat{u}(\lambda). \qquad (47)$$

This property of the integral operator makes the dispersion relation form the linear case derived from Eq. (45) $\omega = |\lambda|$ instead of the characteristic harmonic dispersion relation $\omega = \lambda^2$. Such a dispersion relation is observed in the study of certain metallic lattices, atomic and molecular (KITTEL 1966; ISHIMORI 1982). In that last work is proposed an equation for the study of lattices of particles with long range interactions $1/x^2$ for which traveling wave solutions can be described by Eq. (45).

For nonlinear power $p = 2$ the equation can also be obtained as the equation for a solitary waves solution for the Benjamin-Ono equation. This concrete case solution is known and has a pole in a complex plane. This is not the case for $p \neq 2$.

Localized solutions of Eqs. (42) and (45) can be obtained by using numerical techniques. Equation (45) has been studied in ALFIMOV et al. (2000) and numerical solutions for different values of p have been obtained using a numerical Fourier technique. The numerical algorithm can be extended to the fractional case of (42) (see TURCHETTI et al. 2002; USERO 2014; VÁZQUEZ 2005; VÁZQUEZ and USERO 2005).

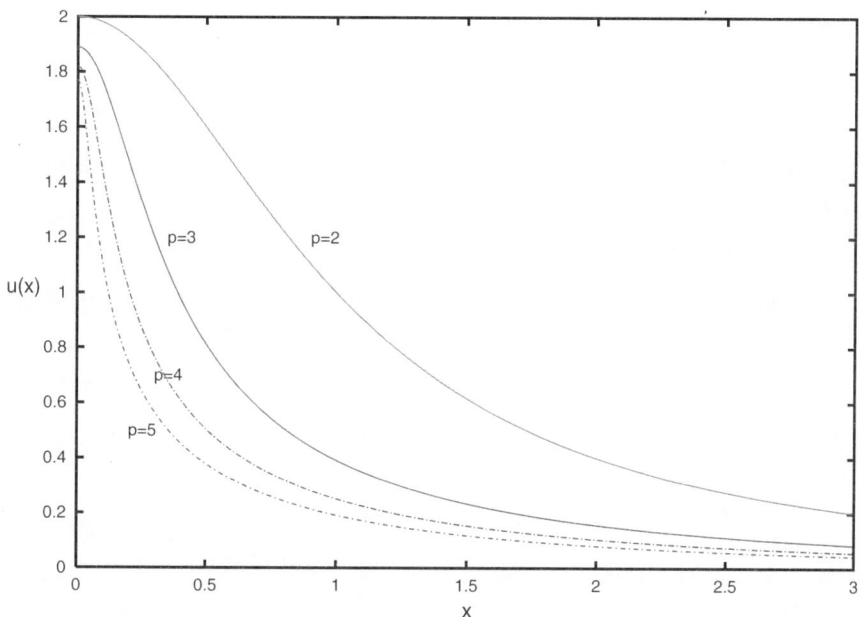

Figure 7
Numerical solution of (45) for different nonlinear powers. Only $x > 0$ is shown since solutions are even

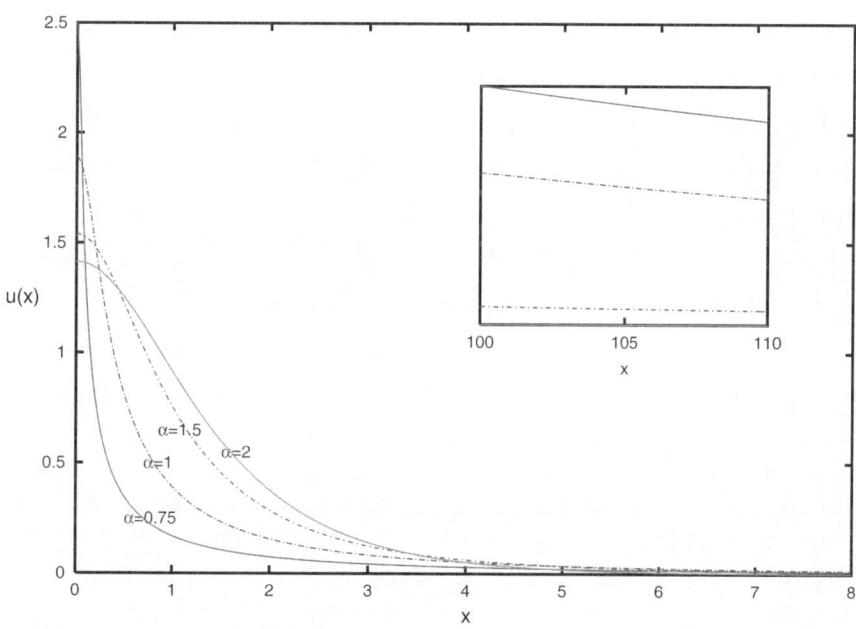

Figure 8
Numerical solution of (42) for nonlinear power $p = 3$ and different values of fractional degree α. The *square* window shows asymptotic decay of the solutions. Only $x > 0$ is shown since solutions are even

Writing Eq. (45) in Fourier space reads

$$(|\lambda| + 1)\widehat{U}(\lambda) = \widehat{U^p}(\lambda). \qquad (48)$$

In order to obtain $U(x)$ an iterative spectral method is used (for details see USERO 2014). Since we are interested in localized solutions, cosine and odd-

Reprinted from the journal

cosine Fourier transforms are used in order to accelerate convergence of the iterations.

In Fig. 7 four solutions are plotted for different values of the nonlinear power ($p = 2, 3, 4, 5$).

This algorithm can be easily extended to fractional degree α and numerical solutions can be found.

In Fig. 8 four solutions are plotted for nonlinear power $p = 3$ and nonlinear degree $\alpha = 2, 1.5, 1, 0.75$. The dotted line $\alpha = 2$ corresponds to the usual nonlinear wave equation and the dashed-dotted line $\alpha = 1$ to the previous Hilbert-transform case. It is also interesting to see that this solutions also exist for values of $\alpha < 1$ (continuous line). These solutions were not considered at the beginning of the study, but numerical simulations can also be extended to those cases.

Non-local problems such as time-fractional or long-range interaction problems are difficult to treat both analytically and numerically.

From an analytical point of view non-local problems give rise to integrodifferential equations that require complicated mathematical tools. Fractional calculus is a relatively modern science and there are many questions to be solved.

From a numerical point of view the integrodifferential equations arising in non-local problems require great computational effort since the solution at one time depends on all the previous values and sometimes on all the future values. With spatial non-locality the situation is similar since integration usually covers both sides.

Computing integrals is time-consuming and error increases quickly, especially when long-term solutions are required. For spatial non-localities the use of fast-Fourier routines simplifies and accelerates integration but this routine can only be used when the problem has a clear equation in Fourier space.

Many efforts must be done in order to develop new numerical algorithms for non-local problems and many open questions are waiting for future students.

Acknowledgments

This work has been partially supported by the Spanish Ministerio de Economía y Competitividad under Grant AYA2011-29967-C05-02.

REFERENCES

ALFIMOV, G.L., USERO, D., and VÁZQUEZ, L. (2000), *On complex singularities of solutions of the equation* $Hu_x - u + u^p = 0$, J. Phys. A: Math. Gen. *33*(38), 6707.

ANGSTROM, A. (1929), *On the Atmospheric Transmission of Sun Radiation and on Dust in the Air*, Geografiska Annaler *11*, 156–166.

ARRUEGO, I., DÍAZ-MICHELENA, M., JIMÉNEZ, J.J., MARTÍNEZ, J., APÉSTIGUE, V., GONZÁLEZ-GUERRERO, M., AZCUE, J., VALVERDE, A., DE MANUEL, V., DOMÍNGUEZ, J.A., MARTÍN, I., MARTÍN, B., ÁLVAREZ, J., ÁLVAREZ, M., HERNANDO, C., CERDÁN, M.F., RUIZ DE GALARRETA, C., SÁNCHEZ, J., MARTÍNEZ, G., VÁZQUEZ, L., and GUERRERO, H. (2010), Development of miniaturized instrumentation for Planetary Exploration and its application to the Mars MetNet Precursor Mission, Proceeding of the European Geosciences Union - 7th General Assembly, Vienna, Austria 02–07 May 2010. Geophysical Research Abstracts 12, EGU2010. http://meetingorganizer.copernicus.org/EGU2010/EGU2010-13330

BONA, J.L., BOSE, D.K., and BENJAMIN, T.B., Solitary-wave solutions for some model equations for waves in nonlinear dispersive media, In Applications of methods of functional analysis to problems in mechanics (Joint Sympos., IUTAM/IMU, Marseille, 1975) Lecture Notes in Math., 503 (Springer, Berlin, 1976) pp. 207–218.

BONA, J.L., and LI, Y.A. (1997), *Decay and analyticity of solitary waves*, J. de Math. Pures et Appl. *76*, 377–430.

CAPUTO, M., and MAINARDI, F. (1971), *Linear models of dissipation in anelastic solids*, Riv. Nuovo Cimento (Ser. 2) *1*, 161–198.

CÓRDOBA-JABONERO, C., and VÁZQUEZ, L. (2003), *Characterization of atmospheric aerosols by an in-situ photometris technique in planetary environments*, SPIE *4878*, 54–58.

Description of the Rover Environmental Monitoring Station (REMS), JPL-NASA web pages: http://msl-scicorner.jpl.nasa.gov/Instruments/REMS/. Accessed 2014.

DEHANT, V., BANERDTB, B., LOGNONNÉC, P., GROTTD, M., ASMARB, S., BIELEE, J., BREUERD, D., FORGETF, F., JAUMANND, R., JOHNSONG, C., KNAPMEYERD, M., LANGLAISI, B., LE FEUVREI, M., MIMOUNJ, D., MOCQUETI, A., READK, P., RIVOLDINIA, A., ROMBERGM, O., SCHUBERTN, G., SMREKARB, S., SPOHND, T., TORTORAO, P., ULAMECE, S., and VENNERSTRØM, S. (2012), *Future Mars geophysical observatories for understanding its internal structure, rotation, and evolution*, Planetary and Space Science 68, 123–145.

DE LUCAS, E., MIGUEL, M.J., MOZOS, D., and VÁZQUEZ, L. (2012), *Martian dust devils detector over FPGA*, Geosci. Instrum. Method. Data Syst. *1*, 23–31.

EL-SAYED, M.A. (1996), *Fractional order diffusion-wave equation*, Internat. J. Theoret. Phys. *35*, 311–322.

ESPOSITO, F., DEBEI, S., BETTANINI, C., MOLFESE, C., ARRUEGO, I., COLOMBATTI, G., HARRI, A.M., MONTMESSIN, F., WILSON, C., ABOUDAN, A., ZACCARIOTTO, M., ABAKI, S., BELLUCCI, G., BERTHELIER, J.J., BRUCATO, J.R., CALCUTT, S.B., CORTECCHIA, F., CUCCIARRÉ, F., DI ACHILLE, G., FERRI, F., FORGET, F., FRISO, E., GENZER, M., GILBERT, P., GOUTAIL, J.P., HAUKKA, H., JIMÉNEZ, J.J., JIMÉNEZ, S., JOSSET, J.L., KARATEKIN, O., LANDIS, G., LORENTZ, R., MARTHY, L., MARTINEZ, J., MENNELLA, V., MÖHLMANN, D., PALOMBA, E., PATEL, M., POMMEREAU, J.P., POPA, C.I., RAFKIN, S., RANNOU, P., RENNO, N.O., SCHIPANI, P., SCHMIDT, W., SEGATO, E., SIMOES, F., SPIGA, A., VALERO, F., VÁZQUEZ, L., VIVAT, F., WITASSE, O., YAHI, S., MUGNUOLO, R., and PIRROTTA, S. (2013),

DREAMS for the ExoMars 2016 mission: a suite of sensors for the characterization of Martian environment, EPSC Abstracts 8.

GÓMEZ-ELVIRA, J., ARMIENS, C., CASTAÑER, L., DOMÍNGUEZ, M., GENZER, M., GÓMEZ, F., HABERLE, R., HARRI, A.M., JIMÉNEZ, V., KAHANPÄÄ, H., KOWALSKI, L., LEPINETTE, A., MARTÍN, J., MARTÍNEZ-FRÍAS, J., McEWAN, I., MORA, L., MORENO, J., NAVARRO, S., DE PABLO, M.A., PEINADO, V., PEÑA, A., POLKKO, J., RAMOS, M., RENNO, N.O., RICART, J., RICHARDSON, M., RODRÍGUEZ-MANFREDI, J., ROMERAL, J., SEBASTIÁN, E., SERRANO, J., DE LA TORRE JUÁREZ, M., TORRES, J., TORRERO, F., URQUÍ, R., VÁZQUEZ, L., VELASCO, T., VERDASCA, J., ZORZANO, M.P., and MARTÍN-TORRES, J. (2012), *REMS: The Environmental Sensor Suite for the Mars Science Laboratory Rover*, Space Sci. Rev. *170*, 583–640.

GORENFLO, R., MAINARDI, F., and SRIVASTAVA, H.M., Special functions in fractional relaxation oscillation and fractional diffusion-wave phenomena, In Proceedings of the Eighth International Colloquium on Differential Equations (Plovdiv, Bulgaria, 18–23 August 1997) (ed. BAINOV D.) (VSP Publishers, Utrecht and Tokyo, 1998), pp. 195–202.

ISHIMORI, Y. (1982), *Solitons in a One-Dimensional Lennard-Jones Lattice*, Prog. Theor. Phys. *68*, 402–410.

JIMÉNEZ, S., and VÁZQUEZ, L. (2014), Elements for the Retrieval of the Solar Spectrum on the Surface of Mars from an Array of Photodiodes, Proceedings: Stochastic and Infinite Dimensional Analysis, Trends of Mathematics, Birkhäuser.

KITTEL, C., Introduction to Solid State Physics (John Wiley & Sons, New York, 1966).

LI, Y.A., and BONA, J.L. (1996), *Analyticity of solitary-wave solutions of model equations for long waves*, SIAM J. Math. Anal. *27*(3), 725–737.

MAGIN, R.L., Fractional Calculus in Bioengineering (Begell House Publishers, Connecticut 2006).

MANKO, M.A., MANKO, V.I., and VILELA-MENDES, R. (2001), *Tomograms and other transforms: a unified view*, J. Phys. A: Math. Gen. *34*, 8321–8332.

MANKO, V.I., and VILELA-MENDES, R. (1999), *Non-commutative time-frequency tomography*, Phys. Let. A *263*, 53–59.

MEIGA project main web page (in English): http://www.meiga-metnet.org//node/87. Accessed 2014.

METNET project main web page (in English): http://metnet.fmi.fi. Accessed 2014.

PIERANTOZZI, T., and VÁZQUEZ, L. (2005), *An interpolation between the wave and diffusion equations through the fractional evolution equations Dirac like*, J. Math. Phys *46*, 1135123.

PIERANTOZZI, T., and VÁZQUEZ, L., A Numerical Study of Fractional Evolution-Diffusion Dirac-like Equations, In Proceedings of the Fifth International Conference on Engineering Computational Technology (ed. TOPPING B.H.V., MONTERO G., and MONTENEGRO

R.) (Civil-Comp Press, Stirlingshire, United Kingdom, 2006) paper 20.

PORTAL of the MARS SCIENCE LABORATORY, Curiosity Rover (2014) JPL-NASA web pages: http://mars.jpl.nasa.gov/msl/. Accessed 2014.

PRESTON, L.J., and DARTNELL, L.R. (2014), *Planetary habitability: Lessons learned from terrestrial analogues*, International Journal of Astrobiology *13*, 81–98.

ROCCO, A., and WEST, B.J. (1999), *Fractional calculus and the evolution of fractal phenomena*, Physica A *265*, 535–546.

SAMKO, S.G., KILBAS, A.A., and MARICHEV, O.I., Fractional Integrals and Derivatives: Theory and Applications (Gordon and Breach Science, Yverdon, Switzerland, 1993).

SEREDINSKA, M., and HANYGA, A. (2000), *Nonlinear Hamiltonian equations with damping*, J. Math. Phys *41*, 2135–2155.

SMITH, P. H., *et al.* (2008), *Introduction to special section on the Phoenix Mission: Landing Site Characterization Experiments, Mission Overviews, and Expected Science*, J. Geophys. Res. *113*, E00A18.

TURCHETTI, G., USERO, D., and VÁZQUEZ, L. (2002), *Hamiltonian systems with fractional time derivative*, Tamsui Oxford Journal of Mathematical Sciences *18*(1), 31–44.

USERO, D., Propagación de ondas no lineales en medios heterogéneos (PhD. Thesis. Universidad Complutense de Madrid).

USERO, D., and VÁZQUEZ, L. (2003), *Fractional derivative: A new formulation for damped systems*, Localization and Energy Transfer in Nonlinear Systems, 296–303.

VÁZQUEZ, L. (2003), *Fractional Diffusion Equations with Internal Degrees of Freedom*, J. Comp. Math. *21*, 491–494.

VÁZQUEZ, L. (2005), *Singularity analysis of a nonlinear fractional differential equation*, Revista de la Real Academia de Ciencias, Serie A, Matemáticas *99*(2), 211–217.

VÁZQUEZ, L., and USERO, D. (2005), *Ecuaciones no locales y modelos fraccionarios*, Revista de la Real Academia de Ciencias Exactas, Físicas y Naturales *99*(2), 203–223.

VÁZQUEZ, L., and VILELA-MENDES, R. (2003), *Fractionally coupled solutions of the diffusion equation*, App. Math. Comp. *141*, 125–130.

VÁZQUEZ, L., ZORZANO, M.P., and JIMÉNEZ, S. (2007), *Spectral information retrieval from integrated broadband photodiode Martian ultraviolet measurements*, Optics Letters *32*, 2596–2598.

WEST, B.J., BOLOGNA, M., and GRIGOLINI, P., Physics of Fractal Operators (Springer-Verlag, New York, 2003).

ZORZANO, M.P., VÁZQUEZ, L., and JIMÉNEZ, S. (2009), *Retrieval of ultraviolet spectral irradiance from filtered photodiode measurements*, Inverse Problems *25*, 115023.

(Received March 11, 2014, accepted May 12, 2014, Published online July 8, 2014)

Pure Appl. Geophys. 172 (2015), 49–56
© 2014 Springer Basel
DOI 10.1007/s00024-014-0892-x

❙ Pure and Applied Geophysics

A Simple Thousand-Year Prognosis for Oceanic and Atmospheric Carbon Change

A. C. Fowler[1,2]

Abstract—A simple carbon-cycle box model allows for very simple quantitative insight into the evolution of climate over the next millennium. While melting ice sheets, rising sea levels, and ocean acidification are well recognised, we show that a further consequence of the repartitioning of ocean carbon is a dramatic rise in atmospheric carbon dioxide on a millennial time scale.

Key words: Climate change, carbon cycle, ocean acidification.

1. Introduction

Climate change is prevalent in the news, and there is a growing consensus on some of the more pertinent facts and hypotheses. It is a simple consequence of radiative transfer theory that increasing CO_2 levels in the atmosphere cause atmospheric warming (Pierrehumbert 2010), with an approximate logarithmic dependence (Houghton 2009). It is well-known that CO_2 levels oscillated between 180 and 280 ppmv during the Pleistocene ice ages (Petit *et al.* 1999) but have risen since 1750 to a level near 400 ppmv. Here, we will preferentially use partial pressures for atmospheric CO_2 levels. Dalton's law of partial pressures states that, for a dilute perfect gas, the ratio of partial CO_2 pressure to air pressure is approximately the molar volume fraction. For an atmospheric pressure of 10^5 Pa and a partial CO_2 pressure of p_{CO_2}, we get a volume fraction $p_{CO_2}/10^5 = 10 p_{CO_2}$ ppmv; thus, roughly, 400 ppmv = 40 Pa.

The Earth has been cooling since the Eocene, some 50 My (million years) ago (Bijl *et al.* 2013). This cooling is generally associated with declining levels of atmospheric carbon (Pearson and Palmer 2000) that are likely due to the increased weathering associated with Himalayan uplift following the collision of India with Asia, as well as the associated subduction of pelagic carbonates as the Tethys Ocean closed (Caldeira 1992).

The Antarctic continent was sub-tropical on its coasts (Pross *et al.* 2012), and the Antarctic Ice Sheet grew slowly thereafter, possibly originating as three separate ice caps (Pollard and DeConto 2005) before these coalesced at the Eocene–Oligocene transition at 34 My (Zachos *et al.* 1992). In model experiments, the hysteretic decay of the Antarctic Ice Sheet (Pollard and DeConto 2005) is associated with the increase of atmospheric CO_2 levels above about 900 ppmv (90 Pa), but one would expect melting to be initiated at levels well below that in conditions of rapid warming. And indeed, the great ice sheets do appear to be melting: outlet glaciers on Greenland and Antarctica have all shown dramatic signs of increased discharge and collapse in recent decades (Holland *et al.* 2008; Rott *et al.* 1996; Rignot *et al.* 2002).

Apart from the atmospheric warming caused by increased atmospheric CO_2 levels, a secondary effect is that of ocean acidification (Zeebe 2012a). The increased CO_2 in the atmosphere causes an increase in dissolved CO_2 in the ocean, and the effect of the bicarbonate buffering system is to cause a decrease in carbonate ion, and hence a decrease in pH (Archer *et al.* 1997; Archer 1999). Emphasis has been placed on the possible consequences for calcifiers in the ocean (Ridgewell and Zeebe 2005; Zeebe *et al.* 2008), with the suggestion that decreased calcium carbonate formation leads to lower oceanic burial rates, and thus, further enhancement of ocean carbon stocks. In addition, simulation models (Zeebe 2012b) indicate an associated increase in temperature of the

---
[1] MACSI, University of Limerick, Limerick, Ireland. E-mail: andrew.fowler@ul.ie
[2] OCIAM, University of Oxford, Oxford, UK.

order of ten degrees on a millennial time scale (ZEEBE and ZACHOS 2013).

Much of the discussion concerning climate prediction focusses on details of recent warming over the last century, and restricts itself to a range of forecasts over the coming century based on ever more detailed numerical weather prediction models, as described in the 2007 Intergovernmental Panel on Climate Change (IPCC) report (SOLOMON et al. 2007), for example. Together with the increasing complexity of these models comes an inevitable uncertainty, due both to chaotic dynamics and model uncertainty (PALMER 2001), with much of the latter being attributable to uncertainty in cloud parameterisation (TIEDTKE 1993).

Arguably, this 'kitchen sink' approach (include every process possible) misses the elephant in the room, which becomes visible when one adopts a 'top-down' approach (include only the essentials of the problem); indeed, this is partly the view of Archer, and of Zeebe and Zachos, who put an emphasis on ocean acidification beyond the century time scale.

There have been a number of recent studies which emphasise the long-term (millennial) evolution of atmospheric and oceanic carbon (GILLETT et al. 2011; SOLOMON et al. 2009, 2010; ARCHER and BROVKIN 2008; EBY et al. 2009). The studies use models of intermediate complexity, suitable for such longer time-scale studies, but still necessary to be solved numerically, so that the mechanisms of the processes are somewhat opaque.

The purpose of the present paper is to illustrate the way in which atmospheric carbon evolves over long time scales using the simplest model that still represents the essential exchange processes in the system, and particularly, to point out a third consequence of carbon injection (beyond warming and acidification). The criticism of over-simplicity must be balanced against the clarity of the insight that emerges.

2. A Simple Model

Our simplest model for climate prediction is based on a box model introduced by FOWLER et al. (2013). Conceptually, the model is a generalised energy balance model of the type introduced by BUDYKO (1969) and SELLERS (1969), and whose progeny are the models of intermediate complexity (GANOPOLSKI et al. 2010), which are applied with some success to climate models, ocean models (JOHNSON et al. 2007; TOGGWEILER 2008), and biogeochemical models (ZEEBE 2012b). Our approach here is to choose the very simplest such model that still encapsulates the essential physics of climate control. Because the energy balance part of the model simply reflects an essentially algebraic dependence of global average temperature on atmospheric CO_2, our focus is on the evolution of carbon within the ocean and atmosphere.

The basic ingredient of the model of FOWLER et al. (2013) is the dependence of mean atmospheric temperature on CO_2 content. To describe the atmospheric CO_2, we must balance anthropogenic or volcanic sources with fluxes to the ocean via dissolution or indirectly through weathering of silicate and carbonate rocks. At the ocean surface, the CO_2 concentration is related to the dissolved CO_2 in the water by Henry's law:

$$p_{CO_2}^s = \frac{[CO_2]}{K_H};\qquad(2.1)$$

K_H is Henry's constant, and has units M Pa^{-1}, where 1 M = 1 mol kg^{-1}. Similarly, we may take the dissolved CO_2 in raindrops to be $K_H p_{CO_2}$, where p_{CO_2} is atmospheric CO_2 partial pressure. If P is the global rate of precipitation, then this leads to a net CO_2 precipitation of $K_H \rho_{H_2O} P p_{CO_2}$, with units of mol y^{-1}. This leads us to a form of FOWLER et al.'s (2013) equation (3.21):

$$\frac{A_E}{M_a g}\dot{p}_{CO_2} = I - K_H \rho_{H_2O} P p_{CO_2} - h_{oc}(p_{CO_2} - p_{CO_2}^s),$$
$$(2.2)$$

which describes conservation of CO_2 in the atmosphere; the units are mol y^{-1}. The constants are as described in Table 1.

In order to determine p_{CO_2}, we need to know the dissolved CO_2 concentration in the ocean, which itself depends on the carbon balance. Particularly, carbon is buffered in the ocean by means of the following reactions:

Table 1

Values and description of the constants in the model. The pre-industrial CO_2 production rate is based on an estimated volcanic production rate of 3×10^{11} kgCO_2 y^{-1} (WALKER et al. 1981), while the anthropogenic value corresponds to a fossil fuel and cement emissions rate of 6 GtC y^{-1}, a value typical of the 1990s (DENMAN et al. 2007; indicating production rates of 5.4, 6.4 and 7.2 GtC y^{-1} in the 1980s, 1990s, and early 2000s, respectively)

Symbol	Meaning	Typical value	Comment
A_E	Earth surface area	5.1×10^{14} m^2	
A_L	Land surface area	1.5×10^{14} m^2	
g	Acceleration due to gravity	9.81 m s^{-2}	
h_{oc}	Interfacial transport coefficient	1.8×10^{13} mol Pa^{-1} y^{-1}	
I	CO_2 production rate	0.7×10^{13} mol y^{-1}	Pre-industrial
		5×10^{14} mol y^{-1}	Anthropogenic
K_H	Henry's law coefficient	4.5×10^{-7} M Pa^{-1}	
M_a	Molecular weight of air	2.88×10^{-2} kg mol^{-1}	
m_{oc}	Ocean mass	1.38×10^{21} kg	
P	Global precipitation	5×10^{14} m^3 y^{-1}	
w	Net carbonate loss	0.5×10^{13} mol y^{-1}	
ρH$_2$O	Density of sea water	1.025×10^3 kg m^{-3}	

$$(\text{H}_2\text{O}) + \text{CO}_2 \underset{k_{-1}}{\overset{k_1}{\rightleftharpoons}} \text{HCO}_3^- + \text{H}^+,$$
$$\text{HCO}_3^- \underset{k_{-2}}{\overset{k_2}{\rightleftharpoons}} \text{CO}_3^{2-} + \text{H}^+. \qquad (2.3)$$

These reactions are very fast (ZEEBE and WOLF-GLADROW 2001), and thus can be taken to be in equilibrium, whence we have

$$X = \frac{K_2 Q}{S}, \quad Y = \frac{K_2 Q^2}{K_1 S}, \qquad (2.4)$$

where we write

$$X = [\text{H}^+], \quad Y = [\text{CO}_2], \quad Q = [\text{HCO}_3^-], \quad S = [\text{CO}_3^{2-}], \qquad (2.5)$$

and the equilibrium constants K_i are defined by

$$K_1 = \frac{k_1}{k_{-1}}, \quad K_2 = \frac{k_2}{k_{-2}}. \qquad (2.6)$$

The total dissolved inorganic carbon concentration in the ocean is

$$C = Q + S + Y, \qquad (2.7)$$

and a conservation law for total carbon is

$$m_{oc}\dot{C} = h_{oc}(p_{\text{CO}_2} - p_{\text{CO}_2}^s) + \left(\frac{A_E - A_L}{A_E}\right)K_H \rho_{\text{H}_2\text{O}} P p_{\text{CO}_2}$$
$$- r_c - r_p + W, \qquad (2.8)$$

in which m_{oc} is ocean mass, A_L is land surface area, r_c is the removal rate of bicarbonate by calcifiers, r_p is the removal rate of carbonate by calcium carbonate precipitation, and W represents the supply of bicarbonate to the ocean through weathering and riverine discharge. FOWLER et al. (2013) provide further individual equations for the separate components, but in the present situation a more succinct discussion is appropriate.

The source and sink terms in (2.8) are relatively small; they cause adjustments to oceanic concentrations on time scales upwards of forty thousand years, which represents the normal adjustment time for carbonate. Because these time scales are much longer than the ocean mixing time, the box model makes sense on such long time scales. On the much shorter time scale associated with anthropogenic input to the atmosphere and thence, the ocean, the dissolved CO_2 adds no charge to the charge-neutral ocean. The concentration of the conservative ions chloride, sodium, etc., do not change at all, and the calcium also will not change, since its time scale for adjustment is so long, even if the calcifier populations were wiped out. Therefore, the net-negative charge $Q + 2S - X$ of the active ions H^+, HCO_3^-, and CO_3^{2-} must remain constant under conditions of rapid change. Since from Table 2 we see that $X \ll Q, S$, this implies

$$Q + 2S \approx \text{constant} \qquad (2.9)$$

over relatively short (millennial) time scales. Since also $Y \ll S$, we see that (2.8) takes the form

$$-m_{oc}\dot{S} = h_{oc}(p_{\text{CO}_2} - p_{\text{CO}_2}^s) + \left(\frac{A_E - A_L}{A_E}\right)K_H \rho_{\text{H}_2\text{O}} P p_{\text{CO}_2}$$
$$- r_c - r_p + W. \qquad (2.10)$$

This completes our millennial-scale model. For such short time scales, we will eventually ignore the small

Table 2

Present estimates of the equilibrium constants K_i and the ionic species concentrations in units of M (1 M = 1 mol kg^{-1}) (EMERSON and HEDGES 2008). C is dissolved inorganic carbon

Species/constant	Typical value (M)
$[Ca^{2+}]$	1.03×10^{-2}
$[CO_2] = Y$	0.8×10^{-5}
$[CO_3^{2-}] = S$	0.24×10^{-3}
$[H^+] = X$	0.63×10^{-8}
$[HCO_3^-] = Q$	1.7×10^{-3}
K_1	1.4×10^{-6}
K_2	1.1×10^{-9}
C	2.0×10^{-3}

source and sink terms, but we retain them initially so that we can sensibly discuss the pre-industrial situation.

3. The Past and the Future

We take our model in the form (2.2) and (2.10), as follows:

$$\frac{A_E}{M_a g} \dot{p}_{CO_2} = I - K_H \rho_{H_2O} P p_{CO_2} - h_{oc}(p_{CO_2} - p^s_{CO_2}),$$

$$m_{oc} \dot{S} = -h_{oc}(p_{CO_2} - p^s_{CO_2})$$
$$- \left(\frac{A_E - A_L}{A_E}\right) K_H \rho_{H_2O} P p_{CO_2} + w, \tag{3.1}$$

where w represents the net loss of dissolved inorganic carbon (DIC) due to precipitation and calcification minus the supply from weathering. Our estimate of its value in Table 1 is based on a river efflux to the ocean of 3.5×10^{13} m^3 y^{-1} and a river bicarbonate concentration of 0.96 mM (EMERSON and HEDGES 2008). This gives a net molar flux of carbon of 3.4×10^{13} mol y^{-1}, and we suppose the loss in the ocean is comparable to this. The value we choose for w is determined by assuming the pre-industrial carbon reservoir in the ocean is approximately in a state of balance between input and output (the balance will not be exact due to transient effects associated with recovery from the last ice age, but is not expected to be wildly out of balance, because the variation of atmospheric CO_2 during the ice ages is not enormous). In addition, we have Henry's law and charge conservation from (2.1), (2.4), and (2.9):

$$p_s = \frac{K_2 Q^2}{K_1 K_H S}, \quad Q + 2S = A, \tag{3.2}$$

where A is, approximately, the carbonate alkalinity.

It is helpful, indeed essential, to non-dimensionalise the equations. We do this by writing

$$p_{CO_2} = p_0 p, \quad p^s_{CO_2} = p_0 p_s, \quad Q = Aq, \quad S = S_0 s,$$
$$t \sim t_0, \tag{3.3}$$

where

$$t_0 = \frac{A_E}{M_a g h_{oc}}, \quad p_0 = \frac{K_2 A^2}{K_1 K_H S_0}, \tag{3.4}$$

and

$$A = 2.2 \text{ mM}, \quad S_0 = 0.24 \text{ mM} \tag{3.5}$$

are the current values in Table 2. This leads us to the dimensionless model

$$\dot{p} = \varepsilon(v - p) - (p - p_s),$$
$$\dot{s} = \delta \left[-\frac{1}{\varepsilon}(p - p_s) - \{(1 - \alpha)p - \Omega\} \right], \tag{3.6}$$
$$p_s = \frac{q^2}{s}, \quad q = 1 - vs,$$

where

$$\varepsilon = \frac{w_0}{h_{oc} p_0}, \quad v = \frac{I}{w_0}, \quad \delta = \frac{w_0 t_0}{m_{oc} S_0}, \quad \alpha = \frac{A_L}{A_E},$$
$$\Omega = \frac{w}{w_0}, \quad v = \frac{2S_0}{A}, \tag{3.7}$$

and where we also define

$$w_0 = K_H \rho_{H_2O} P p_0. \tag{3.8}$$

Based on Table 1, typical values of the parameters are given in Table 3.

We are expecting pre-industrial steady values for p and S to be of $O(1)$. It is easy to ascertain steady states of these equations. These are

$$p = \frac{v - \Omega}{\alpha}, \tag{3.9}$$

which with the values in Table 3, give $p = 0.86$, corresponding to 30.3 Pa (but this motivated the choice of w), and then from $p_s \approx p$, we have $s \approx 0.79$, correponding to 0.19 mM. Obviously, the

Table 3

Values of scales and parameters

Symbol	Meaning	Typical value
A	Carbonate alkalinity	2.2 mM
p_0	CO_2 scale	35.2 Pa
S_0	Carbonate scale	0.24 mM
t_0	Time scale	100 y
v		0.87 (pre-industrial)
		60.3 (anthropogenic)
w_0	Weathering scale	0.81×10^{13} mol y^{-1}
α		0.29
δ		2.48×10^{-3}
ε		1.28×10^{-2}
v		0.22
Ω		0.62

values do not exactly match, but they indicate self-consistency in the parameter choice.

3.1. Pre-Industrial Steady State

More interesting is the transient approach to this steady state, which is easy to read, based on the small values of the parameters ε and δ. On a time scale of $t_0 \sim 100$ y (dimensionlessly $t \sim O(1)$), p approaches an equilibrium in which $p \approx p_s$, but more accurately

$$p - p_s = \varepsilon(v - p) \approx \varepsilon(v - p_s). \qquad (3.10)$$

For longer time scales we substitute this into $(3.6)_2$ to obtain (also using $p \approx p_s$)

$$\dot{s} \approx \delta\left[\Omega - v + \frac{\alpha(1 - vs)^2}{s}\right], \qquad (3.11)$$

from which we can see that the carbonate ion approaches equilibrium on the longer time scale $\frac{t_0}{\delta} = \frac{m_{oc}S_0}{w_0} \sim 40{,}000$ y (dimensionlessly $t \sim \frac{1}{\delta}$).

3.2. Anthropogenic Future

Now, consider the nature of the solutions if we use the anthropogenic value of $v = 60.3$ (corresponding as in Table 1 to a 1990s release rate of 6 GtC y^{-1}). We take this initially as a constant, though in practice the value is steadily rising and is unlikely even to level off in the near future. We define

$$v_A = \varepsilon v \approx 0.77, \qquad (3.12)$$

and note that it is of $O(1)$. The Eq. (3.6) takes the form

$$\dot{p} = v_A - \varepsilon p - (p - p_s),$$
$$\dot{s} = \delta\left[-\frac{1}{\varepsilon}(p - p_s) - \{(1 - \alpha)p - \Omega\}\right], \qquad (3.13)$$

and again, it is simple to trace their solutions. On the same atmospheric time scale of 100 y, p reaches an approximate equilibrium

$$p \approx v_A + p_s. \qquad (3.14)$$

The meaning of this is that the carbon we are pouring into the atmosphere reaches a quasi-steady state in which the influx is balanced by the efflux to the ocean. With the values in Table 3, this quasi-steady state is (taking p_s as a pre-industrial value 0.8, corresponding to 28 Pa) $p \approx 1.57$, corresponding to 55.2 Pa, or 552 ppmv. This is the good news; on a shortish time scale, the ocean can soak up our anthropogenic excess.

Beyond the century time scale, we have $p - p_s \approx v_A - \varepsilon p$, so that s is given by

$$\dot{s} \approx \delta\left[-\frac{v_A}{\varepsilon} + \alpha\left\{v_A + \frac{(1 - vs)^2}{s}\right\} + \Omega\right]. \qquad (3.15)$$

The behaviour of s is altered dramatically. It decreases almost linearly for a dimensional time

$$\frac{\varepsilon t_0}{\delta v_A} = \frac{m_{oc}S_0}{I} \approx 670 \text{ y}, \qquad (3.16)$$

when it fairly abruptly reaches an equilibrium in which

$$s = \frac{\varepsilon\alpha}{v_A}\sigma, \quad p = \frac{v_A}{\varepsilon\alpha}\Pi, \qquad (3.17)$$

so that $s \sim 4.8 \times 10^{-3}$ and $p \sim 208$. Thereafter, $\sigma \approx \frac{1}{\Pi}$, and Π increases toward an eventual equilibrium of $\Pi = 1$ on a time scale of $t \sim \frac{1}{\varepsilon\alpha}$, corresponding to 27,000 years. Illustrations of the short-term (millennial) response are shown in Fig. 1.

In Fig. 2, we show the results of releasing the estimated reserves of 5,000 GtC (ARCHER *et al.* 1997; we take 4,800 GtC) in different ways. The different

53

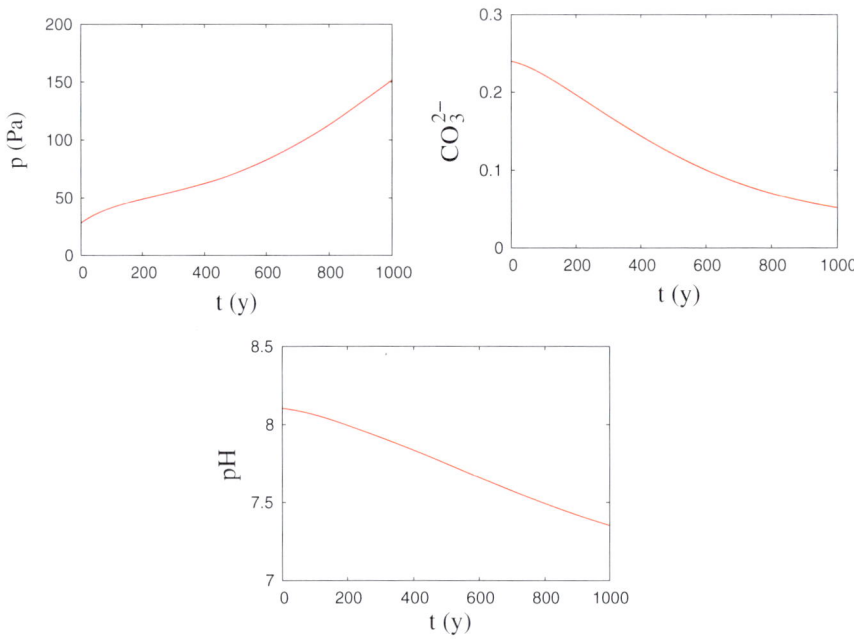

Figure 1
Evolution of CO_2, CO_3^{2-} (mM), and pH over the next millennium under conditions of continuing production of CO_2 at present levels

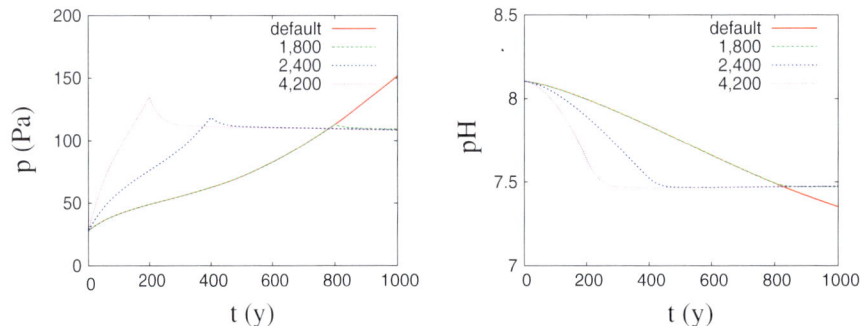

Figure 2
Evolution of p_{CO_2} and pH under different emissions scenarios. The labels have this meaning: 'default' means continued release at 'present' levels (1990s, 6 GtC y^{-1}) indefinitely; 1,800 indicates 6 GtC y^{-1} for 800 years and then zero emissions (switch-off); 2,400 means 12 GtC y^{-1} for 400 years and then a switch-off; 4,200 means 24 GtC y^{-1} for 200 years and then a switch-off. After the switch-off there are a further few decades while the ocean and atmosphere come into quasi-equilibrium, which is then maintained for thousands of years.

release mechanisms yield no ultimate difference, either in atmospheric CO_2 or in ocean pH; the only difference is one of time scale.

The message is simply stated. The carbon that we pour into the atmosphere pours into the ocean, but the maintenance of charge neutrality forces the carbonate ion to decrease dramatically to the level of ≈ 1 µM, a drop of about a factor of 200. Consulting (2.4), we see that the hydrogen ion concentration increases by 200, as does the dissolved CO_2. This causes a drop of

pH of about 2.3 to 6.9, while the atmospheric CO_2, which follows ocean CO_2 via Henry's law in (3.14), jumps by about 200 to a dimensionless value of around 166, correspondng to 5,800 Pa, or 58,000 ppmv: 6 % of the atmosphere! Mankind would be long extinct by then. Of course, this assumes unlimited reserves and their continuing use; the more realistic case of finite reserves results in a saturation at 110 Pa, or 1,100 ppmv, and ocean pH levels at 7.5, as seen in Fig. 2.

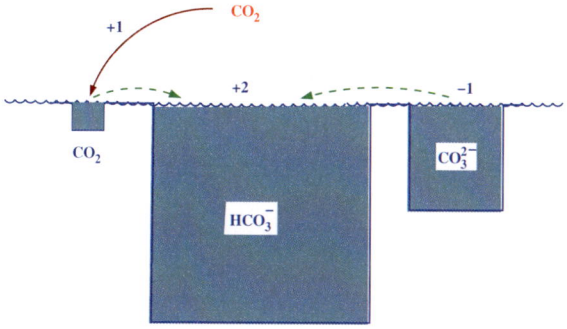

Figure 3
The ocean bicarbonate buffering system

4. Conclusions

The main purpose of this article was to make explicit in simple mathematical terms what the consequences of continued anthropogenic CO_2 output will be. Our model is stripped to its essentials, and consists only of two ordinary differential equations, for atmospheric CO_2 and oceanic CO_2. Yet this model yields essentially the same kind of quantitative result as do more complex simulation models, as used for example by ARCHER et al. (1997) and ZEEBE et al. (2008). Specifically, the same results of rapid CO_2 rise and rapid pH decline are seen, with much slower recovery following the termination of carbon emissions. However, unlike the computational models, the simplicity of our approach enables the mechanisms of warming and acidification to be portrayed more clearly, and the time scales of the responses can be explicitly identified.

In its essence, the message is this. Increasing atmospheric CO_2 causes increased flux to the ocean, until a quasi-equilibrium is reached after a century or two when the ocean uptake balances the discharge to the atmosphere. However, when a mole of CO_2 is dissolved, the bicarbonate buffering system (Fig. 3) immediately empties it into the bicarbonate pool; this would cause charge imbalance, and thus at the same time, a carbonate ion is also transferred to the bicarbonate pool. In effect, the reaction is that described by ARCHER et al. (1997):

$$CO_2 + CO_3^{2-} + H_2O \rightarrow 2HCO_3^-. \quad (4.1)$$

Thus, the flux of CO_2 to the ocean provides also an equivalent flux from the carbonate pool.

Because the acidity is controlled through the buffering relation (2.4)

$$[H^+] \propto \frac{[HCO_3^-]}{[CO_3^{2-}]}, \quad (4.2)$$

$[H^+]$ increases, i. e., pH decreases, resulting in ocean acidification. But at the same time, because

$$[CO_2] \propto \frac{[HCO_3^-]^2}{[CO_3^{2-}]}, \quad (4.3)$$

the dissolved CO_2 increases; not directly because of the flux from the atmosphere, but because of the decrease in carbonate, and the rate of increase is faster than linear. This second consequence of carbonate loss is what causes the accelerated increase of atmospheric CO_2, and its attendant greenhouse effect, following the initial lull.

Our model simulations in Fig. 1 are based on a constant input, while ZEEBE et al. (2008) consider the input of a fixed quantity corresponding to the total reserves, perhaps 5,000 GtC. It is straightforward to read from (3.15) what happens in this case. The carbonate declines until the ocean has absorbed most of the atmospheric load, but the subsequent recovery of the ocean carbonate takes place over a much longer time scale, that being tens of thousands of years.

As with all models, and particularly this one, the results must be tempered with caution. In particular, the neglect of ocean mixing requires consideration, most notably as regards the vertical profile. With a mixing time on the order of 1,000 years, carbonate changes on shorter time scales than this will be restricted to the upper parts of the ocean. The practical effect is to reduce the effective value of m_{oc} in (3.1), and thus increase δ in (3.6). Consulting (3.16), this suggests that the drawdown in carbonate ion content will be even faster, and further study of this issue is warranted.

Acknowledgments

I acknowledge the support of the Mathematics Applications Consortium for Science and Industry (http://www.macsi.ul.ie) funded by the Science Foundation Ireland mathematics initiative grant 12/1A/1683.

REFERENCES

ARCHER, D. 1999 Modeling CO_2 in the ocean: a review. In: Scaling of trace gas fluxes between terrestrial and aquatic ecosystems and the atmosphere, ed. A., F. BOUWMAN. Developments in Atmospheric Science, Vol. 24, pp. 169–184, Elsevier Sciences, Amsterdam.

ARCHER, D. and V. BROVKIN 2008 The millennial atmospheric lifetime of anthropogenic CO_2. Climate Change 90, 283–297.

ARCHER, D., H. KHESHGI and E. MAIER-REIMER 1997 Multiple timescales for neutralization of fossil fuel CO_2. Geophys. Res. Letts. 24, 405–408.

BIJL, P.K., J.A.P. BENDLE, S.M. BOHATY, J. PROSS, S. SCHOUTEN, L. TAUXEG, C. E. STICKLEY, R.M. MCKAY, U. RÖHL, M. OLNEY, A. SLUIJS, C. ESCUTIA, H. BRINKHUIS, and Expedition 318 Scientists 2013 Eocene cooling linked to early flow across the Tasmanian Gateway. Proc. Nat. Acad. Sci. 110, 9,645–9,650.

BUDYKO, M.,I. 1969 The effect of solar radiation variations on the climate of the Earth. Tellus 21, 611–619.

CALDEIRA, K. 1992 Enhanced Cenozoic chemical weathering and the subduction of pelagic carbonate. Nature 357, 578–581.

DENMAN, K.L., G. BRASSEUR, A. CHIDTHAISONG, P. CIAIS, P.M. COX, R.E. DICKINSON, D. HAUGLUSTAINE, C. HEINZE, E. HOLLAND, D. JACOB, U. LOHMANN, S. RAMACHANDRAN, P.L. DA SILVA DIAS, S.C. WOFSY and X. ZHANG 2007 Couplings between changes in the climate system and biogeochemistry. *In: Climate change 2007: the physical science basis. Contribution of Working Group I to the Fourth Assessment Report of the Intergovernmental Panel on Climate Change*, eds. S. SOLOMON, D. QIN, M. MANNING, Z. CHEN, M. MARQUIS, K.B. AVERYT, M. TIGNOR and H.L. MILLER. C.U.P., Cambridge.

EBY, M., K. ZICKFELD, A. MONTENEGRO, D. ARCHER, K.J. MEISSNER and A.J. WEAVER 2009 Lifetime of anthropogenic climate change: millennial time scales of potential CO_2 and surface temperature perturbations. J. Climate 22, 2,501–2,511.

EMERSON, S.R. and J.I. HEDGES 2008 Chemical oceanography and the marine carbon cycle. C.U.P., Cambridge.

FOWLER, A.C., R.E.M. RICKABY and E.W. WOLFF 2013 A simple model of ice ages. Int. J. Geomath. 4, 227–297.

GANOPOLSKI, A., R. CALOV and M. CLAUSSEN 2010 Simulation of the last glacial cycle with a coupled climate ice-sheet model of intermediate complexity. Clim. Past 6, 229–244.

GILLETT, N.P., V.K. ARORA, K. ZICKFELD, S.J. MARSHALL and W.J. MERRYFIELD 2011 Ongoing climate change following a complete cessation of carbon dioxide emissions. Nature Geosci. 4, 83–87.

HOLLAND, D.M, R.H. THOMAS, B. DE YOUNG and M.H. RIBERGAARD 2008 Acceleration of Jakobshavn Isbrae triggered by warm subsurface ocean waters. Nature Geoscience 1, 659–664.

HOUGHTON, J.T. 2009 Global warming: the complete briefing, 4th ed. C.U.P., Cambridge.

JOHNSON, H .L., D.P. MARSHALL and D.A.J. SPROSON 2007 Reconciling theories of a mechanically driven meridional overturning circulation with thermohaline forcing and multiple equilibria. Clim. Dyn. 29, 821–836.

PALMER, T.N. 2001 A nonlinear dynamical perspective on model error: a proposal for non-local stochastic-dynamic parametrization in weather and climate prediction models. Quart. J. R. Met. Soc. 127, 279–304.

PEARSON, P.N. and M.R. PALMER 2000 Atmospheric carbon dioxide concentrations over the past 60 million years. Nature 406, 695–699.

PETIT, J.R., J. JOUZEL, D. RAYNAUD, N.I. BARKOV, J.-M. BARNOLA, I. BASILE, M. BENDER, J. CHAPPELLAZ, M. DAVIS, G. DELAYGUE, M.

DELMOTTE, V.M. KOTLYAKOV, M. LEGRAND, V.Y. LIPENKOV, C. LORIUS, L. PÉPIN, C. RITZ, E. SALTZMAN and M. STIEVENARD 1999 Climate and atmospheric history of the past 420,000 years from the Vostok ice core, Antarctica. Nature 399, 429–436.

PIERREHUMBERT, R.T. 2010 Principles of planetary climate. C.U.P., Cambridge.

POLLARD, D. and R.M. DECONTO 2005 Hysteresis in Cenozoic Antarctic ice-sheet variations. Global Planet. Change 45, 9–21.

PROSS, J., L. CONTRERAS, P.K. BIJL, D.R. GREENWOOD, S.M. BOHATY, S. SCHOUTEN, J.A. BENDLE, U. RÖHL, L. TAUXE, J.I. RAINE, C.E. HUCK, T. VAN DE FLIERDT, S.,S.R. JAMIESON, C.E. STICKLEY, B. VAN DE SCHOOTBRUGGE, C. ESCUTIA, H. BRINKHUIS and Integrated Ocean Drilling Program Expedition 318 Scientists 2012 Persistent near-tropical warmth on the Antarctic continent during the early Eocene epoch. Nature 488, 73–77.

RIDGEWELL, A. and R.E. ZEEBE 2005 The role of the global carbonate cycle in the regulation and evolution of the Earth system. Earth Planet. Sci. Letts. 234, 299–315.

RIGNOT, E., D.G. VAUGHAN, M. SCHMELTZ, T. DUPONT and D. MACAYEAL 2002 Acceleration of Pine Island and Thwaites Glaciers, West Antarctica. Ann. Glaciol. 34, 189–194.

ROTT, H., P. SKVARCA and T. NAGLER 1996 Rapid collapse of Northern Larsen Ice Shelf, Antarctica. Science 271, 788–792.

SELLERS, W.D. 1969 A climate model based on the energy balance of the earth-atmosphere system. J. Appl. Meteorol. 8, 392–400.

SOLOMON, S., D. QIN, M. MANNING, Z. CHEN, M. MARQUIS, K.B. AVERYT, M. TIGNOR and H.L. MILLER (eds.) 2007 Climate Change 2007: The Physical Science Basis. C.U.P., Cambridge.

SOLOMON, S., G.-K. PLATTNER, R. KNUTTI and P. FRIEDLINGSTEIN 2009 Irreversible climate change due to carbon dioxide emissions. Proc. Nat. Acad. Sci. 106, 1,704–1,709.

TIEDTKE, M. 1993 Representation of clouds in large-scale models. Month. Weath. Rev. 121, 3.040–3.061.

TOGGWEILER, J.R. 2008 Origin of the 100,000–year timescale in Antarctic temperatures and atmospheric CO_2. Paleoceanogr. 23, PA2211, doi:10.1029/2006PA001405.

WALKER, J.C.G., P.B. HAYS and J.F. KASTING 1981 A negative feedback mechanism for the long-term stabilization of Earth's surface temperature. J. Geophys. Res. 86 (C10), 9,776–9,782.

ZACHOS, J.C., J.R. BREZA and S.W. WISE 1992 Early Oligocene ice-sheet expansion on Antarctica: stable isotope and sedimentological evidence from Kerguelen Plateau, southern Indian Ocean. Geology 20, 569–573.

ZEEBE, R.E. 2012a History of seawater carbonate chemistry, atmospheric CO_2, and ocean acidification. Ann. Rev. Earth Planet. Sci. 40, 141–65.

ZEEBE, R.E. 2012b 'LOSCAR': Long-term Ocean-atmosphere-Sediment CArbon cycle Reservoir model v2.0.4. Geosci. Model Dev. 5, 149–166.

ZEEBE, R.E. and D. WOLF-GLADROW 2001 CO_2 in seawater: equilibrium, kinetics, isotopes. Elsevier, Amsterdam.

ZEEBE, R.E., J.C. ZACHOS, K. CALDEIRA and T. TYRELL 2008 Carbon Emissions and Acidification. Science 321, 51–52.

ZEEBE, R.E. and J.C. ZACHOS 2013 Long-term legacy of massive carbon input to the Earth system: Anthropocene vs. Eocene. Phil. Trans. R. Soc. A, in press.

(Received March 5, 2014, revised June 26, 2014, accepted June 27, 2014, Published online July 25, 2014)

Pure Appl. Geophys. 172 (2015), 57–74
© 2014 The Author(s)
This article is published with open access at Springerlink.com
DOI 10.1007/s00024-014-0879-7

Earth's Rotation: A Challenging Problem in Mathematics and Physics

José M. Ferrándiz,[1] Juan F. Navarro,[1] Alberto Escapa,[1] and Juan Getino[2]

Abstract—A suitable knowledge of the orientation and motion of the Earth in space is a common need in various fields. That knowledge has been ever necessary to carry out astronomical observations, but with the advent of the space age, it became essential for making observations of satellites and predicting and determining their orbits, and for observing the Earth from space as well. Given the relevant role it plays in Space Geodesy, Earth rotation is considered as one of the three pillars of Geodesy, the other two being geometry and gravity. Besides, research on Earth rotation has fostered advances in many fields, such as Mathematics, Astronomy and Geophysics, for centuries. One remarkable feature of the problem is in the extreme requirements of accuracy that must be fulfilled in the near future, about a millimetre on the tangent plane to the planet surface, roughly speaking. That challenges all of the theories that have been devised and used to-date; the paper makes a short review of some of the most relevant methods, which can be envisaged as milestones in Earth rotation research, emphasizing the Hamiltonian approach developed by the authors. Some contemporary problems are presented, as well as the main lines of future research prospected by the International Astronomical Union/International Association of Geodesy Joint Working Group on Theory of Earth Rotation, created in 2013.

Key words: Earth rotation, nutation, precession, polar motion, UT1.

1. Relevance and Features of the Earth Rotation Problem

The accurate determination and prediction of the orientation and the motion of the Earth in the space is needed in various fields, especially since the advent of the space age. Direct examples in which that knowledge is essential are: carrying out astronomical observations from an observatory located on the Earth's surface, making observations of spacecrafts from ground-located tracking stations, observing the Earth from the space, determination and prediction of satellite orbits, etc.

A good knowledge of the Earth's orientation is necessary for any applications related to pinpointing of points or objects with respect to the Earth at a global scale. There is a very broad set of such applications, ranging from popular handy simple navigation devices to the most sophisticated investigations of Space Geodesy that address the quantification of the physical effects of climate change. The most popular issue is the determination of sea level variation, whose magnitude is typically of a few millimetres per year. Besides, there is a variety of geodetic research aimed at finding the fingerprints of different geophysical processes: mass movements in oceans, ice sheets, terrestrial water storages, displacement fields associated with earthquakes, etc. (PLAG *et al.* 2009a, 2010) All of those geodetic studies have very demanding requirements of accuracy. The GGOS (Global Geodetic Observing System) initiative developed by the International Association of Geodesy (IAG) targeted the requirements of accuracy on the level of 1 mm in position and 1 mm/year in stability (PLAG *et al.* 2009b).

The IAG considers Earth rotation as one of the three pillars of Geodesy, because of the relevant role it plays in Space Geodesy, the other two being Earth geometry and gravity. Those "three pillars" provide the basis for the realization of the reference systems required to assign time-dependent coordinates to points and objects, and to describe the Earth's motion in space. This is not at all new: a quick look to the table of contents of some classic treatises like Tisserand (1891) would suffice to appreciate how the interaction of those pillars have fostered theoretical

[1] Department of Applied Mathematics, University of Alicante, P.O. Box 99, 03080 Alicante, Spain. E-mail: jm.ferrandiz@ua.es; jf.navarro@ua.es; alberto.escapa@ua.es

[2] Department of Applied Mathematics, Faculty of Sciences, University of Valladolid, 47011 Valladolid, Spain. E-mail: getino@maf.uva.es

advances in many fields, such as Mathematics, Physics, Astronomy, Geodesy or Geophysics.

The solution to the Earth rotation problem consists mainly in the determination of the rotation matrix linking the celestial and the terrestrial reference frames. Nowadays, one of its most remarkable features is the extreme requirements of accuracy that must be fulfilled in the near future, at the level of a millimetre on the tangent plane to the planet surface, which corresponds to an angle about 30 µ as from the Earth centre, roughly speaking. Due to its relevance and the broad range of its applications, there is an international service in charge of monitoring and predicting the Earth rotation, the International Earth Rotation and Reference Systems Service (IERS). It was established in 1987 by the International Astronomical Union (IAU) and the International Union of Geodesy and Geophysics (IUGG). IERS is also responsible of the realization and maintenance of the celestial and terrestrial reference frames associated with Earth rotation, namely the International Celestial Reference Frame (ICRF) (FEY et al. 2004) and the International Terrestrial Reference Frame (ITRF) (ALTAMIMI et al. 2011). More information appears in the Annual Reports yearly published by IERS (DICK 2011). This service provides the international community with combined solutions for the EOP (Earth Orientation Parameters) (BIZOUARD and GAMBIS 2009) and also publishes the IERS Conventions, which are widely used not only in the field of Earth rotation, but in satellite orbit determination and many other geodetic or geophysical applications (PETIT and LUZUM 2010).

The Earth rotation is affected by many factors that must be accounted for to obtain solutions suitable to meet the present needs. Apart from the mathematical methods used to derive solutions, the main physical influences come from:

- Lunisolar gravitational attraction and planetary attraction.
- Earth figure and tensor of inertia (really not constant but time-varying).
- Earth internal structure: fluid outer core (FOC), solid inner core (SIC), etc.
- Effects at the boundaries of the inner layers, with dissipations and topography.

- Deformations (which produce geometric and dynamical effects).
- Tides: solid earth tides, ocean tides.
- Many other geophysical influences: redistribution of ice–water-vapor masses, currents, winds, hydrology, magnetism, post-glacial rebound, earthquakes, etc.

2. The Rigid-Earth Model: A First Step Towards the Solution

Assuming the Earth is a rigid body is a logic first step, which has fulfilled the practical needs of accuracy for centuries. The solutions for nutations are close enough to the actual non-rigid Earth nutations, since the maximum differences between them (for each frequency) are below 30 mas (milliarcseconds), about 1 m on the Earth surface, and therefore irrelevant in ancient observations. The accuracy was thus satisfactory for applications until the development of highly accurate space geodetic techniques. Besides, the rigid model allows a great simplification for several reasons:

- In this case, perturbations only arise from the gravitational attraction of celestial bodies on an Earth with a constant tensor of inertia.
- The theoretical definition of the terrestrial frame is trivial, since there are no intricacies associated to deformations.
- Rigid body rotations have been widely studied for centuries, and there are lots of well-known topics easily found in the literature: Euler equations for rigid body rotation, systems of variables, integrability issues, etc.
- There are several well-established approaches at hand: Newtonian, Eulerian, Lagrangian, Hamiltonian.
- The unperturbed motion is essentially the Euler-Poinsot problem; therefore, it is integrable (in the Liouville sense) and convenient to derive asymptotic solutions by means of perturbation methods.

Eulerian formulation The formulation using Euler equations (1749) is the most extended and well-known in rotational dynamics. In the Newtonian

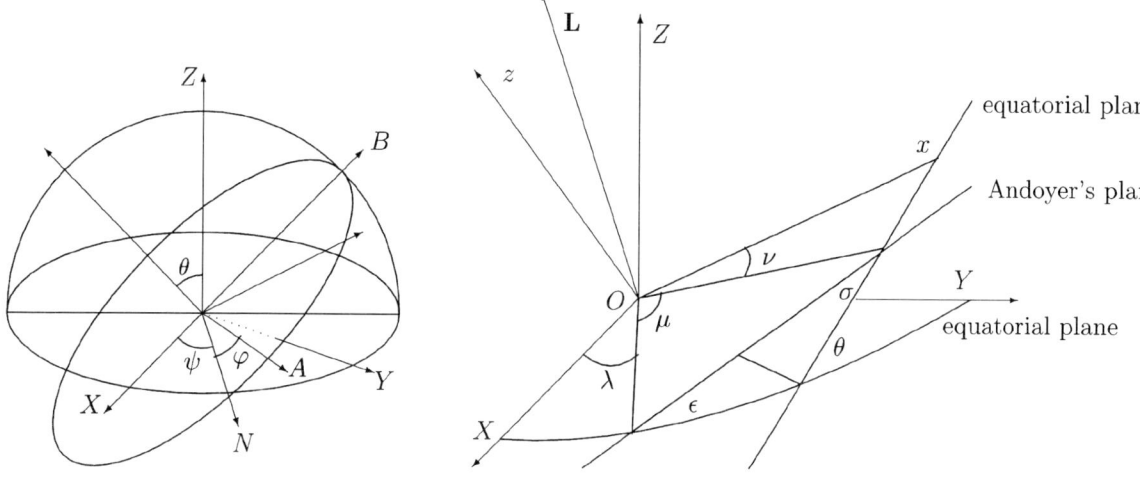

Figure 1
Euler (*left*) and Andoyer (*right*) variables

framework, we consider an inertial reference system $\mathcal{F}$ centered at the barycenter O of the body and a system $\mathcal{B}$ attached to the body, moving with angular velocity ω. The velocity of a body particle P with respect to both frames holds $v_{\mathcal{F}} = v_{\mathcal{B}} + \omega \times r = \omega \times r$, since P is in rest in $\mathcal{B}$. Therefore, the absolute angular momentum M can be expressed in the body frame as $M = \Pi\omega$, Π being the matrix of inertia, constant in this case. If L stands for the external torque, the basic Newtonian equation in the inertial frame, $dM_{\mathcal{F}}/dt = L_{\mathcal{F}}$, writes in the body frame as

$$\frac{dM}{dt} + \omega \times M = L, \quad \text{or} \quad \Pi \frac{d\omega}{dt} + \omega \times \Pi\omega = L.$$

$$(1)$$

It is usual to choose the body axes of $\mathcal{B}$ aligned with the principal axes of inertia of the body, and then Euler equation (1) reduces to the most familiar form

$$A\dot{\omega}_1 + (C - B)\omega_2\omega_3 = L_1,$$
$$B\dot{\omega}_2 + (A - C)\omega_1\omega_3 = L_2,$$
$$C\dot{\omega}_3 + (B - A)\omega_1\omega_2 = L_3.$$

Let us notice that those equations provide the derivatives of the angular velocities, but not the angles specifying the orientation of the body system $\mathcal{B}$ relative to the inertial one $\mathcal{F}$. This information is necessary to determine the rotational motion, since usually the torque L depend on the attitude of the body. Therefore, Eq. (1) must be complemented with other equations. One choice in Astronomy is to describe the attitude by means of the Euler angles ψ, θ, φ, by performing a sequence of three consecutive rotations with respect to the 3, 1, and 3 axes, respectively (Fig. 1, left). The time derivatives of the Euler angles are related to the components of the angular velocity vector ω by the kinematical equations (WOOLARD 1953a; LEIMANIS 1965)

$$\frac{d\varphi}{dt} = \omega_1 \frac{\sin\psi}{\sin\theta} + \omega_2 \frac{\cos\psi}{\sin\theta},$$
$$\frac{d\theta}{dt} = \omega_1 \cos\psi - \omega_2 \sin\psi,$$
$$\frac{d\psi}{dt} = \omega_3 - \omega_1 \cot\theta \sin\psi - \omega_2 \cot\theta \cos\psi.$$

$$(2)$$

Variational formulations Eulerian formulations are seldom used nowadays in Earth rotation studies. An exception is the solution by ROOSBEEK and DEHANT (1998), that followed the Eulerian method, with direct computation of torques. WOOLARD (1953a) used a Lagrangian approach to derive the most accurate solution of his epoch, which was adopted by the IAU for a period. BRETAGNON *et al.* (1988) later computed a highly accurate solution with that formalism. Nevertheless, the most successful variational approach has been the Hamiltonian one. It allowed systematic derivations of accurate solutions with the concourse of perturbation methods based on the Lie series. KINOSHITA (1977) established such an approach to rigid Earth rotation that became a model to follow. His solution was the best of its class and was the base of the non-rigid solution by Wahr, adopted by the IAU in 1980. Accuracy was improved later by KINOSHITA and SOUCHAY (1990) and SOUCHAY and KINOSHITA (1996, 1997). The fundamentals of this method are presented in the next section.

3. The Hamiltonian Treatment of Rigid Earth Rotation

Variables. A key of the success of the Hamiltonian method application to rotation problems is the use of Andoyer variables, also named after Serret (TISSERAND 1891; DEPRIT and ELIPE 1993). The fixed frame $\mathcal{F}$ of coordinates $(OXYZ)$ is transformed into the moving frame $\mathcal{F}$ $(Oxyz)$ by means of five rotations, since the plane orthogonal to the angular momentum vector (often called Andoyer plane) is used as an intermediate step to go from the equinoctial plane OXY to the equator Oxy. The scheme of rotations is 3–1–3–1–3, and the corresponding angles are $(\lambda, I, \mu, \sigma, v)$. They are shown in Fig. 1, to the right.

Three of those five angles (λ, μ, v) are canonical coordinates and the other two (I, σ) are auxiliary angles related to the conjugate momenta through $\cos I = \Lambda/M$, $\cos \sigma = N/M$. The canonical momenta are denoted as Λ, M, N, and they have a clear dynamical meaning: M is the modulus of the angular momentum vector $\boldsymbol{M}$, N its component along the polar or figure axis Oz of the body, and Λ its component along the OZ axis of the fixed frame. Let us point that the notation $(\lambda, \mu, v; \Lambda, M, N)$ corresponds to $(h, g, l; H, G, L)$ used by Kinoshita; the auxiliary angle I (obliquity of the Andoyer plane) has the same notation and σ stands for Kinoshita's J, the angle between $\boldsymbol{M}$ and the body figure axis. Further analyses of the Andoyer set appear in EFROIMSKY and ESCAPA (2009).

The components of $\boldsymbol{M}$ in the body system have the expressions

$$M_1 = M \sin \sigma \sin v, \quad M_2 = M \sin \sigma \cos v, \quad M_3 = M \cos \sigma.$$
(3)

Unperturbed problem In Andoyer variables, the kinetic energy is

$$T = \frac{1}{2} \left(M^2 - N^2 \right) \left(\frac{\sin^2 v}{A} + \frac{\cos^2 v}{B} \right) + \frac{1}{2} \frac{N^2}{C}. \quad (4)$$

It is especially simple in the case of axial symmetry, since all the coordinates are cyclic:

$$T = \frac{1}{2} \frac{M^2 - N^2}{A} + \frac{1}{2} \frac{N^2}{C}. \quad (5)$$

The free motion of a symmetric body is easily described in Andoyer variables. As the potential $V = 0$, the Hamiltonian $\mathcal{H} = T + V = T$. Therefore, the moments M, N, Λ, are constant and angles I, σ are as well. Coordinate λ (representing an ecliptic longitude) is also constant. The remaining two angles, μ, v, are linear functions of time. The sum $\mu + v$ corresponds to the diurnal rotation with angular velocity Ω_E and v provides the rotation of the angular momentum around the figure (or polar) axis. Its frequency is proportional to the dynamical ellipticity $H = (C - A)/A$. That is the simplest case of the free polar motion, known as Euler free oscillation.

Let us remark that the solution to the angular variables is computed directly and no previous solution to the angular velocity is needed, unlike in the approaches based on Euler–Liouville equations. The longitude and obliquity of the Earth equator, λ_f, I_f can be computed from the approximate relationships

$$\lambda_f - \lambda = \sigma \frac{\sin \mu}{\sin I}, \quad I_f - I = \sigma \cos \mu \quad (6)$$

which hold up to first order in σ, of the order of 10^{-6} rad. Those differences are known as Oppolzer terms. The classic precession-nutation angles used in Astronomy are $\psi = -\lambda_f$ and $\epsilon = -I_f$, since they are reckoned in the opposite direction.

Perturbing potential The free rotation of a rigid earth is perturbed by the gravitational attraction of the Sun, Moon and planets. The gravitational potential due to a body of mass m^* and coordinates (r^*, α, δ) in the fixed frame has a known expansion as a series of spherical harmonics (SH) multiplied by the corresponding Stokes coefficients. In the case of the Earth, the zonal term due to the Earth oblateness (associated to P_{20} and also named MacCullagh's term) is at least 1,000 times larger than the others, and thus provides a good approximation to the potential

$$V \simeq V_0 = \frac{Gm^*}{a^{*3}}(C - A)\left(\frac{a^*}{r^*}\right)^3 P_2(\sin \delta), \qquad (7)$$

It must be expressed in terms of the canonical variables and the coordinates of the disturbing body, assumed to be known functions of time provided by some ephemeris, usually referred either to an inertial (or quasi-inertial) system or to a non-inertial system with known motion (e.g., ecliptic of date). Nutation theories have relied on analytical or semi-analytical ephemeris since a long time ago (NEWCOMBE 1898; WOOLARD 1953b, etc.), and they still use such ephemeris, instead of numerical ones as in the JPL series (e.g., DE432, FOLKNER *et al.* 1994, 2014). The main reason is that nutation theories intend to derive semi-analytical solutions in which the constituent frequencies are explicit, and resorting to numerical ephemeris would introduce additional complications. The ephemeris commonly used are ephemeris VSOP87 (BRETAGNON 1982) and ELP2000 (CHAPRONT-TOUZÉ 1980), respectively. The transformation of the standard expansion in SH is a difficult task, since it requires performing five rotations. Kinoshita successively applied Wigner's theorem of transformation of SH under rotation and calculated the second degree SH of Moon and Sun to obtain up to the order of σ

$$V = \sum_{p=S,M} k'_p \sum_i \left[\frac{1}{2}(3\cos^2 \sigma - 1)B_i \cos \Theta_i\right.$$
$$\left. - \frac{1}{2}\sin 2\sigma \sum_{\tau=\pm 1} C_{i,\tau} \cos(\mu - \tau\Theta_i)\right] \qquad (8)$$

where subindex p stands for the perturbing body (S = Sun, M = Moon), the parameter $k'_p = 3Gm_p(C - A)/a_p^3$ factorises the main terms of the potential generated by body p and the coefficients B_i, $C_{i,\tau}$, depending on the variable I, are

$$B_i = -\frac{1}{6}(3\cos^2 I - 1)A_i^{(0)} - \frac{1}{2}\sin 2I A_i^{(1)} - \frac{1}{4}\sin^2 I A_i^{(2)},$$

$$C_{i,\tau} = -\frac{1}{4}\sin 2I A_i^{(0)} + \frac{\tau}{4}\sin I(1 + \tau\cos I)A_i^{(1)}$$
$$+ \frac{1}{2}(1 + \tau\cos I)(-1 + 2\tau\cos I)A_i^{(2)}. \qquad (9)$$

As for arguments Θ_i, we have that $\Theta_i = m_{1i}l_M + m_{2i}l_S + m_{3i}F + m_{4i}D + m_{5i}\Omega$, where l_M, l_S, F, D, and Ω are the *Delaunay* arguments of the

Moon and the Sun. Within our level of approximation, we can assume that $d\Theta_i/dt = n_i$, the mean motion n_i being constant. Let us stress that the canonical variable λ is implicitly contained in Ω through $\Omega = \Omega_0 - \lambda$, where Ω_0 is the mean longitude of the Moon referred to the origin of longitude on the ecliptic of date. The numerical values of coefficients $A_i^{(j)}$, Θ_i and n_i, as well as of the list of the five integer numbers $(m_{1i}, m_{2i}, m_{3i}, m_{4i}, m_{5i})$ associated to each value of the index i, depend on the orbital theories of the Moon and the Sun. They were first computed by KINOSHITA (1977) and updated by KINOSHITA and SOUCHAY (1990) and NAVARRO (2002).

3.1. Note on the Efficient Expansion of the Potential

A main difficulty of the Hamiltonian theory in the rigid case is the handling of terms of the gravitational potential due to lunisolar attraction. The expansion of the potential contains spherical harmonics (SH) of the perturbing bodies (Sun, Moon and Planets) beyond McCullagh's approximation. Their spherical coordinates are given by numerical or semi-analytical ephemeris, which provide them as multiple Fourier series whose arguments are linear combinations of the orbital variables of the relevant body. The semi-analytical expansion of those SH is a difficult task because of the large number of terms (thousands), and the help of computer algebra is essential.

The best option is designing and using special purpose symbolic manipulators to handle the so-called Poisson series. More advanced processors exist, capable of manipulating the full expansions of the potential, including canonical variables (Kinoshita series) and performing transformations of SH, even rotations applying Wigner's Theorem (NAVARRO and FERRÁNDIZ 2002).

Since the early 1960s, investigators have used computers to generate analytical expressions. The first symbolic processors were developed to work with Poisson series, that is, multivariate Fourier series whose coefficients are multivariate Laurent series,

$$\sum_{i_1,\ldots,i_n} \sum_{j_1,\ldots,j_m} C_{i_1,\ldots,i_n}^{j_1,\ldots,j_m} x_1^{i_1}\ldots x_n^{i_n} \frac{\cos}{\sin}(j_1\phi_1 + \cdots + j_m\phi_m),$$

where $C_{i_1,\ldots,i_n}^{j_1,\ldots,j_m} \in \mathbb{R}$, $i_1,\ldots,i_n,j_1,\ldots,j_m \in \mathbb{Z}$, and $x_1,\ldots,x_n$ and $\phi_1,\ldots,\phi_m$ are called polynomial and angular variables, respectively. These processors were applied to problems in non-linear mechanics or non-linear differential equations in the field of celestial mechanics. One of their first applications was concerned with the orbital motion of the Moon. Delaunay devised his perturbation method to treat the lunar problem and spent 20 years doing algebraic calculations by hand to solve it. DEPRIT *et al.* (1971) extended the solution of Delaunay's work with the help of a special purpose symbolic processor, and HENRARD (1979) pushed it to order 25. This solution was improved by iteration by CHAPRONT-TOUZÉ (1980), and planetary perturbations were also introduced by CHAPRONT-TOUZÉ (1980). Later, analytical theories for the rotation of the Earth (KINOSHITA 1977) were treated with the help of symbolic computation packages. Nowadays, there are many open problems that require massive symbolic computation. To cite one example, we will refer to the analytical theory of the resonant motion of Mercury. Motivated by the projects of space missions like BepiColombo and MESSENGER, D'HOEDT and LEMAITRE (2004) developed a spin–orbit model for the rotation of Mercury. The computation of the spherical harmonics of Mercury are performed with the use of the planetary theory "Variations Séculaires des Orbites Plantaires" (VSOP) (BRETAGNON 1988). The VSOP87 analytical solution of the motion of Mercury contains trigonometric series that represent the coordinates of the body (elliptic, rectangular or spherical coordinates according to the version). For instance, the solution for the distance Sun–Mercury (r) is given as a Poisson series containing 2,371 terms, and so, the calculus of $1/r$ through a Taylor expansion requires high accuracy symbolic computation with Poisson series containing hundreds of thousands of terms.

Many Poisson series processors have been developed until now, as PSP (BROUCKE 1970), mechanised algebraic operations (MAO) (ROM 1969), Trigonometric Manipulator (TRIGMAN) (JEFFERYS 1970), MSNam (HENRARD 1986), PARSEC (RICHARDSON 1989), and others.

3.2. *Analytical Solutions up to the Second Order*

The Hamiltonian method allows the derivation of highly accurate asymptotic solutions, depending analytically on the canonical variables and the arguments of lunisolar and planetary orbits, considered as known functions of time. Accurate solutions need to derive perturbations up to the second order, using, e.g., Hori's perturbation method (1966). The Hamiltonian can be cast in the form

$$\mathcal{H}(p,q) = \mathcal{H}_0(p,q) + \mathcal{H}_1(p,q) + \mathcal{H}_2(p,q), \quad (10)$$

$\mathcal{H}_i$ being of order $O(\varepsilon^i)$, where i is a nonnegative integer and ε is a small parameter measuring the perturbation. We will sketch this procedure at the second order in ε following a similar method to that of KINOSHITA (1977).

The algorithm consists in performing a canonical transformation from the actual canonical set (p,q) to a new one (p^*, q^*). This transformation is given at the second order by the generating function $\mathcal{W} = \mathcal{W}_1 + \mathcal{W}_2$, with $\mathcal{W}_i = O(\varepsilon^i)$, which depends on the transformed set (p^*, q^*) of canonical variables. The transformed Hamiltonian at the second order has a similar form

$$\mathcal{H}^*(p^*, q^*) = \mathcal{H}_0^*(p^*, q^*) + \mathcal{H}_1^*(p^*, q^*) + \mathcal{H}_2^*(p^*, q^*), \quad (11)$$

with $\mathcal{H}_i^*(p^*, q^*) = O(\varepsilon^i)$. In addition, some extra conditions are imposed on $\mathcal{H}_i^*$ in order to ensure that $\mathcal{H}^*$ is easier to integrate than $\mathcal{H}$. In particular, we force $\mathcal{H}^*$ to be free from periodic terms, that is to say, we combine the Lie transformation with an averaging method. By so doing, the transformed Hamiltonian $\mathcal{H}^*$ and the generating function $\mathcal{W}$ are determined by the so–called equations of the method (HORI 1966), which can be written up to the second order as

$$\mathcal{H}_0^* = \mathcal{H}_0, \quad \mathcal{H}_1^* = \mathcal{H}_{1sec},$$
$$\mathcal{H}_2^* = \mathcal{H}_{2sec} + \frac{1}{2}\{\mathcal{H}_1 + \mathcal{H}_{1sec}; \mathcal{W}_1\}_{sec},$$
$$\mathcal{W}_1 = \int_{UP} \mathcal{H}_{1per}\, dt, \quad \mathcal{W}_2 = \int_{UP} \mathcal{H}_{2per}\, dt \quad (12)$$
$$+ \frac{1}{2}\{\mathcal{H}_1 + \mathcal{H}_{1sec}; \mathcal{W}_1\}_{per},$$

where the subscripts per and sec denote the periodic or secular part of the corresponding function, and the Poisson brackets are computed in the (p^*, q^*) canonical set. The integrals are evaluated along the solutions to the unperturbed problem generated by the

Hamiltonian $\mathcal{H}_0^*$, obtained by literal substitution of the variables (p, q) by the variables (p^*, q^*) in $\mathcal{H}_0$. The time evolution of the transformed canonical variables (p^*, q^*) is determined by solving the Hamiltonian equations

$$\frac{dp^*}{dt} = -\frac{\partial \mathcal{H}^*}{\partial q^*}, \frac{dq^*}{dt} = \frac{\partial \mathcal{H}^*}{\partial p^*}. \qquad (13)$$

The variation of a function $f(p, q)$ of the canonical variables can be computed at the second order by the expression $f(p, q) = f^*(p^*, q^*) + \Delta f(p^*, q^*)$, with

$$f^*(p^*, q^*) = f(p^*, q^*); \Delta f = \Delta_1 f + \Delta_2 f + \Delta_3 f$$

$$\rightarrow \begin{cases} \Delta_1 f = \{f^*; \mathcal{W}_1\} \\ \Delta_2 f = \{f^*; \mathcal{W}_2\} \\ \Delta_3 f = \frac{1}{2}\{\{f^*; \mathcal{W}_1\}; \mathcal{W}_1\}. \end{cases}$$

The determination of the transformed Hamiltonian $\mathcal{H}^*$ and the generating function $\mathcal{W}$ allows to describe the time evolution of any variable of the Earth rotation up to the second order in the perturbation parameter ε. For the sake of brevity, only a few expressions corresponding to a first order integration are displayed, following GETINO and FERRÁNDIZ (1995). The first order generating function $\mathcal{W}_1$ is

$$\mathcal{W}_1 = K_0' \left\{ \frac{1}{2}\left(3\cos^2 \sigma - 1\right) W_a - \frac{1}{2}\sin 2\sigma W_b \right\}, \qquad (14)$$

with

$$W_a = \sum_i \frac{B_i}{n_i}\sin \Theta_i, W_b = \sum_{\tau=\pm 1}\sum_i \frac{C_i(\tau)}{n_\mu - n_i}\sin(\mu - \tau\Theta_i), \qquad (15)$$

n_μ, n_i being the mean motions of μ and τ_i, i.e., $\left(\frac{d\mu}{dt}, \frac{d\Theta_i}{dt}\right)$ respectively.

The perturbations of all the canonical variables can be obtained in a straight forward manner by taking derivates before being simplified doing $\sigma = 0$, in short as

$$\Delta(\Lambda, M, N) = -\frac{\partial W}{\partial(\lambda, \mu, v)}, \qquad (16)$$

$$\Delta(\lambda, \mu, v) = \frac{\partial W}{\partial(\Lambda, M, N)}$$

The first order nutations of the angular momentum axis (or Andoyer plane), are

$$\Delta\lambda = -K_0 \frac{1}{\sin I}\frac{\partial W_a}{\partial I} = -\frac{K_0}{\sin I}\sum_i \frac{\partial}{\partial I}\left(\frac{B_i}{n_i}\right)\sin \Theta_i,$$

$$\Delta I = K_0 \frac{1}{\sin I}\frac{\partial W_a}{\partial \lambda} = \frac{K_0}{\sin I}\sum_i (-m_5)\frac{B_i}{n_i}\cos \Theta_i, \qquad (17)$$

with $K_0 = K_0'/M$. The nutations of the figure axis are obtained by adding the *Oppolzer* terms

$$\Delta(\lambda_f - \lambda) = \frac{K_0}{\sin I}\sum_{\tau=\pm 1}\sum_i \frac{\tau C_i(\tau)}{n_\mu - \tau n_i}\sin \Theta_i,$$

$$\Delta(I_f - I) = K_0 \sum_{\tau=\pm 1}\sum_i \frac{C_i(\tau)}{n_\mu - \tau n_i}\cos \Theta_i. \qquad (18)$$

No other approach but the Hamiltonian succeeded in computing the nutations up to the second order. Solutions provide the longitude of the equinox and the obliquity of the equator as Poisson series of the arguments Θ_i, with coefficients depending analytically of Andoyer variables and of numbers $A_i^{j)}$. Final series result after numerical evaluation. Solutions can be computed for any of the three axes of interest: axis of figure, angular momentum and angular velocity. The number of accounted terms is very high:

- REN 2000 solution (SOUCHAY *et al.* 1999) contains several 1,000 terms of lunisolar and planetary origin; the truncation level is approximately 0.1 µas.
- FGN 2000 solution (Ferrándiz, Navarro and Getino) fully derived by computer algebra has a similar number of terms. A detailed second order solution showing the origin of the various terms was published by GETINO *et al.* (2010).

Solutions only include the perturbations due to the external potential ("forced nutations" and precession). The secular part of the solution, arising from Eq. (13) provides the precession. The non-rigidity effects on precession are so small (WILLIAMS 1994) that is not difficult to read that the precession is independent of the considered Earth model, which is not really true (FERRÁNDIZ *et al.* 2004, 2007).

4. Effect of the Liquid Core

Poincaré equations In (1891), Chandler detected variations of latitude in astrometric observations that pointed to a pole wobble, with a period of about 430 days, far from the Euler period for a rigid earth of about 305 days. The discovery of the so-called Chandler wobble (CW) stimulated the research on the effects of elasticity and the potential existence of a liquid core on Earth rotation. POINCARÉ (1901, 1910) developed the first satisfactory model for an Earth model consisting of a rigid mantle and a liquid core undergoing certain simple motion, often denoted as Poincaré model. He used two differentiated approaches to derive a set of equations quite similar to those of Euler,

$$\dot{M} + \omega \times M = L, \quad \dot{M}_c - \delta\omega \times M_c = 0. \quad (19)$$

Here, M and L are the total angular momentum and torque acting on the whole earth, ω the angular velocity of the frame linked to the mantle and $\delta\omega$ the relative angular velocity of the fluid core with respect to the mantle. M_c is the total angular momentum of the core, given by $M_c = \Pi_c(\omega + \delta\omega)$, Π_c being the tensor of inertia of the core in the mantle frame. Assuming axial symmetry and after neglecting the second and higher order terms in $\omega_1, \omega_2, \tilde{\omega}_3 = \omega_3 - \Omega_E$ and $\delta\omega$, the equations for $\tilde{\omega}_3$ and $\delta\omega$ happen to be uncoupled and the problem reduces to four linear equations. In this approximation, the complex variables $u = \omega_1 + i\omega_2$, and $v = \delta\omega_1 + i\delta\omega_2$ oscillate with two free frequencies

$$\sigma_1 = \frac{C-A}{A_m}\Omega, \quad \sigma_2 = -\Omega\left(1 + \frac{A}{A_m}\frac{C_c - A_c}{A_c}\right), \quad (20)$$

whilst the solution of the free polar motion (PM) is a linear combination of $e^{i\sigma_1 t}$, $e^{i\sigma_2 t}$. The frequency σ_1 corresponds to CW, which replaces the Euler free oscillation of the rigid case. As $A_m < A$, frequency σ_1 is larger for a Poincaré Earth than for a rigid Earth, so that the period of the polar oscillation is shortened by the fluid core. The lengthening of the period is mainly due to the elastic yielding of the earth, as already explained by NEWCOMBE (1892)—see also GETINO and FERRÁNDIZ (1995) who performed more detailed calculations.

Besides, a second new free frequency σ_2 emerges due to the presence of the liquid core, that is known as NDFW (nearly diurnal free wobble) or RFCN

(retrograde free core nutation), because it does not contribute solely to PM, but also gives rise to an observable nutation, named free core nutation (FCN). It was predicted by theory in early times (VICENTE and JEFFREYS 1964), but its observation remained elusive for a long period and only could be evidenced after some years of very long baseline interferometry (VLBI) observations.

POINCARÉ (1910) also found the ratio of the amplitudes of the nutations of a rigid planet with and without a liquid core (in the linear approximation). That established the basis of the transfer function approach, which has been followed in most of the research on non-rigid earth nutations. Besides, he included in that paper a section treating a body with a fluid core contained in an elastic shell.

Hamiltonian approach to Poincaré's earth model Nevertheless, Poincaré did not perform any numerical evaluation of his solution to obtain values of the main nutations. Accurate solutions for a Poincaré model were computed much later by GETINO (1995) and GETINO and FERRÁNDIZ (1997), but using their Hamiltonian method. Let the tensors of inertia be Π_m for the mantle, Π_c for the core and $\Pi = \Pi_m + \Pi_c$ for the whole Earth, which are assumed constant in a frame attached to the (rigid) mantle. If M, M_m and M_c are, respectively, the angular momenta of the total Earth, the mantle and the core, they satisfy:

$$\begin{aligned} M = M_m + M_c &= \Pi_m\omega + \Pi_c(\omega + \delta\omega) \\ &= \Pi\omega + \Pi_c\delta\omega. \end{aligned} \quad (21)$$

Notice that setting $M_c = \Pi_c(\omega + \delta\omega)$ means that an appropriate definition of the core rotation (MORITZ 1982) has been made, so that it is referred to a Tisserand frame (MORITZ 1982), as detailed in GETINO (1995). The kinetic energy is thus

$$T = \frac{1}{2}(M - M_c)^t \Pi_m^{-1}(M - M_c) + \frac{1}{2}M_c^t \Pi_c^{-1}M_c. \quad (22)$$

This expression is canonically formulated by means of a set of canonical variables, $\lambda, \mu, \nu, \Lambda, M, N$ for the whole Earth, and $\lambda_c, \mu_c, \nu_c, \Lambda_c, M_c, N_c$ for the core, with the help of the auxiliary angles σ, I, σ_c, I_c described by GETINO (1995). The angular momenta M and M_c are given by

$$\mathbf{M} = \begin{pmatrix} \mathbf{K}\sin v \\ \mathbf{K}\cos v \\ \mathbf{N} = \mathbf{M}\cos\sigma \end{pmatrix}, \quad \mathbf{M}_c = \begin{pmatrix} \mathbf{K}_c\sin v_c \\ -\mathbf{K}_c\cos v_c \\ \mathbf{N}_c = \mathbf{M}_c\cos\sigma_c \end{pmatrix},$$

$$(23)$$

where $K = M\sin\sigma$, $K_c = M_c\sin\sigma_c$. Note that σ and σ_c are small quantities, of the order of 10^{-6} rad. The kinetic energy can be written as

$$T_0 = \frac{1}{2A_m}\left[K^2 + \frac{A}{A_c}K_c^2\right] + \frac{KK_c}{A_m}\cos(v + v_c)$$
$$+ \frac{1}{2C_m}\left[N^2 - 2NN_c + \frac{C}{C_c}N_c^2\right],$$

$A = A_m + A_c$, $C = C_m + C_c$, being the principal moments of the total Earth. Let us remark that these hypotheses pose no problem related to the terrestrial frame, since the mantle is rigid and its principal axes are well defined. However, the number of canonical variables has been doubled, which increases the difficulty of the treatments. But there is an additional, essential difference with respect to the rigid case: resonance phenomena occur, which amplify the amplitudes of some Oppolzer terms, hence of nutations. That fact helps to constrain the values of certain geophysical parameters. Besides, the unperturbed problem is not integrable any more, irrespective of the axial symmetry of the body. The integrability issues in this model were studied by FERRÁNDIZ and BARKIN (2001).

5. Theories of Non-Rigid Earth Nutations

5.1. Two-Layer Earth Models

As we pointed out above, explaining the observed CW period requires taking into account elasticity besides the liquid core. A number of solutions were developed between about 1950 and 1990 by considering Earth models composed of an elastic mantle and a liquid core, the standard two-layer model. They made use of the theory of elasticity, developed by Cauchy, Green, Poisson, Stokes, Lord Kelvin, etc. These approaches are very different, but they share some features:

- Kelvin solutions of the Laplace equations in terms of SH are used, as well as generalizations,

including the assumption of variability for some parameters, such as density or Lamé parameters.
- Some simplifying hypotheses are usually made, as radial dependence of parameters or certain equilibrium conditions.
- In general, this procedure allows the reduction of the original continuous problem of elasticity to a discrete one, with the relevant parameters determined by quadratures assuming certain rheological models.

JEFFREYS and VICENTE (1957) proposed a variational formulation of Lagrangian type; therefore, the computation of the internal dissipative moments is avoided. In Molodenski's model (1961), the elastic equations for the mantle are approximated by spherical functions, and the fluid core is treated using hydrodynamical equations. SOS equations (SASAO et al. 1980) had deep impact on later research. They are a simple generalization of Poincaré's, including elasticity and dissipations at the core–mantle boundary (CMB) due to friction and electromagnetic coupling. The original derivation was carried out by direct methods (Euler–Liouville). Using variational methods allowed a drastic simplification (Moritz).

The IAU 1980 nutation theory Wahr's solution WAHR (1981) was obtained by applying a certain transfer function to the rigid earth solution by KINOSHITA (1977). An IAU Working Group proposed its adoption (SEIDELMANN 1982) and the theory was endorsed by IAU as its first non-rigid Earth nutation theory in 1981. This solution follows the method of Smith and Whar: The partial differential equations of the elastic problem are transformed into an infinite system of ordinary differential equations through a series expansion of spheroidal and toroidal harmonics. A drastic truncation produces a finite system. The resulting equations are integrated numerically over the Earth volume, assuming a certain rheological model. This solution gives the nutations of an oceanless, elastic solid Earth with a fluid core. In the framework of IAU 1980, other effects not addressed in the official theory (oceanic, atmospheric, anelastic, etc.) are treated in the moving (terrestrial) reference frame, mainly using some versions of the Euler–Liouville equations, and are usually classified as "polar motion" terms (see Sect. 6).

Earth elasticity in the Hamiltonian method The Hamiltonian method contributed with a series of papers by GETINO and FERRÁNDIZ who introduced the Hamiltonian formalism to study an elastic Earth (1990, 1991, 1995). Let us notice that the definition of the body frame has no special difficulties under the assumption of linear elasticity, since the deformations have known expressions depending on constant Love numbers, and the variations of the principal axes and moments of inertia can be derived analytically (BARKIN and FERRÁNDIZ 2000). More properties of the rotation of weakly deformable bodies are given by BARKIN (1998, 2000a, b).

GETINO and FERRÁNDIZ (2000, 2001) also combined their previous results to derive an accurate Hamiltonian solution for a two-layered earth made of a liquid core and an inelastic mantle, and accounted for dissipation at CMB. That way of proceeding guarantees consistency of the new considered effects with the former pieces of theory. For instance, the main change when elasticity is put into the Poincaré model is the addition of a new term T_t to the Hamiltonian, which represents the increment of the kinetic energy due to the tidal deformation and is given by

$$T_t = \frac{N - N_c}{A_m C_m} D_m^t \left[K_c \left(t_{13} \sin v_c - t_{23} \cos v_c \right) \right.$$
$$\left. - .-K \left(t_{13} \sin v + t_{23} \cos v \right) \right] \qquad (24)$$
$$- \frac{N_c}{A_c C_c} D_c^t K_c \left(t_{13} \sin v_c - t_{23} \cos v_c \right),$$

where $D_{m,c}^t$ are constants related to the Love number k_2 and functions $t_{i,j}$ have expansions similar to the components of the potential.

5.2. Three-Layer Earth Models

The improvements of the space geodetic observation techniques since the late 1980s revealed that IAU1980 was not accurate enough. Besides, the launch and operation of new geodetic satellites improved the observational possibilities and contributed to obtaining more insight into matters such as bodily tides and other geophysical properties of the Earth. New investigations aimed at explaining the new results, among them DEHANT *et al.* (1999). MATHEWS *et al.* (1991a, b) introduced a solid inner core in the basic Earth structure. An empirical

nutation model was adopted in the IERS Conventions 1996 (McCARTHY 1996). In this context, an IAU Working Group on non-rigid Earth nutation theory started in 1994 and recommended that theories be based on geophysical models closer to the actual Earth (DEHANT *et al.* 1999).

Most of the theories developed in that epoch assumed a three-layered Earth made of elastic mantle, fluid outer core (FOC) and solid inner core (SIC). They had to rely upon a pre-existent rigid Earth solution, since they used a transfer function approach. Depending on theories, elasticity might be extended to deal with in-elastic or an-elastic assumptions, include dissipations in the inner layers boundaries or consider oceanic and atmospheric effects to some extent. Among those theories, we can cite first DD97 (DEFRAIGNE and DEHANT 1998), Sch97 (SCHASTOK 1997), Hg2000 (HUANG *et al.* 2001) among the main differentiated approaches.

In March 2000, three theories were selected as candidates to become the IAU 2000 nutation model (DEHANT 2002). They were:

- MHB2000 (MATHEWS *et al.* 2002): a transfer function derived from a generalization of Poincaré–SOS equations was applied to REN2000. It was complemented with the Kinoshita–Souchay–Folgueira (1999) planetary perturbations for the rigid Earth.
- SF2000 (SHIRAI and FUKUSHIMA 2000): applied a numerical convolution in the time domain to adjust parameters of Herrings transfer function.
- GF2000 (GETINO and FERRÁNDIZ 2000): Hamiltonian, analytical theory for the Earth rotation, extending Kinoshita and Souchay's rigid Earth theory. It was complemented with the planetary non–rigid perturbations by Ferrándiz–Navarro–Getino and Huang et al. oceanic corrections, the final series being named FGHN.

All of them fit a low number of basic Earth parameters to observations and got similar accuracy, about 150 μas in terms of wrms (weighted root mean squared) observations-model differences (if an empirical model for FCN is used). The accuracy of IAU 1980 was thus improved in more than one order of magnitude. MHB2000 was preferred and selected as IAU2000, and it is in force since 2003. In that year

FUKUSHIMA (2003) published a new precession theory. The IAU1976 model of the precession (LIESKE et al. 1977) was changed 6 years later and the P03 model by CAPITAINE et al. (2003) was adopted as the IAU2006 precession model (HILTON et al. 2006).

5.3. The Hamiltonian Method

Main features The Hamiltonian or global approach is the only one that allows the direct derivation of a non-rigid solution up to the second order of perturbation, in a fully consistent manner, since it is independent of any previous rigid Earth solution. That is because transfer function approaches are intrinsically linear. The rigid solution can be recovered when some parameters vanish. The calculation of some poorly known internal torques is avoided, since the approach is variational. The effect of geophysical Earth models is concentrated in a reduced set of parameters. Analytical solutions are convenient for several reasons, like fitting parameters, allowing the identification of resonances and providing more insight into the Earths interior and geophysical properties. Besides numerical methods have failed to provide good solutions in the non-rigid case so far: the attempts which have been successful within a fitting time interval (KRASINSKY 2006) showed a quick degradation when extrapolated beyond that interval (CAPITAINE et al. 2009).

Free motion of a three layers Earth in the Hamiltonian approach The definition of the Andoyer variables for FOC and SIC takes into account the relations among a frame fixed to the mantle, $Ox_m y_m z_m$, the Andoyer planes defined by the angular momentums of FOC and SIC, and frames "attached" to the FOC or SIC, $Ox_f y_f z_f$ or $Ox_s y_s z_s$. It originates a set of 18 canonical variables, λ, μ, ν, Λ, M, N for the total Earth, λ_f, μ_f, ν_f, Λ_f, M_f, N_f for the fluid outer core, and λ_s, μ_s, ν_s, Λ_s, M_s, N_s for the solid inner core, with auxiliary angles σ, I, σ_f, I_f, σ_s and I_s. Their geometrical meaning is displayed in Fig. 2. Denoting by $\mathbf{M}$, $\mathbf{M_f}$ and $\mathbf{M_s}$ the absolute angular momenta of the whole earth, FOC and SIC, respectively; the canonical moments satisfy

$$M = |\mathbf{M}|, \qquad M_f = |\mathbf{M_f}| \qquad M_s = |\mathbf{M_f}|,$$
$$N = M\cos\sigma, \quad N_f = M_f\cos\sigma_f, \quad N_s = M_s\cos\sigma_s,$$
$$\Lambda = M\cos I, \quad \Lambda_f = M_f\cos I_f, \quad \Lambda_s = M_s\cos I_s.$$

The three layers Earth kinetic energy T is written as

$$T = \frac{1}{2}(\mathbf{M} - \mathbf{M_f} - \mathbf{M_s})^t \Pi_m^{-1}(\mathbf{M} - \mathbf{M_f} - \mathbf{M_s})$$
$$+ \frac{1}{2}\mathbf{M_f}^t \Pi_f^{-1}\mathbf{M_f} + \frac{1}{2}\mathbf{M_s}^t \Pi_s^{-1}\mathbf{M_s},$$

Π, Π_f and Π_s being the respective inertia matrices. The angular momenta hold

$$\mathbf{M} = \begin{pmatrix} \mathbf{K}\sin\nu \\ \mathbf{K}\cos\nu \\ \mathbf{N} = \mathbf{M}\cos\sigma \end{pmatrix}, \quad \mathbf{M_f} = \begin{pmatrix} \mathbf{K_f}\sin\nu_f \\ -\mathbf{K_f}\cos\nu_f \\ \mathbf{N_f} = \mathbf{M_f}\cos\sigma_f \end{pmatrix},$$

$$\mathbf{M_s} = \begin{pmatrix} \mathbf{K_s}\sin\nu_f \\ -\mathbf{K_s}\cos\nu_f \\ \mathbf{N_s} = \mathbf{M_s}\cos\sigma_s \end{pmatrix},$$

with $K = M\sin\sigma$, $K_f = M_f\sin\sigma_f$, $K_s = M_s\sin\sigma_s$.

The explicit expression of the Hamiltonian is involved even for the unperturbed motion, especially if no restrictive hypothesis on the SIC attitude is made. ESCAPA et al. (2001) derived a solution to the linearised equations, which gives the frequencies of the four normal modes or free harmonic oscillations of the rotation pole in terms of the ellipticities and an additional small parameter δ:

$$m_1 = \frac{A}{A_m}e \qquad \rightarrow \quad \text{CW or Chandler wobble,}$$

$$m_2 = -1 - \frac{A_f + A_m}{A_m}e_f \quad \rightarrow \quad \text{RFCN or retrograde free core nutation,}$$

$$m_3 = -1 + \delta \qquad \rightarrow \quad \text{PFCN or prograde free core nutation,}$$

$$m_4 = e_s - \delta \qquad \rightarrow \quad \text{ICW or inner core wobble.}$$

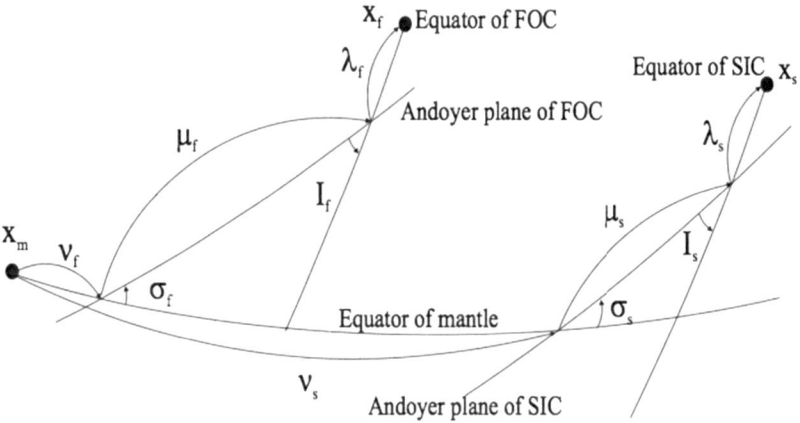

Figure 2
The Andoyer variables for FOC and SIC relate a fixed frame of mantle, $Ox_m y_m z_m$, the plane defined by the angular momentum of FOC and SIC, Andoyer plane of FOC and SIC, and a fixed frame of FOC or SIC, $Ox_f y_f z_f$ or $Ox_s y_s z_s$

Recent progress in the Hamiltonian theory of non-rigid Earth Since the year 2000, many effects have been investigated by different authors and under various approaches. Several effects have been found to contribute to the nutations with direct or indirect terms reaching the magnitude of some tens of μas. Most of them are not included in the current IAU or IERS models. Those terms are often referred to as of second order, although they can be cast in at least two distinct groups. The first group is made of second order terms in the sense of perturbation theory (crossing of the ordinary first order precession–nutation terms with themselves). They are part of a solution that is non-linear with respect to the dynamical ellipticity *H*. The other group gathers small terms of various physical origins but sharing some properties: arising from unaccounted terms of the potential, like high frequency nutations (ESCAPA *et al.* 2002) and indirect effect of sectorial and tesseral third order harmonics (FERRÁNDIZ *et al.* 2003), effect of fluid core on the precession (FERRÁNDIZ *et al.* 2004, 2007), direct effects of the actual rotation of the inner core (ESCAPA *et al.* 2012), effects on nutations (FERRÁNDIZ *et al.* 2011) of the observed J_2 variation (CHENG and TAPLEY 2004; CHENG *et al.* 2011; COX and CHAO 2002), and other time variations of the geopotential as unaccounted effects of tidal models (FERRÁNDIZ *et al.* 2011), etc. The Hamiltonian method provided a systematic, consistent procedure to approach all of them in the non-steady, non-rigid case.

6. The Solution for Polar Motion

Theories of earth rotation usually devote a part to calculate the frequencies of the unperturbed or free polar motion (PM), corresponding to the oscillations or wobbles of the axis of angular velocity or angular momentum around the figure axis or vice versa. Let us note that the differences between free periods in the rigid and non-rigid cases are more marked than those in the corresponding forced motions (nutation amplitudes), whose main components are the so-called Poisson terms, practically independent of the Earth model. Conversely, the amplitudes and phases of the polar wobbles are highly dependent on the Earth physics. Woolard already mentioned the relevance of geophysical effects on nutations and what he called diurnal nutations, although his terminology differs from the currently used.

The main components of nutations have long periods in the "inertial" frame. However, the main components of PM have long periods in the terrestrial or body-fixed frame, therefore they are in the diurnal band when seen from the inertial frame. However, the terrestrial frame is more convenient for their study as well as their determination since the advent of radio-interferometric techniques like VLBI.

In fact, the actual motion of the Earths pole, displayed in Fig. 3, has not been fully explained by any theory yet. It includes noticeable changes of amplitudes and phases of its main components (the

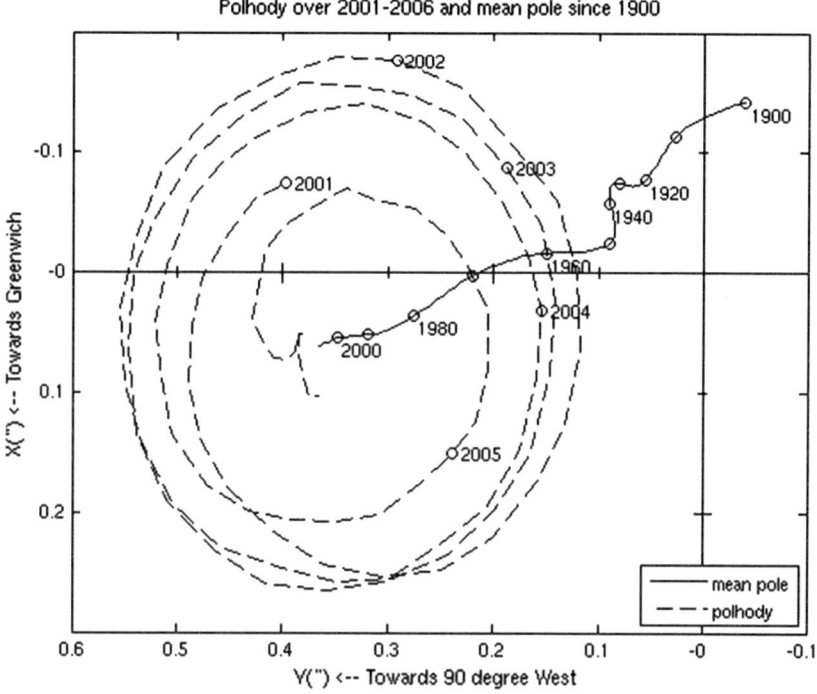

Figure 3
Motion of the Earth pole. Source: IERS

Chandler wobble with an amplitude usually ranging from 100 to 200 μas and the annual term with an amplitude nearby 100 μas) as well as a long-term drift (Barkin 2000a, b; Schuh *et al.* 2001). A thorough review can be found in, e.g., Gross (2007).

Whilst nutations arise from a mainly astronomical forcing, the free wobbles of the pole are excited mainly by geophysical processes, and are difficult to predict (Chao and Gross 1987; Dickey *et al.* 2002, Gross *et al.* 2005). For that reason, the solution for the forced PM is not derived analytically along with the nutations, but is computed from different equations using empirical time series providing the relevant excitation functions (Gross 1992; Brzeziński 1992).

That behaviour was essential for the definition of the set of Earth orientation parameters (EOP) currently in use. In 1982, an IAU Working Group on Nutation (Seidelmann 1982) recommended the adoption of five EOP, namely: the precession/nutation angles ϵ, ψ, referred to the equinox and equator; UT1 (universal time 1), corresponding to the sidereal diurnal revolution and GMST or GAST (Greenwich

Mean Sidereal Time or Greenwich Apparent Sidereal Time); the polar motion angles x, and y.

This set provides the transformation of coordinates from the celestial to the terrestrial frame (or vice-versa) by performing five rotations. The transformation is mathematically redundant, since the relative orientation of two reference systems can be specified by only three independent parameters. Nevertheless, that redundancy was convenient for the analysis of the VLBI observations of Earth rotation, which started in the early 1980s. It proceeded by fitting one set of five EOP to each observation session spanning a whole day.

7. Present State of the Earth Rotation Modelling and Outlook

Since the IERS establishment, EOP solutions are provided by IERS along with several Analysis Centres. Besides VLBI, other techniques contribute to determine a subset of EOP (UT1 and PM), namely satellite laser ranging (SLR) and GNSS (Global

Navigation Satellite Systems). Time series of daily EOP values are produced by IERS (BIZOUARD and GAMBIS 2009), by a combination of individual solutions computed by various associated Analysis Centres for each technique. Nowadays, IERS releases two sets of EOP related by a known transformation, since the nutation offsets dX, dY and the Earth rotation angle (ERA) were recommended to replace the former three equinox-based EOP after a new paradigm was adopted by IAU in 2000, based in the use of the celestial intermediate origin and pole (CIO and CIP, respectively). Precise definitions of the main and auxiliary parameters and frames can be found in, e.g., the IERS Conventions 2010, Supplement to the Nautical Almanac (URBAN and SEIDELMANN 2013) or standards of fundamental astronomy (SOFA) documentation (HOHENKERK 2010).

Current accuracy of EOP series is difficult to assess. Comparisons between combined solutions and individual solutions corresponding to different techniques and analysis centres provide some insight into their accuracy or uncertainty. Following the IERS Annual Report 2011 (DICK 2011), the uncertainty of VLBI solutions may be near 90 μas for nutations in average and in about 170 μas for PM. The accuracy of precession/nutation models, when used to make forward predictions, is stabilised at around 150 μas, in terms of wrms of the observation-model differences, and the figures are larger for PM prediction. The remarkable efforts made in the last years provided a better insight into the problem and unveiled new potential sources of error, but have not been compensated yet by a significant reduction of the residuals. Let us recall that the IAG's Global Geodetic Observing System (GGOS) initiative demands an accuracy of 1 mm to the systems of reference, besides a stability in time of 0.1 mm/year. That corresponds roughly to a value of 30 μas for angular EOP (2 μs for time).

From the observational side, the accuracy and performance of the major techniques is increasing. Therefore, series of more accurate EOP will be available in a few years. Besides, higher time resolution is expected. There are still many difficult open problems, such as magnetic effects (HUANG et al. 2011), motions of inner layers (BARKIN and VILKE 2004), relativistic effects (KLIONER et al. 2009),

consistent and comprehensive treatment of a more realistic time-varying earth model, etc. Clearer separation of nutations and polar motion is also sought as we approach the EOP determination at a sub-diurnal rate (NILSSON et al. 2010) and non-predictable constituents are accounted in the nutation angles like the free core nutation (FCN) (LAMBERT 2007; KRASNA et al. 2013), whereas some short periodic predictable astronomical effects are included into PM (GETINO et al. 2001; ESCAPA et al. 2002, BRZEZINSKI 2001).

In this context, the International Association of Geodesy (IAG) and the International Astronomical Union (IAU) set up a new Joint Working Group on Theory of Earth Rotation (or JWG ThER) in April 2013. The purpose of the new JWG is: "To promote the development of theories of Earth rotation that are fully consistent and that agree with observations and provide predictions of the Earth rotation parameters (ERP) with the accuracy required to meet the needs of the near future as recommended by, e.g., GGOS, the Global Geodetic Observing System of the IAG. Its structure is more complex than usual and adapts to the characteristics of the current EOP, as well as the specialised fields of research. The people in charge are:

- Chair: José M. Ferrándiz (IAU)
- Vice-Chair: Richard S. Gross (IAG)

The JWG is composed of three Sub-Working Groups (SWG):

1. Precession/Nutation (Chair: Juan Getino)
2. Polar Motion and UT1 (Chair: Aleksander Brzezinski)
3. Numerical Solutions and Validation (Chair: Robert Heinkelmann)

These SWG should work independently but in parallel for the sake of efficiency, and they must be linked together as closely as the needs of consistency demand. More information is available in FERRÁNDIZ and GROSS (2014) and on the JWG website: http://web.ua.es/en/wgther/.

7.1. Future Prospects of the Hamiltonian Method

Meeting the stringent GGOS accuracy and stability goals is a challenging task, whose fulfilment requires a joint cooperative effort of the scientific community involved in the determination, modelling and prediction of Earth rotation. In the authors' opinion, the Hamiltonian approach can provide a valuable contribution to the theoretical modelling because of some of its features. First, the treatment addresses the Earth rotation globally, as a whole problem, and its previous results, described in former sections, show that the theory can incorporate any kind of geophysical models or effects that have been considered up to date, like the Earth division in solid and fluid layers, the various assumptions on its elastic behaviour, the dissipations at the layers boundaries, the time variation of the geopotential, etc. It also allows the incorporation of small corrections obtained independently by other theories. The inclusion of all the components in a sole Hamiltonian function (or more precisely formalism, to distinguish the generalised forces) helps to assess the magnitude of any neglected effect and ensures the self-consistency of the developments, so that there is no need to introduce corrections aimed at restoring consistency when some background models are updated. But the essential characteristic of the method is its capability to derive solutions with a prescribed level of accuracy in a systematic way, by calculating the approximate solution up to the suitable order of perturbation (usually first or second, depending on the magnitude of each group of terms), as well as to identify the contributions of the different effects included in the chosen geophysical model. This last property is not shared by any solution derived by numerical integration that also can reach high accuracy, but cannot separate the free motion component of the solution to the Earth attitude from the forced one, which is a difficulty according to the current conventions and EOP definitions.

Acknowledgments

The authors acknowledge the valuable suggestions of the anonymous referees. This work has been partially supported by the Spanish government under Grants AYA2010-22039-C02-01 and AYA2010-22039-C02-02 from Ministerio de Economía y Competitividad (MINECO), the University of Alicante under Grant GRE11-08 and the Generalitat Valenciana, Grant GV/2014/072.

References

ALTAMIMI, Z., COLLILIEUX, X. and METIVIER, L. (2011), *ITRF2008, an improved solution of the International Terrestrial Reference Frame*, J. Geod. *85*(8), 457–473.

BARKIN, T. V. (1998), *Unperturbed chandler motion and perturbation theory of the rotation motion of deformable celestial bodies*, Astron. Astrophys. Trans. *17*(3), 179–219.

BARKIN, Y. V. (2000a), *Towards on explanation of the secular motion of the earth's rotation axis pole*, Astron. Astrophys. Trans. *19*(1), 13–18.

BARKIN, Y. V. (2000b), *Perturbated rotational motion of weakly deformable celestial bodies*, Astron. Astrophys. Trans. *19*(1), 19–65.

BARKIN, Y. V. and FERRÁNDIZ, J. M. (2000), *The motion of the Earth's principal axes of inertia caused by tidal and rotational deformations*, Astron. Astrophys. Trans. *18*, 605–620.

BARKIN, Y. V. and VILKE, V. G. (2004), *Celestial mechanics of planet shells*, Astron. Astrophys. Trans. *23*(6), 533–553.

BIZOUARD, C. and GAMBIS, D. (2009), *The Combined Solution C04 for Earth Orientation Parameters consistent with International Terrestrial Reference Frame 2005*, IAG Symp *134*, 265–270.

BRETAGNON, P. (1982), *Theory for the motion of all the planets—The VSOP82 solution*, Astron. Astrophys. *114*, 278.

BRETAGNON, P. (1988), *Planetary theories in rectangular and spherical variables. VSOP 87 solution*, Astron. Astrophys. *202*, 304–315.

BRETAGNON, P., ROCHER, P., and SIMON, J.-L. (1997), *Theory of the rotation of the rigid Earth*, Astron. Astrophys. *319*, 305–317.

BROUCKE, R. (1970), *How to assemble a Keplerian processor*, Celest. Mech. *2*, 9–20.

BRZEZIŃSKI, A. (1992), *Polar motion excitation by variations of the effective angular momentum function: considerations concerning deconvolution problem.* Manuscr. Geod. *17*, 3–20.

BRZEZIŃSKI, A. (2001), Diurnal and sub-diurnal terms of nutation: a simple theoretical model for a nonrigid Earth, In N. CAPITAINE (ed.), Proc. of the Journées 2000—Systèmes de Référence Spatio-temporels, Observatoire de Paris, pp. 243–251.

CAPITAINE, N., WALLACE, P. T. and CHAPRONT, J. (2003), *Expressions for IAU 2000 precession quantities*, Astron. Astrophys. *412*, 567–586.

CAPITAINE, N., MATHEWS, P. M., DEHANT, V., WALLACE, P. T. and LAMBERT, S. B. (2009), *On the IAU 2000/2006 precession nutation and comparison with other models and VLBI observations*, Celest. Mech. Dyn. Astron. *103*, 179–190.

CHAO, B. F. and R. S. GROSS (1987), *Changes in the Earths rotation and low-degree gravitational field induced by earthquakes*, Geophys. J. Roy. Astr. Soc. *91*, 569–596.

CHANDLER, S.C. (1891) *On the variation of latitude.* Astron. J. *11*, 59–61.

CHAPRONT-TOUZÉ, M. (1980), *La solution ELP du problème central de la Lune*, Astron. Astrophys. 83–86.

CHAPRONT-TOUZÉ, M. (1982), *Progress in the analytical theories for the orbital motion of the Moon*, Celest. Mech. *26*, 53–62.

CHENG, M. and TAPLEY, B. D. (2004), *Variations in the Earth's oblateness during the past 28 years*, J. Geophys. Res. *109*, B09402.

CHENG, M. K., RIES, J. C. and TAPLEY, B. D. (2011), *Variations of the Earth's Figure Axis from Satellite Laser Ranging and GRACE*, J. Geophys. Res. *116*, B01409.

COX, C. M. and CHAO, B. F. (2002), *Detection of a large-scale mass redistribution in the terrestrial system since 1998*, Science *297*, 831–833.

DEFRAIGNE, P. and DEHANT, V. (1998), New theoretical model for nutations and comparison with VLBI observations. In: CAPITAINE, N. (ed) Proc. Journées 1997—Systèmes de Référence Spatio-Temporels, Observatoire de Paris, pp 69–72.

DEHANT, V., DEFRAIGNE, P. and WAHR, J. M. (1999a), *Tides for a convective Earth*, J. Geophys. Res. *104*, 1035–1058.

DEHANT V. *et al.* (1999b), *Considerations concerning the non-rigid Earth nutation theory*, Celest. Mech. Dyn. Astron. *72*, 245–310.

DEHANT, V. (2002), *Report of IAU Working Group on 'Non-rigid Earth rotation theory'*, Highlights of Astronomy *12*, 117–119.

DEPRIT, A., HENRARD, J. and ROM, A. (1971), *Analytical Lunar Ephemeris: Delaunay's Theory*, Astron. J. *76*, 269–272.

DEPRIT, A. and ELIPE, A. (1993), *Complete reduction of the Euler-Poinsot problem*, J. Astronaut. Sci. *41*, 603–628.

DICKEY, J. O. *et al.* (2002), *Recent Earth Oblateness Variations: Unraveling Climate and Postglacial Rebound Effects*, Science, *298*, 1975–1977.

D'HOEDT, S. and LEMAITRE, A. (2004), *The spin-orbit resonant rotation of Mercury: a two degree of freedom Hamiltonian model*, Celest. Mech. Dyn. Astron. *89*, 267–283.

DICK, W. R. (ed) (2011), IERS Annual Report 2011. Verlag des Bundesamts fr Kartographie und Geodsie, Frankfurt AM.

EFROIMSKY, M. and ESCAPA, A. (2009), *The theory of canonical perturbations applied to attitude dynamics and to the Earth rotation. Osculating and nonosculating Andoyer variables*, Celest. Mech. Dyn. Astron. *98*, Issue 4, 251–283.

ESCAPA, A., GETINO, J. and FERRÁNDIZ, J. M. (2001), *Canonical approach to the free nutations of a three-layer Earth model*, J. Geophys. Res. *106*, 11387–11397.

ESCAPA, A., GETINO, J. and FERRÁNDIZ, J. M. (2002), *Indirect effect of the triaxiality in the Hamiltonian theory for the rigid Earth nutations*, Astron. Astrophys. *389*, 1047–1054.

ESCAPA, A., FERRÁNDIZ, J. M. and GETINO, J. (2012), Influence of the inner core on the rotation of the Earth revisited, IAU Joint Discussion 7 "Space-time reference systems for future research", XXVIIIth General Assembly of the International Astronomical Union.

FERRÁNDIZ, J. and BARKIN, Y. (2001), *On integrable cases of the Poincaré problem*, Astron. Astrophys. Trans. *19*, 769–780.

FERRÁNDIZ, J. M., ESCAPA, A., NAVARRO, J. F., and GETINO, J. (2003), Recent work on theoretical modelling of nutation. In: RICHTER, B., SCHWEGMANN, W. and DICK, W.R. (eds) Proceedings of the IERS Workshop on Combination Research and Global Geophysical Fluids, IERS Technical Note 30, pp 163–167.

FERRÁNDIZ, J. M., NAVARRO, J. F., ESCAPA, A. and GETINO, J. (2004), *Precession of the Nonrigid Earth: Effect of the Fluid Outer Core*, Astron. J. *128*, 1407–1411.

FERRÁNDIZ, J. M., NAVARRO, J. F., ESCAPA, A., GETINO, J. and BAENAS, T. (2007), Influence of the mantle elasticity on the precessional motion of a two-layer Earth model, In: LEMAÍTRE, A.

(ed) The rotation of celestial bodies, Press. Universitaires de Namur, pp 9–14.

FERRÁNDIZ, J. M., MARTÍNEZ-ORTIZ, P. A. and GARCÍA, D. (2011), *Effects of time gravity changes on the Earth nutations*, Geophysical Research Abstracts *13*, EGU2011-4981.

FERRÁNDIZ, J. M., BAENAS, T. and ESCAPA, A. (2012), *Effect of the potential due to lunisolar deformations on the Earth precession*, Geophysical Research Abstracts 14, EGU2012-6175.

FERRÁNDIZ, J. M. and GROSS, R. S. (2014), The New IAU/IAG Joint Working Group on Theory of Earth Rotation, IAG Symp 143 (to appear).

FEY, A. L., ARIAS, E. F., CHARLOT, P., FEISSEL-VERNIER, M., GONTIER, A. M., JACOBS, C. S., LI, J. and MACMILLAN, D. S. (2004), *The second extension of the International Celestial Reference Frame: ICRF-EXT. 1*, Astron. J. *127*, 3587–3608.

FOLKNER, W. M., CHARLOT, P., FINGER, M. H., WILLIAMS, J. G., SOVERS, O. J., NEWHALL, X., STANDISH, E. M. Jr. (1994), *Determination of the extragalactic-planetary frame tie from joint analysis of radio interferometric and lunar laser ranging measurements*, Astron. Astroph. *287*, 279–289.

FOLKNER, W. M *et al.* (2014), JPL Interplanetary Network Progress Report 42–196, (2014) Available at http://ipnpr.jpl.nasa.gov/progress_report/42-196/196C.

FUKUSHIMA, T. (2003) *A new precession formula*, Astron. J. *126*, 494–534.

GETINO, J. and FERRÁNDIZ, J. M. (1990), *A Hamiltonian theory for an elastic earth: Canonical variables and kinetic energy*, Celest. Mech. Dyn. Astron. *49*, 303–326.

GETINO, J. and FERRÁNDIZ, J. M. (1991), *A Hamiltonian Theory for an Elastic Earth—First Order Analytical Integration*, Celest. Mech. Dyn. Astron. *51*, 35–65.

GETINO, J. and FERRÁNDIZ, J. M. (1995), *On the effect of the mantle elasticity on the Earth's rotation*, Celest. Mech. Dyn. Astron. *61*, 117–180.

GETINO, J. and FERRÁNDIZ, J. M. (1997), *A Hamiltonian approach to dissipative phenomena between the Earth's mantle and core, and effects on free nutations*, Geophys. J. Int. *130*, 326–334.

GETINO, J. and FERRÁNDIZ, J. M. (2000), *Effects of dissipation and a liquid core on forced nutations in Hamiltonian theory*, Geophys. J. Int. *142*, 703–715.

GETINO, J. and FERRÁNDIZ, J. M. (2000b), Advances in the Unified Theory of the Rotation of the Nonrigid Earth. In: JHONSTON, T. *et al.* (ed) Towards models and constants for sub-microarcsecond astrometry, Proc. IAU Col. *180*, pp 236–241 Geophys. J. Int. *142*, 703–715.

GETINO, J. and FERRÁNDIZ, J. M. (2001), *Forced nutations of a two-layer Earth model*, Mon. Not. R. Astron. Soc. *322*, 785–799.

GETINO, J., FERRÁNDIZ, J. M. and ESCAPA, A. (2001), *Hamiltonian theory for the non-rigid Earth: semidiurnal terms*, Astron. Astroph. *370*, 330–341

GETINO, J., ESCAPA, A. and MIGUEL, D. (2010), *General theory of the rotation of the non-rigid Earth at the second order. I. The rigid model in Andoyer variables*, Astron. J. *139*, 1916–1934.

GROSS, R. S. (1992), *Correspondence between theory and observations of polar motion*, Geophys. J. Int. *109*, 162–170.

GROSS, R. S., FUKUMORI, I. and MENEMENLIS, D. (2005), *Atmospheric and oceanic excitation of decadal-scale Earth orientation variations*, J. Geophys. Res. *110*, B09405.

GROSS, R. S. (2007), Earth rotation variations long period, In: HERRING TA (ed) Physical Geodesy. Treatise on Geophysics vol *3*, Elsevier, Oxford, *239*–294.

HENRARD, J. (1979), *A New Solution to the Main Problem of Lunar Theory*, Celest. Mech. *19*, 337–355.

HENRARD, J. (1986), *Algebraic manipulation on computers for lunar and planetary theories*. In: KOVALEVSKY, J. and BRUMBERG, V. (eds.) Proceedings IAU Symposium, 114, Reidel , pp 59–62.

HILTON, J. L., CAPITAINE, N., CHAPRONT, J., FERRNDIZ, J. M., FIENGA, A., FUKUSHIMA, T., GETINO, J., MATHEWS, P., SIMON, J. L., SOFFEL, M., VONDRAK, J., WALLACE, P. and WILLIAMS, J. (2006), *Report of the Internacional Astronomical Union Division I Working Group on precession and the ecliptic*, Celest. Mech. Dyn. Astron. *94*, 351–367.

HOHENKERK, C., and the IAU SOFA BOARD (2010), SOFA Tools for Earth Attitude. IAU. Available at http://www.iausofa.org

HORI, G. (1966), *Theory of General Perturbation with Unspecified Canonical Variable*, Publ. Astron. Soc. Jpn. *18*, 287–296.

HUANG, C. L., JIN, W. J. and LIAO, X. H. (2001), *A new nutation model of a non-rigid earth with ocean and atmosphere*, Geophys. J. Int. *146*, 126–133.

HUANG, C. L., DEHANT, V., LIAO, X. H., VAN HOOLST, T. and ROCHESTER, M. G. (2011), *On the coupling between magnetic field and nutation in a numerical integration approach*, J. Geophys. Res. *116*, B03403, doi:10.1029/2010JB007713.

JEFFERYS, W. H. (1970), *A Fortran-based list processor for Poisson series*. Celest. Mech. *2*, 474–480.

JEFFREYS, H. and VICENTE, RO. (1957), *The theory of nutation and the variation of latitude: the Roche model core*, Month. Not. Roy. Astron. Soc. *117*, 162–173.

KINOSHITA, H. (1977), *Theory of the rotation of the rigid Earth*, Celest. Mech. Dyn. Astron. *15*, 277–326.

KINOSHITA, H. and SOUCHAY, J. (1990), *The theory of the nutation for the rigid earth model at the second order*, Celest. Mech. Dyn. Astron. *48*, 187–265.

KLIONER, S. A., GERLACH, E., and SOFFEL, M. (2009), Relativistic aspects of rotational motion of celestial bodies, In: S. KLIONER, K. SEIDELMANN, M. SOFFEL (eds.) Relativity in Fundamental Astronomy, Proc. of the IAU Symposium 261, Cambridge University Press, Cambridge, pp 112–123.

KRASINSKI, G.A. (2006), *Numerical theory of rotation of the deformable Earth with the two-layer fluid core. Part 1: Mathematical model*, Celest. Mech. Dyn. Astron. *96*, 169–217.

KRÁSNÁ, H., BÖHM, K. and SCHUH, H. (2013), *Free core nutation observed by VLBI*, Astron. Astrophys. *555*, A29.

LAMBERT, S. (2007), Empirical model of the Free Core Nutation, Technical note, available at http://syrte.obspm.fr/lambert/fcn/

LEIMANIS, E. (1965), The general problem of the motion of coupled rigid bodies about a fixed point. Springer-Verlag, Berlin-Heidelberg-New York.

LIESKE, J. H., LEDERLE, T., FRICKE, W., and MORANDO, B. (1977), *Expressions for the Precession Quantities Based upon the IAU (1976) System of Astronomical Constants*, Astron. Astrophys. *58*, 1–16.

MATHEWS, P. M., HERRING, T. A., BUFFET, B. A. and SHAPIRO, I. I. (1991a), *Forced nutations of the Earth: Influence of inner core dynamics 1. Theory*, J. Geophys. Res. *96*, 8291–8242.

MATHEWS, P. M., HERRING, T. A., BUFFET, B. A. and SHAPIRO, I. I. (1991b), *Forced nutations of the Earth: Influence of inner core dynamics 2. Numerical results and comparisons*, J. Geophys. Res. *96*, 8243–8257.

MATHEWS, P. M., HERRING, T. A. and BUFFET, B. A. (2002), *Modelling of nutation and precession: New nutation series for nonrigid Earth and insights into the Earth's interior*, J. Geophys. Res. *107 B4*, 2068–2094.

MCCARTHY, D. D. (ed) (1996), IERS Conventions, IERS Technical Note, 21, Observatoire de Paris, Paris, available at http://www.iers.org/TN21

MORITZ, H. (1982), *A variational principle for Moledensky's liquid-core problem*, Bull. Geod. *56*, 381–400.

NAVARRO, J. F. and FERRÁNDIZ, J. M. (2002), *A new symbolic processor for the Earth rotation theory*, Celest. Mech. Dyn. Astron. *82*, 243–263.

NAVARRO, J. F. (2002), Teoría analítica de la rotación de la tierra rígida mediante manipulación simbólica, Doctoral Dissertation.

NEWCOMBE, S. (1892), *On the dynamics of the Earth's rotation, with respect to the periodic variations of latitude*, Mon. Not. R. Astron. Soc. 248–249, 336–341.

NEWCOMBE, S. (1898), *Sur les formules de nutation basées sur les décisions de la conférence de 1896*, Bull. Astron. *15*, 241–246.

NILSSON, T., BÖHM, J. and SCHUH, H. (2010), *Sub-diurnal Earth rotation variations observed by VLBI*, Artificial Satellites, *45*, No. 2.

PETIT, G. and LUZUM, B., IERS Conventions (2010), IERS Technical Note 36, Verlag des Bundesamtes für Kartographie und Geodäsie, Frankfurt am Main.

PLAG, H. P., Gross, R. S. and ROTAHACHER, M. (2009a), *Global geodetic observing system for geohazards and global change*, Geosciences, BRGM's Journal for a Sustainable Earth *9*, 96–103.

PLAG, H. P. and PEARLMAN, M. (eds), Global Geodetic Observing System: Meeting the Requirements of a Global Society on a Changing Planet in 2020, Springer-Verlag, Berlin-Heidelberg, (2009b).

PLAG, H. P., RIZOS, C., ROTHACHER, M. and NEILAN, R., The Global Geodetic Observing System (GGOS): Detecting the Fingerprints of Global Change in Geodetic Quantities. In: Advances in Earth Observation of Global Change, CHUVIECO, E., LI, J., YANG, X. (eds.), Springer (2010).

POINCARÉ, H. (1901), *Sur une forme nouvelle des équations de la mécanique*, C.R. Acad. Sci. Paris, *132*, 369–371.

POINCARÉ, H. (1910), *Sur la précesion des corps déformables*, Bull. Astronom. *27*, 321–356.

RICHARDSON, D. L. (1989), *PARSEC: An interactive Poisson series processor for personal computing systems*, Celest. Mech. Dyn. Astron. *45*, 267–274.

ROM, A. (*1969*), *Mechanized algebraic operations (MAO)*, Celest. Mech. *1*, 301–319.

ROOSBEEK, F. and DEHANT, V. (1998), *RDAN97: An analytical development of rigid Earth nutations series using the torque approach*, Celest. Mech. Dyn. Astron. *70*, 215–253.

SASAO T., OKUBO S., SAITO M. (1980), A simple theory on the dynamical effects of a stratified fluid core upon nutational motion of the Earth In FEDOROV, E. P., SMITH, M. L., BENDER, P. L. (eds) Nutation and the Earth's rotation, Proc. IAU Symp. 78, pp 165–183.

SCHASTOK J. (1997), *A new nutation series for a more realistic model Earth*, Geophys. J. Int., *130*, 137–150.

SCHUH, H., NAGEL, S. and SEITZ, T. (2001), *Linear drift and periodic variations observed in long time series of polar motion*, J. Geod. *74*: 701–710.

SEIDELMANN, P. K. (1982), *1980 IAU theory of nutation—the final report of the IAU Working Group on Nutation*, Celest. Mech. *27*, 79–106.

SHIRAI, T. and FUKUSHIMA, T. (2000), *Numerical Convolution in the Time Domain and Its Application to the Nonrigid-Earth Nutation Theory*, Astron. J., *119*, 2475–2480.

SOUCHAY, J. and KINOSHITA, H. (1996), *Corrections and new developments in rigid earth nutation theory. I. Lunisolar influence including indirect planetary effects*, Astron. Astrophys. *312*, 1017–1030.

SOUCHAY, J. and KINOSHITA, H. (1997), *Corrections and new developments in rigid-Earth nutation theory. II. Influence of second-order geopotential and direct planetary effect*, Astron. Astrophys. *318*, 639–652.

SOUCHAY, J., LOSLEY, B., KINOSHITA, H. and FOLGUEIRA, M. (1999), *Corrections and new developments in rigid Earth nutation theory III. Final tables REN-2000 including crossed-nutation and spin-orbit coupling effects*, Astron. Astrophys. Suppl. Ser. *135*, 111–131.

TISSERAND, F.F. (1891), Traité de Mécanique Céleste, T. II Théorie de la figure des corps célestes et de leur mouvement de rotation. Gauthier Villars, Paris. Reprinted by Jacques Gabay, Paris, 1990.

URBAN, S. E. and SEIDELMANN P. K. (eds) (2013), The Explanatory Supplement to the Astronomical Almanac. University Science Books, Mill Valley.

VICENTE, R. O. and JEFFREYS, H. (1964), *Nearly diurnal nutation of the Earth*, Nature *204*, 120–121.

WAHR, J. M. (1981), *The forced nutations of an elliptical, rotating, elastic and oceanless Earth*. Geophys. J. Roy. Astron. Soc. *64*, 705–727.

WILLIAMS, J. G. (1994), *Contributions to the Earth's obliquity rate, precession, and nutation*, Astron. J. *108*, 711–724.

WOOLARD, E. W. (1953a) Theory of the rotation of the Earth around its center of mass, Goddard Space Flight Center.

WOOLARD, E. W. (1953b) *A revedelopment of the theory of nutation*, Astron. J. *58*, 1–3.

(Received April 7, 2014, revised June 4, 2014, accepted June 9, 2014, Published online July 17, 2014)

Pure Appl. Geophys. 172 (2015), 75–89
© 2014 Springer Basel
DOI 10.1007/s00024-014-0912-x

Pure and Applied Geophysics

Geosystemics: A Systemic View of the Earth's Magnetic Field and the Possibilities for an Imminent Geomagnetic Transition

ANGELO DE SANTIS[1,2] and ENKELEJDA QAMILI[1]

Abstract—Geosystemics is a way to see and study the Earth in its wholeness together with the eventual couplings among the subsystems composing our planet. This paper will provide this view for the Earth's magnetic field, reviewing most of the results obtained in our recent works. The main tools used by geosystemics are some nonlinear quantities, such as some kinds of entropy. Through them, it is possible to: (a) establish the chaoticity and ergodicity of the recent geomagnetic field in a direct and simple way; and (b) indentify the most extreme events in its history, as the most rapid and the slowest ones, i.e., jerks and polarity changes (reversals or excursions). In particular, regarding the latter phenomena, with the help of these entropic concepts and together with the use of the theory of critical transitions, some clues can be given for a possible imminent change of the geomagnetic field dynamical regime.

Key words: Geosystemics, Geomagnetic field, Chaos, Ergodicity, Global transition.

1. Introduction

Earth is an ever-changing planet. This statement is obvious if we take into due account some important processes happening in our planet, i.e., a growing population and, consequently, pollution; climate change, biodiversity and resources reduction, and a greater weakness of the present society against disasters caused by geohazards such as earthquakes, volcanic eruptions, hurricanes, etc. Our planet is a complex system constituted by numerous subsystems interacting with each other (SKINNER and PORTER, 1995). For this reason, the understanding of our planet in its whole complexity is a challenging task.

Here, we will deepen the study of an important property of Earth, i.e., the geomagnetic field, and we will then try to prove or to reject the hypothesis that this field is going toward a global transition. In addition, we will try to find out whether and how other processes interact with this field. More in detail, recently a lot of interest has been dedicated to how the present geomagnetic field is quite distinct from the field of the recent past. GUBBINS (1987) found that the southern hemisphere gives the largest contribution to the present decrease in the dipole moment, which is directly related to the intensification and southward movement of a pair of patches of reverse flux under South Africa. This state could eventually lead to a polarity change in terms of a geomagnetic reversal or excursion. Through an inversion of MAGSAT and Ørsted satellite magnetic data, HULOT et al. (2002) confirmed GUBBINS (1987) results, identifying a reverse magnetic flux under the southern hemisphere. RAJARAM et al. (2002) and DE SANTIS (2007) found a rapid fall of the recent geomagnetic field in Antarctica. JACKSON (2003) identified some intense equatorial flux spots on the top of the Earth's core, as manifestation of a high variability in the core. BLOXHAM et al. (2002) explained the frequency increase of the number of jerks, which are very rapid variations in the change of the slope of the secular variation (SV) with a time scale of around one year, as likely due to an increased excitation of torsional waves towards the end of the last century. GUBBINS et al. (2006) evidenced that the recent geomagnetic field increased its rate of decay from 1840 by about 5 % per century even though they attributed this fact to an erratic aspect of the present field. DE SANTIS et al. (2004) proposed that the present geomagnetic field could be in a chaotic state next to a geomagnetic

---
[1] Istituto Nazionale di Geofisica e Vulcanologia, Sezione Roma 2, Rome, Italy. E-mail: angelo.desantis@ingv.it
[2] Università "G. D'Annunzio", Campus Universitario, Chieti, Italy.

Reprinted from the journal

reversal or excursion, with significant possible implications in the biosphere, in the atmosphere, and in some other components of the Earth system (CONSTABLE and KORTE, 2006). All the above papers express some evidence for an irregular, likely chaotic, state of the present geomagnetic field, with some possibilities for an imminent change of magnetic polarity.

A problem that we can meet when we investigate in detail some aspects of the geomagnetic field is a concern with the conventional multipolar approach, i.e., dipolar/non-dipolar representation of the geomagnetic field potential. This approach may not provide clear information regarding the dynamical and configurational properties of the whole field, because the huge dipolar contribution mostly obscures the results taken from the other multipoles. Therefore, other kinds of analyses must be exploited by following a more holistic approach to the problem. This approach looks at the field for its wholeness rather than at each specific and minute part of it (e.g., DE SANTIS, 2009, 2014). This is exactly what we will do in this paper. We will try to use the concepts of *Geosystemics* with the purpose of having a more complete view of the geomagnetic field system. To do it, we will establish and compare the variation of the Shannon Information and some other quantities in order to affirm that the present geomagnetic field behaviour is consistent with a possible current planetary transition in terms of a significant change of its main characteristics (energy, dipole moment, related core dynamics, etc.) possibly going toward an excursion or even a reversal.

This article is a review of our recent contributions to the understanding of the recent geomagnetic field. In the next section we will introduce the concept of Geosystemics, and then we will define some useful mathematical tools like the Shannon entropy (and Information) and its application to the geomagnetic field. We will also illustrate a new technique for the detection of geomagnetic jerks. Then, we will study in detail an important feature of the geomagnetic field, i.e., the South Atlantic Anomaly (SAA) and will relate this feature with another physical quantity, the global sea level (GSL), strictly connected with climate.

2. Geosystemics

Geosystemics studies the Earth system from a holistic point of view, looking with particular attention at self-regulation phenomena and relations among the parts composing the Earth, together with the possible trends of change or persistence of the specific system or subsystem under study (DE SANTIS, 2009, 2014). This approach puts at its centre the concepts of Entropy and Information content. To characterise the world, not only energy and matter are important, but also information is important (BEKENSTEIN, 2003). In particular self-regulation, nonlinear coupling, emergent behaviour, and irreversibility have to be taken into due account since these are important constituents of the planet, and so they must be matter of study for Geosystemics. In particular, the information exchanged and the increased entropy allow us to better understand those irreversible processes occurring in the Earth's interior. Geosystemics has the objective to observe, study, represent and interpret those aspects of geophysics that determine the structural characteristics and dynamics of our planet and the complex interactions of the elements that compose it. Some universal nonlinear tools are fundamental for Geosystemics: among many, we will focus on information and entropy. There will be also important an approach based on multi-scale/parameter/platform observations in order to cover and monitor the particular subsystem of Earth under study as much as possible. Although this latter aspect will not be considered in this review, it is a fundamental issue of geosystemics, because there is no better way to understand the behaviour of a complex system than looking at it from as many perspectives as possible.

3. Shannon Entropy and Shannon Information

The concept of Shannon entropy $H(t)$ (SHANNON, 1948) is an important tool which can be used for the space–time characterization of a dynamical system. In the case of a system characterized by N possible independent states, this entropy is defined in a certain time t as follows:

$$H(t) = -\sum_{n=1}^{N} p_n(t) \cdot \log p_n(t), \qquad (1)$$

where $p_n(t)$ represents the probability of the system to be at the nth state. For convenience, we impose $\sum_n p_n = 1$ and $\log p_n = 0$ if $p_n = 0$ to remove the corresponding singularity.

In the literature we can find a wide number of physical interpretations of the Shannon entropy. Among these, we choose the simplest one. That is, it is a non-negative measure of our ignorance about the state of the system of concern. The Shannon entropy has a great importance in studying and interpreting the behaviour of complex systems like Earth in general, and the geomagnetic field, in particular. On the other hand, we find in the literature also Shannon Information, $I(t)$, which is simply related to $H(t)$ as $I(t) = -H(t)$. It is a negative quantity that measures our knowledge on the state of the system when we know only the distribution of probability $p(t)$ (BECK and SCHLÖGL, 1993). In practice, this quantity denotes our decreasing ability to predict the evolution of the system into the future.

4. Shannon Information and Entropy of the Geomagnetic Field

The Shannon Information has been already applied to the present (DE SANTIS et al., 2004) and recent past (DE SANTIS and QAMILI, 2010) geomagnetic field, $\mathbf{B}(t)$, that can be defined at and above the Earth's surface as the negative gradient of a scalar potential $V(t)$. In turn, this potential can be expressed, at a given time t, by a spherical harmonic expansion in space characterized by a set of Gaussian coefficients, $g_n^m(t)$, $h_n^m(t)$, with $n = 1, \ldots N$ degrees and $m = 0, \ldots n$ orders of the potential field expansion. We can define the Shannon Information $I(t)$ of the geomagnetic field as:

$$I(t) = -H(t) = \sum_{n=1}^{N} p_n(t) \cdot \log p_n(t) \qquad (2)$$

$p_n(t)$ is the probability of having a particular nth multipole rather than another, and it is calculated as (DE SANTIS et al., 2004):

$$p_n = \frac{\langle B_n^2 \rangle}{\langle B^2 \rangle} = \frac{(n+1)\left(\frac{a}{r}\right)^{2n+4} \sum_{m=0}^{n} (c_n^m)^2}{\sum_{n'=1}^{N} (n'+1)\left(\frac{a}{r}\right)^{2n'+4} \sum_{m=0}^{n'} (c_{n'}^m)^2}. \qquad (3)$$

In this formula $(c_n^m)^2 = (g_n^m)^2 + (h_n^m)^2$ and $a = 6{,}371.2$ km. $\langle B^2 \rangle$ and $\langle B_n^2 \rangle$ are the mean squared amplitudes over the sphere with radius r of the total field and of the field due to the nth multipole, respectively (LOWES, 1966). Although the brackets $\langle \ldots \rangle$ denote strictly spatial averages of the squared field strength over the terrestrial sphere, actually they are also time averages because any global model of the geomagnetic field, for the way it is constructed, is a smoothed averaged model in time and space. Therefore, the definition of the Shannon Information (2) with the probabilities (3) assumes an ergodic geomagnetic field. In the next section we will confirm this property of the field. It is important to underline the fact that the time behaviour of the Shannon Information for the geomagnetic field can help us in understanding a possible chaotic scenario for the dynamical system that generates and sustains the field, in terms of geodynamo models (CHILLINGWORTH and HOLMES, 1980). The slow temporal variation of the geomagnetic field with time scales from years to thousand years, is called SV: mathematically, it is defined as the time derivative of the field. Therefore, we can introduce also an analogous Shannon Information for the SV, i.e., $I(SV)$ (we relax the dependence with time, that is implicit) where the corresponding probability $\tilde{p}_n$ will be similar to that of Eq. (3), but $(c_n^m)^2$ will be replaced by $(\dot{c}_n^m)^2 = (\dot{g}_n^m)^2 + (\dot{h}_n^m)^2$:

$$\tilde{p}_n = \frac{\langle \dot{B}_n^2 \rangle}{\langle \dot{B}^2 \rangle} = \frac{(n+1)\left(\frac{a}{r}\right)^{2n+4} \sum_{m=0}^{n} (\dot{c}_n^m)^2}{\sum_{n'=1}^{N} (n'+1)\left(\frac{a}{r}\right)^{2n'+4} \sum_{m=0}^{n'} (\dot{c}_{n'}^m)^2}. \qquad (4)$$

DE SANTIS et al. (2004) showed that the recent geomagnetic field is in the particular situation that $p_n \approx \tilde{p}_n$, which is a necessary (but not sufficient) condition for a geomagnetic reversal or excursion.

In Fig. 1 some synthetic examples are given for different Shannon entropy H values (DE SANTIS and QAMILI, 2008). These configurations are given in terms of the normalised Shannon entropy H^* (with value between 0 and 1) as:

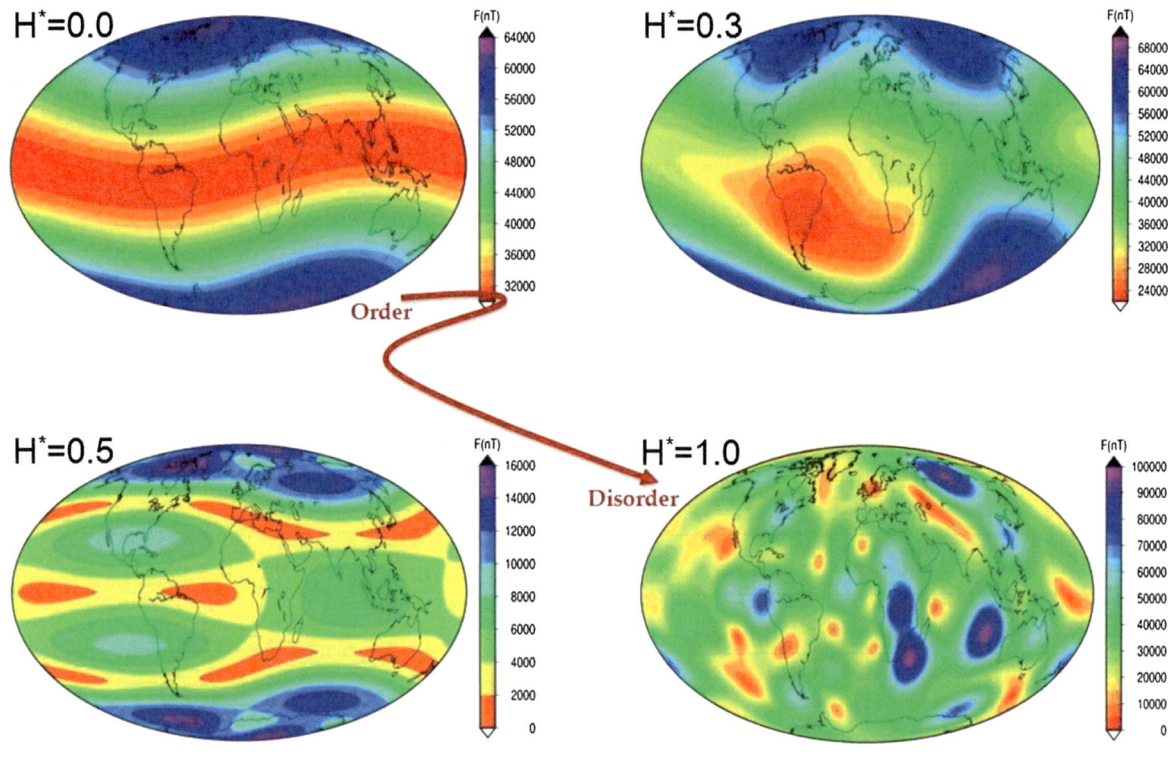

Figure 1

Four geomagnetic total intensity configurations with different increasing (decreasing) normalised Shannon entropy (Information), H^* from 0 to 1 (from 0 to −1). The *upper right* case is real and represents the present geomagnetic field, while the other three cases are synthetic examples (from DE SANTIS and QAMILI, 2008)

$$H^* = \frac{H}{H_{max}} = \frac{I}{I_{min}} \qquad (5)$$

with $H_{max} = -I_{min} = \log N$. In this figure, the example with entropy $H^* = 0.3$ represents the real case of the present geomagnetic field deduced from the IGRF-11 model in 2010 (FINLAY *et al.*, 2010), while the others are synthetic cases. From these examples it is clear that the Shannon entropy provides a way to measure the degree of complexity of the field spatial configuration. The higher the Shannon entropy (or the lower Shannon Information), the more all probabilities are equally possible, and, then, the more complex the derived spatial configuration of the field will be. The interpretation of this is that when the Shannon Information is low there is a lower degree of organization for the system under study (see for instance the recent review by Balasis *et al.*, 2013). For this reason the corresponding Shannon entropy is also called spatial or configuration entropy (e.g., RODRIGUEZ-ITURBE *et al.*, 1998). Of course, when

we apply these concepts to the real geomagnetic field, we recognise that the value of the information quantity must be referred to a specific reference radial distance, *r*. For instance, while the normalised entropy of the present real geomagnetic field is around 0.3 at the Earth's surface, it becomes 0.8 at the core mantle boundary (CMB). This is normal because the field is more complex going toward the sources, and the increase of entropy denotes an increase of complexity.

The temporal trends of $I(t)$ of the real geomagnetic field for the last 7,000 years, at the Earth's surface and at CMB, by using CALS7K (KORTE and CONSTABLE, 2005), CALS3K (KORTE *et al.*, 2009) and IGRF-11 (FINLAY *et al.*, 2010) global models, are shown in Fig. 2 (DE SANTIS and QAMILI, 2010). All these global models are based on a spherical harmonic representation of the field. The trends in this figure show also the estimated error bars associated to CALS7K moving from a maximum of 33 % at 5000

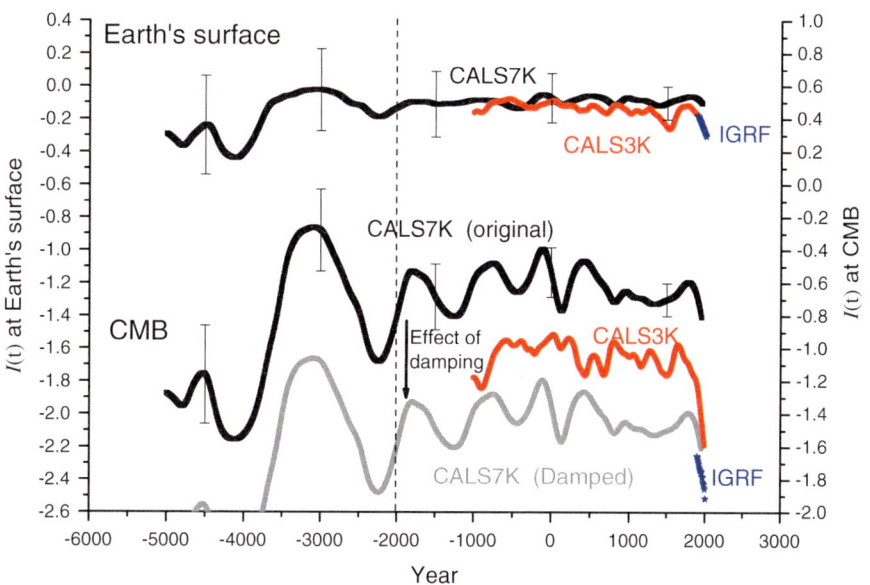

Figure 2

Shannon Information $I(t)$ of the geomagnetic field from 5000 BC to present from CALS7K (*black*), CALS3K (*red*) and IGRF-11 (*blue*) models, at the Earth's surface and at the CMB. For visual convenience, the estimated error vertical bars are shown every 1,500 years for CALS7K only (from DE SANTIS and QAMILI, 2010)

BC to 3 % at 1950 AD. In order to allow the "contact" of the $I(t)$ of CALS7K at the CMB with that of IGRF-11, here we show the possible effect of the spectral damping typical of CALS7K model (at the Earth's surface $I(t)$ is practically unaffected). Our results indicate that the present Shannon Information is much lower than the Shannon Information of the past, i.e., the present field is much more chaotic than the field of the past.

When a process is ergodic and chaotic, $I(t)$ can be related to the Kolmogorov entropy or K-entropy that represents the rate of loss of information, by (WALES, 1991):

$$K = -\frac{dI}{dt}. \qquad (6)$$

This quantity measures the degree of unpredictability of the future evolution of the system between successive points on the trajectory in the phase space (BECK and SCHLOGL, 1993; BUCHNER and ZEBROWSKI, 1998). An alternative definition of the K-entropy can be given also as the sum of all positive Lyapunov exponents of the dynamical system (SCHUSTER and JUST, 2005). A practical consequence is that after a characteristic time $\tau = 1/K$, the

system's behaviour can no longer be predicted. Figure 3 shows the K-entropy of the geomagnetic field as derived from Eq. (6) from 5000 BC to the present at the Earth's surface and at CMB. Here, the K-entropy is derived as a linear fit to each 100-year interval at the Earth's surface and at CMB; each single IGRF-11 value is given in the plots as a star. The K-entropy of the present field is rather high with respect to the past (DE SANTIS and QAMILI, 2010). Although we cannot exclude that some marginal contribution to this feature could be due to insufficient data in the past when compared with the present "abundance", we think that the effect is mostly real, because our definition of the K-entropy is based on the dynamical variations of the Shannon Information of global models with realistic spatial power spectra, and so it is with a reliable repartition between dipolar and non-dipolar parts.

5. Ergodicity of the Recent Geomagnetic Field

In a general case, the information quantities introduced in the previous section by Eqs. (3) and

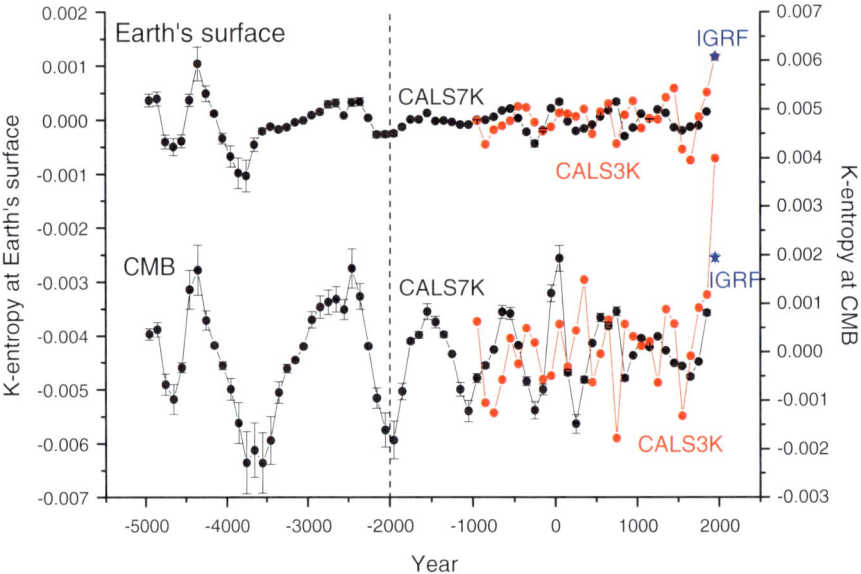

Figure 3

K-entropy of the field from 5000 BC to present from CALS7K (*black*), CALS3K (*red*) and IGRF-11 (*blue*) models, at the Earth's surface and at the CMB. The present value of *K*-entropy of the field based on the IGRF-11 model is the highest over all the investigated period (from DE SANTIS and QAMILI, 2010)

(6), should be estimated in the phase space. However, if we prove that, apart from being chaotic, the geomagnetic field is also ergodic, i.e., time average of the original signal is equal to the density average in the phase space (ECKMANN and RUELLE, 1985), then the phase space reconstruction will not be necessary, and we can perform all the analysis in the time domain. In that case, if we want to investigate the nonlinearities present in a system, we can simply perform a nonlinear forecasting approach in the time domain. Taking into account the chaotic properties of the geomagnetic field, any small change ε of the initial orbit in the phase space propagates exponentially with time, i.e., $\varepsilon(t) = \varepsilon_0 \exp(K \cdot t)$, where K is the above defined *K*-entropy.

Let us consider a generic measure ρ of the dynamical system moving in the phase space Ω. For every continuous function φ, a dynamics f is called ergodic if it has the same behaviour averaged over time as averaged over phase space and the space average is weighted by the invariant measure ρ (ECKMANN and RUELLE, 1985). Under general assumptions, this can be expressed mathematically with the following equation:

$$\lim_{T \to \infty} \frac{1}{T} \int_0^T \varphi[f^t(x_0)] \mathrm{d}t = \int \rho(\mathrm{d}x)\varphi(x). \quad (7)$$

This means that if the system is ergodic, after a certain time evolution, the system is no longer dependent on its initial state x_0 (EGOLF, 2000). In the case of the geomagnetic field, the invariant measures are the *K*-entropy and its inverse value $\langle \tau \rangle = 1/K$, i.e., the limiting mean time of prediction.

Considering all the global geomagnetic models present in the literature, we could have large errors if we extrapolate the geomagnetic field outside their typical time of validity (DE SANTIS *et al.*, 2011, 2013a). These errors can be estimated from a comparison between the predicted and definitive parts of each model. More precisely, these errors can be calculated by means of Gauss coefficients from the formula (MAUS *et al.*, 2008):

$$\varepsilon = \sqrt{\sum_{n=1}^{N} (n+1) \sum_{m=0}^{n} \left[(c_n^m)_{\text{pred}} - (c_n^m)_{\text{def}} \right]^2}. \quad (8)$$

An example of the trend of these errors is given in Fig. 4, where one 10-year segment from 1965 to 1975 and seven 5-year segments from 1975 to 2010 taken

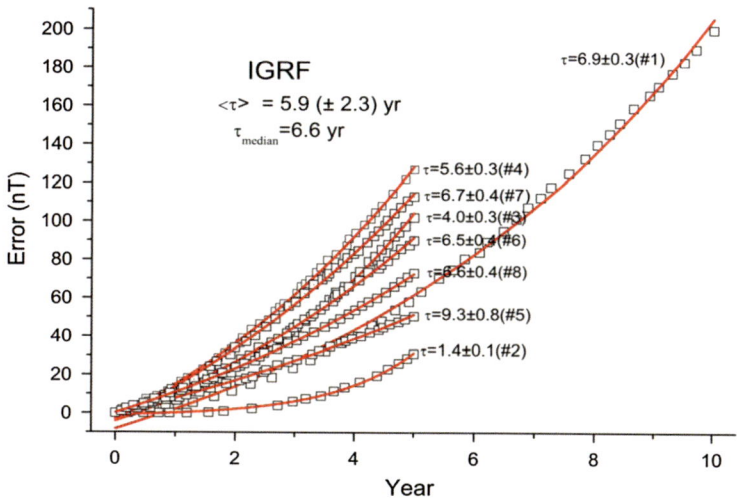

Figure 4

Time evolution of errors between the predicted and observed global IGRF-11 model (from DE SANTIS *et al.* 2011). The exponential increase in time is a symptom of a chaotic geomagnetic field

from the IGRF-11 global model have been considered. DE SANTIS *et al.* (2011) have analyzed also other global models like CHAOS (OLSEN *et al.*, 2014), CM4 (SABAKA *et al.*, 2004), GUFM1 (JACKSON *et al.*, 2000), WMM (MAUS *et al.*, 2010), and POMME (MAUS *et al.*, 2005), and obtained the same results (but they are not shown here). Since the IGRF-11 model gives a constant predictive field, in this analysis we have used the CM4 model as predictive part in order to avoid this problem. For visual convenience, we have imposed the same initial time. This means that each exponential growth will have an offset of $-\varepsilon_0$. In this case the formula for the calculation of the errors, substituting the K-entropy with its inverse τ, will be:

$$\varepsilon(t) = \varepsilon_0 \exp(t/\tau) - \varepsilon_0 = \varepsilon_0(\exp(t/\tau) - 1). \quad (9)$$

In this figure, for each segment we indicate the corresponding τ values together with their associated errors. As a result, all the segments show a clear exponential growth with characteristic mean time $\langle\tau\rangle = 5.9 \pm 2.3$ years. We can conclude that the exponential temporal divergence of the errors between several couples of predictive and definitive global geomagnetic models supports a chaotic state of the present geomagnetic field with no reliable prediction after around six years (see DE SANTIS *et al.*, 2011). This result confirms the nonlinear analysis

performed in the phase space by DE SANTIS *et al.* (2002) and has direct consequences in repeating magnetic surveys and updating global and regional models of the geomagnetic field (DE SANTIS *et al.*, 2013a). The total agreement of these analyses made in the phase space and in the time domain, confirms the ergodicity of the geomagnetic field (DE SANTIS *et al.*, 2011).

6. *Jerks as Chaotic Fluctuations of the Geomagnetic Field*

Geomagnetic jerks have been generally identified in geomagnetic observatory time series (e.g., COURTILLOT *et al.*, 1978; MANDEA *et al.*, 2010). The fact that some timescales of jerk occurrences are overlapping with those of the solar activity (e.g., sunspots cycle of almost 11 years) complicates the clear identification of jerks in the geomagnetic field time series. A better alternative was presented recently. In this section we will describe the results obtained by QAMILI *et al.* (2013) who extended the nonlinear forecasting approach in the time domain over the last 400 years, which is a period covered by GUFM1 (JACKSON *et al.*, 2000) global model with the objective to reinterpret the geomagnetic jerks. We analyze the temporal behaviour of the differences between predicted and definitive values of the geomagnetic field calculated from this model in order to find periods more or less chaotic than others. The *predicted/definitive* comparison is made over successive 10-year segments, moving at steps of one year. The problem is that the GUFM1 model does not give a predictive field. This part was calculated by extrapolating the prior 10-year SV, into the subsequent ten years and comparing them with the real GUFM1 field values for the same period of time. Each of the analyzed segments shows an evident exponential growth with a characteristic time of predictability $\langle \tau \rangle \approx 6 \pm 2.5$ years. The temporal fluctuations of τ value around its mean linear trend from 1600 to 1980 are shown in Fig. 5. What is clear from this figure is that the past field is less predictive than the recent one, because the number of accentuated negative fluctuations of $\langle \tau \rangle$ increases with time. A simple explanation of this result could be the progressive improvement of the data quality used to build the GUFM1 model. Around this general trend, we have found some interesting fluctuations, i.e., periods where the geomagnetic field is more chaotic (smaller time of predictability τ) and also periods where the field is less chaotic (greater time of predictability τ). Checking carefully all the epochs where the τ value becomes suddenly lower with respect to the values that surround it (a sort of V-shape in the temporal behaviour of τ), we find that most of these epochs corresponds to already known geomagnetic jerks (epochs evidenced by arrows but considering an uncertainty of a few years for each event), detected by other authors (MANDEA *et al.*, 2010 and references therein). But not all the chaotic fluctuations correspond to already known geomagnetic jerks. This could be because the techniques introduced till now for the identification of geomagnetic jerks, where most of them are applied to direct measurements, had not been able to detect all these features produced by the geomagnetic field. For this reason, here, we detect a number of new (undetected till now) geomagnetic jerks (events evidenced by blue arrows in Fig. 5). As a conclusion, we can say that geomagnetic jerks appear in those epochs where the geomagnetic field is more chaotic. It is interesting to notice that the more recent field is characterised by more frequent jerks. During the applied analysis, we have identified also some short periods where the field appears less chaotic than usual but this aspect will need more investigation to understand the corresponding origin.

7. *Toward a Global Geomagnetic Transition?*

Some previous (e.g., GUBBINS, 1987; HULOT *et al.*, 2002; DE SANTIS 2007) and more recent (DE SANTIS *et al.*, 2013b) results show that it is very important to investigate the geomagnetic field of the southern hemisphere, since it contributes more to the overall decaying trend of the geomagnetic field. Could this be considered as a symptom that the Earth's magnetic field is going toward a global transition? We will see that this hypothesis is fostered by the presence of one of the most important features of the present geomagnetic field, i.e., the SAA (figure with $H^* = 0.3$ in Fig. 5). We follow here the reasoning of the recent article by DE SANTIS *et al.* (2013b).

The SAA is a significant depression in the total intensity of the present geomagnetic field that has been persisting at least for the last 400 years. It is generally interpreted as the Earth's surface expression of a magnetic vortex present in the outer core, as a component of a strong reversed magnetic flux (OLSON and AMIT, 2006). During the last 400 years, the SAA has changed in space and in time. If we consider its extension from 1590 to present using GUFM1 and IGRF-11 models (we have considered the extent of the 32,000 nT isoline because it is the lowest value in the oldest epoch), we obtain the trend shown as a thick curve in Fig. 6. The continuous and accelerating growth of this anomaly is evident, especially during the last 250 years. We could ask whether this acceleration happens just by chance or not. Figure 7 shows the real acceleration of the SAA in the last 400 years (red curve) compared with 10,000 simulations (blue curves) where all SAA increments have been randomly shuffled. Green curves represent the maximum acceleration (lower green curve) and deceleration (upper green curve). The real acceleration of the SAA stands clearly at the lower limit of the possibilities (maximum acceleration), supporting the case that the present situation is not occurring just by chance.

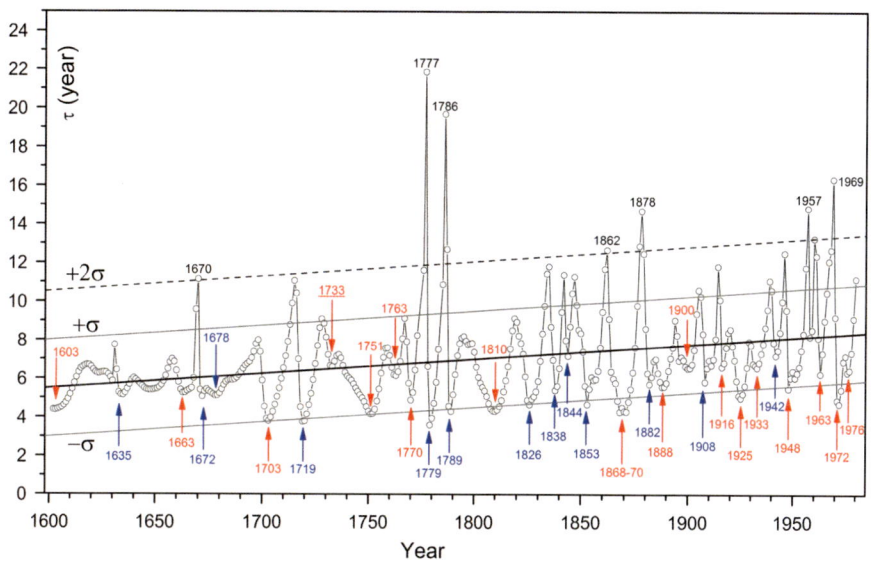

Figure 5

Estimation of the time of predictability $\langle\tau\rangle$ every year over the period 1600–1980 from the GUFM1 model. The epochs of already noted geomagnetic jerks are indicated by *red arrows*, and those for which new possible events are suggested by *blue arrows*. The mean trend is indicated by the best fit line across the data points: the linear increase in time can be ascribed by a better quality of data and model with time (from QAMILI *et al.*, 2013). The most recent times are characterised by more frequent jerks

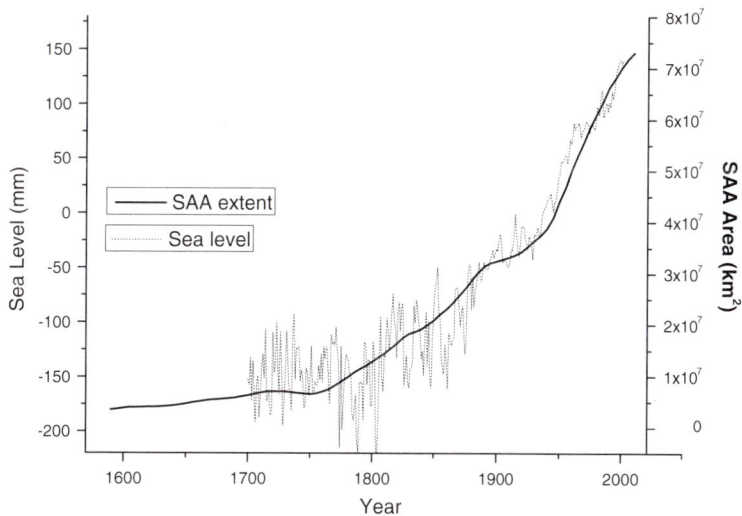

Figure 6

Extent of the SAA surface area obtained from GUFM1 and IGRF-11 (1590–2010) from 1590 to 2010, together with the global sea level rise (original data set) from 1700 to 2002 (redrawn from DE SANTIS *et al.*, 2012)

DE SANTIS *et al.* (2012) found that also another, apparently unrelated quantity, the GSL rise (JEVREJ-EVA *et al.*, 2008) has followed the same growing trend during the last three centuries (thin curve in Fig. 6). To assess a real correlation between the two time series, some statistical tests have been performed, i.e., the Spearman correlation test (DAVIS, 1986) and the Kullback–Leibler entropy (KULLBACK and LEIBLER, 1951). The results taken from the statistics (both with or without a trend removal), confirm the high correlation between the SAA extension area and GSL (see DE SANTIS *et al.*, 2012, 2013b). Although correlation

83

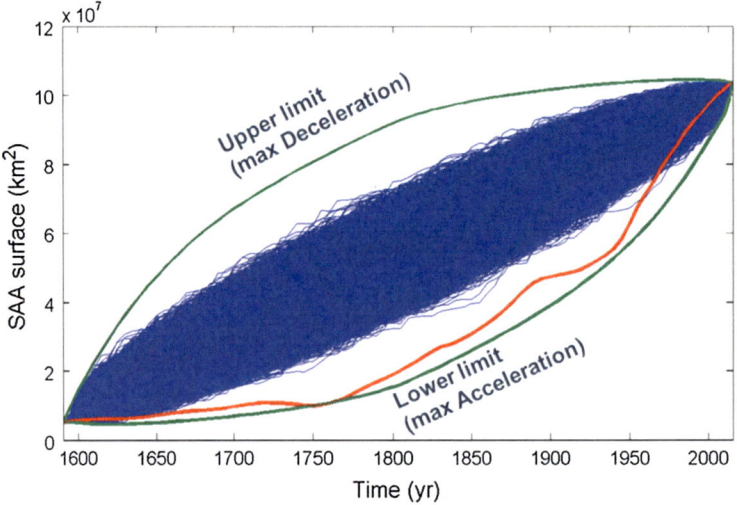

Figure 7
The real acceleration of the SAA in the last 400 years (*red curve*) compared with simulated data (*blue curves*) where all SAA increments have been randomly shuffled (using Matlab routines). *Green curves* represent the maximum acceleration (*lower green curve*) and deceleration (*upper green curve*)

does not always mean causation, we should consider this possibility as a serious hypothesis. In that case, what physical mechanism could be behind the observed correlation? DE SANTIS *et al.* (2012) proposed three possible mechanisms (two external and one internal):

1. An increase of the SAA area facilitates the entrance of charged particles from space. As result we have a warmer atmosphere, which implies a consequent melting of major ice caps (Antarctica and Greenland) that finally causes a global increase of sea level;

2. A possible reduction of the ozone layer in the upper stratosphere over the South Atlantic region can modify the radiative flux at the top of the atmosphere and, hence, can cause changes in the weather and climate patterns, including cloud coverage;

3. Both SAA and GSL time variations could share the same common internal cause, i.e., a convective dynamism in the outer core causes a variation of the magnetic field and an elastic deformation at the Earth's surface (GREFF-LEFFTZ *et al.*, 2004).

An interesting question concerns the best temporal function that fits the SAA surface area change

in time. We will see that this function follows the typical behaviour in time of critical systems, i.e., those complex systems approaching a critical transition.

The deformation (or energy release) $y(t)$ of a material that approaches a failure satisfies the following empirical equation (VOIGHT, 1989):

$$\ddot{y} = a\dot{y}^{\alpha}, \qquad (12)$$

where a and α are two empirical constants. The latter is an exponent that measures the degree of nonlinearity and normally takes values between 1 and 2. We can extend the concept of the failure of a material to critical systems approaching their tipping point, i.e., the time when the system undergoes a dramatic (usually abrupt) change of its dynamical properties. In this way, Eq. (12) assumes a more universal importance. Indeed, the solutions of Eq. (12) have been largely applied for the prediction of different critical systems like volcanic eruptions (VOIGHT, 1988), earthquake main failure (BUFE and VARNES, 1993), financial crashes (SORNETTE, 2003), and magnetic storms (BALASIS *et al.*, 2011), etc. Integrating Eq. (12) for $\alpha \neq 1$, we obtain a first-order equation whose solution takes the form of a power-law increase with time:

$$y = A + B(t_c - t)^p, \qquad (13)$$

where t_c represents the time to failure or critical time of the system under study, $p = [(2 - \alpha)/(\alpha - 1)]$ is a power-law exponent (usually <1); $A > 0$ and $B < 0$ are parameters to be found from the experimental data. VANDEWALLE et al. (1998) introduced a logarithmic function in alternative of the power-law form, i.e., $y = A + B \ln(t_c - t)$. SORNETTE and SAMMIS (1995) propose a more generalised solution which is decorated by a log-periodic function. Thus, we can write:

$$y(t) = A + B \ln(t_c - t) \\ \cdot \{1 + D \cdot \cos[2\pi f \ln(t_c - t) + \phi]\}. \quad (14)$$

D is the magnitude of the log-periodic fluctuations around the acceleration growth, f the frequency of the fluctuations, and ϕ is the phase shift.

We applied Eq. (14) over SAA (Fig. 8) and GSL (Fig. 9) trends (DE SANTIS et al., 2013b). Equation (14) fits very well both the considered time series, with a very high correlation coefficient r (in both cases $r > 0.98$). Since original GSL data set is very noisy (especially the oldest values), before the fit we averaged the data every five years (black circles in Fig. 9). Regarding the estimation of the critical time t_c, we find practically the same time, i.e., $t_c \approx 2,034 \pm 3$ years for SAA and $t_c \approx 2,033 \pm 11$ years for GSL (estimated errors are just statistical, as they could be even larger than indicated; see the end of this section). Also the D and f parameters are very similar in both SAA and GSL fits, indicating that the fluctuations affect the acceleration in almost the same way in both physical quantities. In practice, the A parameter of the SAA fit corresponds to the value of the SAA area at the critical time (actually at $t_c - 1$ year), that, in this case, will reach more than 50 % of the whole Earth surface. Alternative fits with functions with a comparable number of coefficients (such as, for instance, a 5°-polynomial in time), are equally possible; however, they are much more unrealistic outside the data they use. All these results suggest that the same trend of these quantities is not a mere coincidence, and, probably, both these systems are behaving as dynamical systems close to a critical point.

However, some words of caution are necessary. Any fit made with the function of (14) is rather instable when the critical point is far, so it needs as much data as possible before it can provide some stable result: in general, the fit converges to a stable result as the data approach the critical transition (e.g., see SORNETTE et al., 2004 for landslide predictions). To have some quantitative idea of the stability (or instability) of this kind of analysis, we apply the fit in subsequent segments of the SAA dataset, in particular from 1590 to 1960, then from 1590 to 1965, and so on, till the last epoch of 2010, and for every fit we keep note of the prediction, t_{pred}, of the critical transition. This means that in every analysis we truncate the data at some time $t_{max} < t_c$ and use only the data up to t_{max}. We choose the SAA data instead of GSL data because the latter are much noisier and reducing the points of the fit would provide very unstable results. We can then consider the difference $\Delta t = t_{pred} - t_{max}$, i.e., between prediction (t_{pred}) and epoch (t_{max}) at which the prediction is made for the last 50 years (Fig. 10). A linear fit over Δt would predict $\Delta t = 0$ at around 2,060, indicating that the most recent prediction (made at $t_{max} = 2,010$) of $t_c \approx 2,034$ years (or, alternatively, the error of ± 3 years) is probably underestimated. However, in the last 15 years, Δt almost stopped decreasing, so also this linear prediction must be taken with some caution (Fig. 10). What we can affirm is that the analysed time series has been so far behaving as a critical system, but we will need more time and more data before to completely confirm this result. This is also due to the fact that a critical system reveals its criticality as it approaches more closely to the critical point. It is for this reason that log-periodic functions are called "sloppy" functions (e.g. BRÉE et al., 2013).

8. Conclusions

In this paper we made an overview of the complex characteristics of the present Earth's magnetic field by means of the Geosystemics approach, reviewing most of the results obtained in our recent works. More precisely, useful tools like entropy and Information have been applied to the present and to the recent past geomagnetic field in

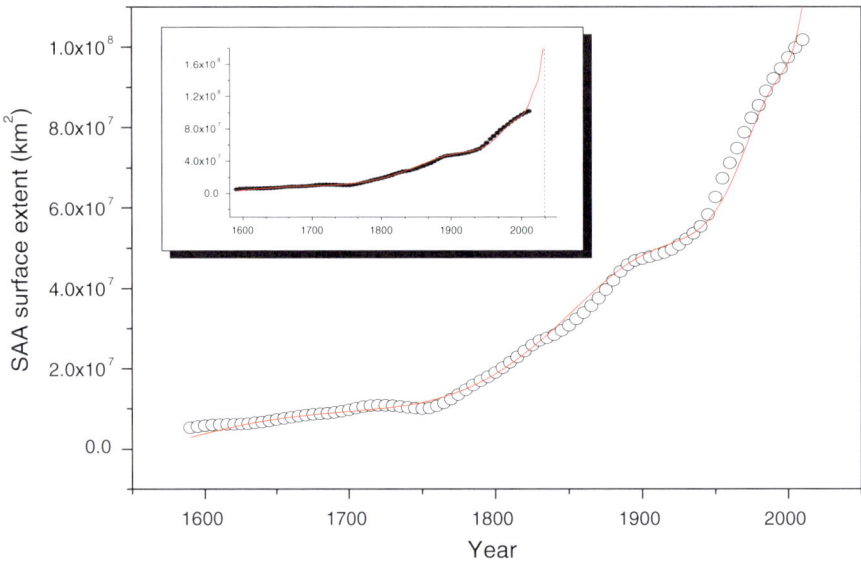

Figure 8
Extent of the SAA surface area over the last 400 years and the best nonlinear fit with the function indicated in the text as Eq. (14). The "critical time" $t_c \approx 2{,}034 \pm 3$ years, where the curve will approach a singularity. This time could represent the time of no return for a great change of the geomagnetic field, possibly going toward a reversal or excursion (redrawn from DE SANTIS et al., 2013b)

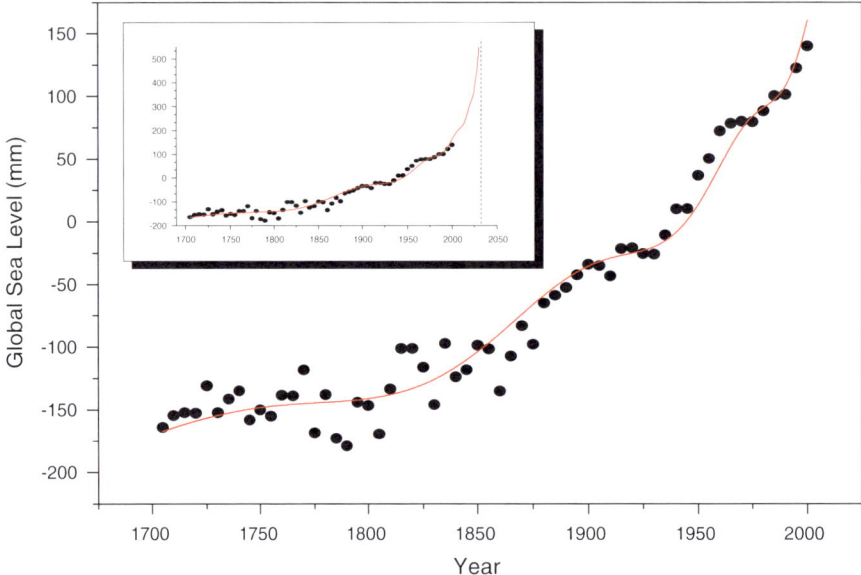

Figure 9
Global sea level rise (averaged every five years) and its best log-periodic fit with the critical time $t_c \approx 2{,}033 \pm 11$ years (redrawn from DE SANTIS et al., 2013b)

order to derive important information regarding the corresponding dynamical system originating in the outer terrestrial core. Moreover, the temporal evolution of an important feature of the present geomagnetic field, the SAA, has been deeply investigated. Since the main objective of Geosystemics is to study the Earth system from the holistic point of view, together with the SAA we have also analysed another physical quantity, the GSL, and studied its possible correlation with the

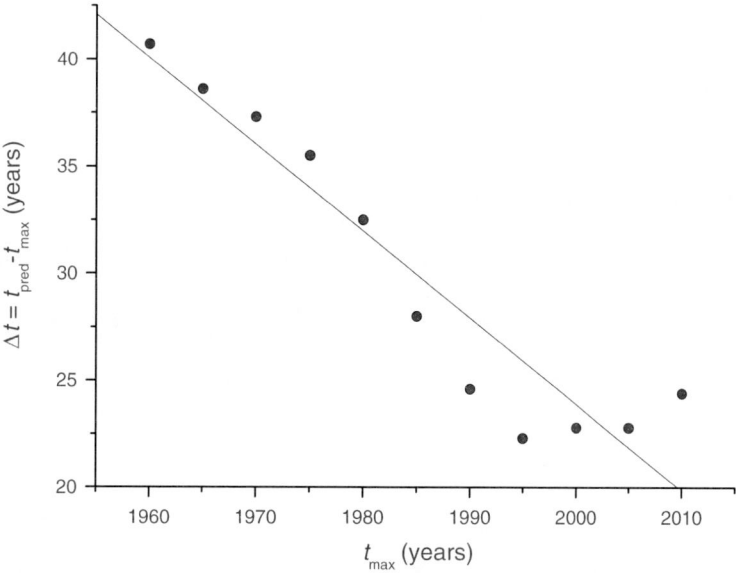

Figure 10

Behavior in time of the difference Δt between prediction (t_{pred}) and epoch (t_{max}) at which the prediction is made for the last 50 years. A linear fit predicts $\Delta t = 0$ at around 2,060, confirming that the most recent prediction (made at $t_{max} = 2,010$) of $t_c \approx 2,034 \pm 3$ years is probably underestimated, or, alternatively, it is the error of ± 3 years that is underestimated. However, in the last 15 years Δt almost stopped decreasing, so also this linear prediction must be taken with some caution

geomagnetic field. After all these analysis, we can conclude that the present geomagnetic field is more chaotic than the one of the past. In addition, there are many intriguing aspects that encourage us to suggest that the present geomagnetic field is rather special and a possible imminent change of geomagnetic polarity could be not so unexpected. We find the interesting result that both SAA and GSL can be described by a critical system evolution, with similar critical time that can be interpreted as "the point of no return" for both the whole geomagnetic field and GSL. We do not interpret the critical time as the exact moment of a geomagnetic reversal, because the typical diffusion time of the Earth's core would require a few thousand years, We rather consider t_c as the time when the irreversible process, that will drive the magnetic field to change its polarity, will start. To make a simple figurative analogy, consider the case in which we play with a ball near a deep well: if the ball falls into the well, the critical point will be the moment at which its centre of gravity is beyond the border of the well and not the time the

ball will touch the bottom. About the exact time of the critical transition, we warned about its uncertainty, mostly due to the instability of the fit of expression (14) over any experimental data. For this reason, further investigation is still needed in the near future to confirm or confute the found results. In particular, the recent ESA Swarm mission (OLSEN and HAAGMANS, 2006) of three twin satellites (launched on 22 November 2013) with precise magnetic sensors aboard will be an unprecedented occasion to verify the present results with great accuracy.

Acknowledgments

We thank three referees (Vladimir Kossobokov and two anonymous) for their useful comments that improved the quality of the paper. The first author (ADS) would like to thank also the organisers of the conference "Mathematics and Geosciences: Global and Local Perspectives", in particular Maria Luisa Osete, for the kind invitation for a keynote presentation and for her enthusiastic encouragement. The simulations shown in Fig. 7 (together with the figure) have been kindly performed by Francisco Javier Pavón-Carrasco using Matlab© routines.

References

Balasis, G., C. Papadimitriou, I. A. Daglis, A. Anastasiadis, L. Athanasopoulou, and K. Eftaxias (2011), *Signatures of discrete scale invariance in Dst time series*, Geophys. Res. Lett., *38*, L13103, doi:10.1029/2011GL048019.

Balasis, G., R. V. Donner, S. M. Potirakis, J. Runge, C. Papadimitriou, I. A. Daglis, K. Eftaxias, and J. Kurths (2013), *Statistical mechanics and information-theoretic perspectives on complexity in the Earth system*, Entropy, *15* (11), 4844–4888; doi:10.3390/e15114844.

Beck, C., and Schlögl, F. (1993), Thermodynamics of Chaotic Systems (Cambridge University Press, Cambridge) p. 306.

Bekenstein, J.D. (2003), *Information in the Holographic Universe*, Scientific American, *289*, 2, August 2003, p. 61.

Bloxham, J., Zatman, S., and Dumberry, M. (2002), *The origin of geomagnetic jerks*, Nature *420* (6911), 65–68.

Brée D., Challet D. and Peirano P.P. (2013) *Prediction accuracy and sloppiness of log-periodic functions*, Quantitative Finance, *13*, 275–280.

Buchner, T., and Zebrowski, J. (1998), *Local entropies as a measure of ordering in discrete maps*, Chaos Soliton. Fract. *9* (1/2), 19–28.

Bufe, C.G. and Varnes, D. J. (1993), *Predictive modelling of the seismic cycle of the Greater San Francisco Bay region*, J. Geophys. Res. *98*, 9871–9883.

Chillingworth, D.R.J. and Holmes, P.J. (1980), *Dynamical systems and models for the reversals of the Earth's magnetic field*, Math. Geol. *12*, 41–59.

Constable, C.G., and Korte, M. (2006), *Is Earth's magnetic field reversing?*, Earth Planet. Sci. Lett. *246*, 1–16.

Courtillot, V., J. Ducruix, and J.L. LeMouël (1978), *Sur une accélération récente de la variation séculaire du champ magnétique terrestre*, C. R. Acad. Sci. Paris Ser. D, *287*, 1095–1098.

Davis, J.C. (1986), Statistics and Data Analysis in Geology. Wiley & Son, New York.

De Santis, A. (2007), *How persistent is the present trend of the geomagnetic field to decay and, possibly, to reverse?*, Phys. Earth Planet. Int. *162*, 217–226.

De Santis, A. (2009), Geosystemics, Proceedings of WSEAS Conference on Geology and Seismology, GES'09, Feb. 2009 Cambridge, 36–40.

De Santis, A. (2014), Geosystemics, entropy and criticality of earthquakes: a vision of our planet and a key of access, in "Nonlinear phenomena in Complex Systems: from Nano to Macro Scale" ed. E. Stanley and D. Matrasulov, NATO Science for Peace and Security Series—C: Environmental Security, 3–20.

De Santis, A., and Qamili, E. (2008), *Are we going towards a global planetary magnetic change?* 1st WSEAS International Conference on Environmental and Geological Science and Engineering (EG'08), 149–152.

De Santis, A., and Qamili, E. (2010), *Shannon Information of the geomagnetic field for the past 7000 years*, Nonlinear Proc. Geoph. *17*, 77–84.

De Santis, A., Barraclough, D.R., and Tozzi, R. (2002), *Nonlinear variability of the Recent Geomagnetic Field*, Fractals *10*, 297–303.

De Santis, A., Tozzi, R., and Gaya-Piqué, L.R. (2004), *Information content and K-Entropy of the present geomagnetic field*, Earth Planet. Science Lett. *218*, 269–275.

De Santis, A., Qamili, E., and Cianchini, G. (2011), *Ergodicity of the recent geomagnetic field*, Phys. Earth Plan. Int. *186*, 103–110.

De Santis, A., Qamili, E., Spada, G., and Gasperini, P. (2012), *Geomagnetic South Atlantic Anomaly and global sea level rise: a direct connection?*, J. Atmos. Sol. Terr. Phys. *74*, 129–135.

De Santis A., Qamili E. and Cianchini G. (2013a), *Repeat-station surveys: implications from chaos and ergodicity of the recent geomagnetic field*, Annals of Geophysics *56* (1), R0103, doi:10.4401/ag-5491.

De Santis A., Qamili E. and Wu L.X. (2013b), *Toward a possible next geomagnetic transition?* Nat. Haz. Earth Syst. Sc. *13*, 3395–3403.

Eckmann, J.-P., and Ruelle, D. (1985), *Ergodic theory of chaos and strange attractors, Part 1*, Rev. Mod. Phys. *57*(0.3), 617–654.

Egolf, D. (2000), *Equilibrium regained: from nonequilibrium chaos to statistical mechanics*, Science *287*, 101–104.

Finlay, C.C., et al. (2010), *International Geomagnetic Reference Field: The eleventh generation*, Geophys. J. Int. *183*, 1216–1230, doi:10.1111/j.1365-246X.2010.04804.x.

Greff-Lefftz, M., Pais, M.A., Le Mouel, J.-L. (2004), *Surface gravitational field and topography changes induced by the Earth's fluid core motions*. Journal of Geodesy *78*, 386–392.

Gubbins, D. (1987), *Mechanism for geomagnetic polarity reversals*, Nature *326*, 167–169.

Gubbins, D., Jones, A.L., and Finlay, C.C. (2006), *Fall in Earth's Magnetic Field is erratic*, Science *312*, 900–902.

Hulot, G., Eymin, C., Langlais, B., Mandea, M., and Olsen, N. (2002), *Small-scale structure of the geodynamo inferred from Ørsted and Magsat satellite data*, Nature *416*, 620–623.

Jackson, A. (2003), *Intense equatorial flux spots on the surface of Earth's core*, Nature *424*, 760–763.

Jackson, A., Jonkers, A.R.T. and Walker, M.R. (2000), *Four centuries of geomagnetic secular variation from historical records*, Phil. Trans. R. Soc. Lond. A, *358*, 957–990.

Jevrejeva, S., Moore, J.C., Grinsted, A., and Woodworth, P.L. (2008), *Recent global sea level acceleration started over 200 years ago?*, Geophys. Res. Lett. *35*, L08715, doi:10.1029/2008GL033611.

Korte, M. and Constable, C.G. (2005), *Continuous geomagnetic field models for the past 7 millennia: 2. CALS7K*, Geochem. Geophy. Geosy. *6*, Q02H16, doi:10.1029/2004GC000801.

Korte, M., Donadini, F. and Constable, C.G. (2009), *Geomagnetic field for 0–3 ka: 2. A new series of time-varying global models*, Geochem. Geophy. Geosy. *10*, Q06008, doi:10.1029/2008GC002297.

Kullback, S., and Leibler, R.A. (1951), *On information and sufficiency*. The Annals of Mathematical Statistics 22 (1), 79–86.

Lowes, F. J. (1966), *Mean-square values on sphere of spherical harmonic vector fields*, J. Geophys. Res. *71*, 2179.

MANDEA, M., HOLME, R., PAIS, A., PINHEIRO, K., JACKSON, A., and VERBANAC, G. (2010), *Geomagnetic Jerks: Rapid Core Field Variations and Core Dynamics*, Space Sci Rev. *155*, 147–175.

MAUS S., LÜHR H., BALASIS G., ROTHER M. and MANDEA M. (2005), Introducing POMME, the POtsdam Magnetic Model of the Earth, in "Earth Observation with CHAMP", ed. by C. REIGBER, H. LÜHR, P. SCHWINTZER and J. WICKERT, pp. 293–298, Springer, New York.

MAUS, S., SILVA, L., and HULOT, G. (2008), *Can core-surface flow models be used to improve the forecast of the Earth's main magnetic field?* J. Geophys. Res. *113*, B08102, doi:10.1029/2007JB005199.

MAUS, S., MACMILLAN S., MCLEAN S., HAMILTON B., THOMSON A., NAIR M., and ROLLINS C. (2010) The US/UK World Magnetic Model for 2010–2015, NOAA Technical Report NESDIS/NGDC.

OLSEN, N., and HAAGMANS, R., (Guest Editors) (2006) *Swarm—The Earth's Magnetic Field and Environment Explorers*. Special Issue, Earth Planets Space, *58*, 349–496.

OLSEN, N., LÜHR, H., FINLAY, C.C., SABAKA, T.J., MICHAELIS, I., RAUBERG, J., and TOFFNER-CLAUSEN, L. (2014) *The CHAOS-4 geomagnetic field model*. Geophys. J. Int., doi:10.1093/gji/ggu033.

OLSON, P., and AMIT, H. (2006), *Changes in earth's dipole*, Naturwissenschaften *93*, 519–542.

QAMILI, E., DE SANTIS, A., ISAC, A., MANDEA, M., and DUKA, B. (2013), *Geomagnetic jerks as chaotic fluctuations of the Earth's magnetic field*, Geochem. Geophys. Geosyst. *14*, 839–850, doi:10.1029/2012GC004398.

RAJARAM, G., ARUN, T. DHAR, A., and PATIL, G. (2002), *Rapid decrease in total magnetic field F at Antarctic stations—its*

relationship to core-mantle features, Antarctic Science *14*, 61–68.

RODRIGUEZ-ITURBE, I., D'ODORICO, P., and RINALDO, A. (1998), *Configuration entropy of fractal landscapes*, Geophys. Res. Lett. *25*(7), 1015–1018.

SABAKA T.J., OLSEN N., PURUCKER M.E. (2004) *Extending comprehensive models of the Earth's magnetic field with Ørsted and CHAMP data*, Geophys. J. Int., *159*:521–547.

SCHUSTER H.G., and JUST W. (2005) Deterministic chaos. An introduction. Wiley-VCH Verlag GmbH & Co.KGaA, Weinbein, pp. 287.

SHANNON, C. (1948), *A mathematical theory of communication*, Bell System Technical Journal *27*, 379–423, 623–656.

SKINNER, B.J., and PORTER, S.C., The blue planet: an introduction to earth system science (Wiley and Sons, 1995).

SORNETTE, D., Why stock markets crash. Critical events in complex financial systems (Princeton Univ. Press, Oxford, 2003).

SORNETTE, D., and SAMMIS, C. (1995) *Complex critical exponents from renormalization group theory of earthquakes: implications for earthquake predictions*, J. Phys. I France *5*, 607–619.

SORNETTE D. HELMSTETTER A. ANDERSEN J.V. GLUZMAN S., GRASSO J.-R. and PISARENKO V.F. (2004) *Towards Landslide Predictions: Two Case Studies*. Physica A *338*, 605–632.

VANDEWALLE, N., AUSOLOS, M., BOVERAUS, P., and MINGUET, A. (1998), *How the financial crash of October 1997 could have been predicted*, Eur. Phys. J. B., *4*, 139–141.

VOIGHT, B. (1988), *A method for prediction of volcanic eruptions*, Nature *332*, 125–130.

VOIGHT, B. (1989), *A relation to describe rate-dependent material failure*, Science *243*, 200–203.

WALES, D.J. (1991), *Calculating the rate of loss of information from chaotic time series by forecasting*, Nature *350*, 485–488.

(Received March 28, 2014, revised July 14, 2014, accepted July 19, 2014, Published online August 3, 2014)

Pure Appl. Geophys. 172 (2015), 91–107
© 2014 Springer Basel
DOI 10.1007/s00024-014-0919-3

Non-Dipole and Regional Effects on the Geomagnetic Dipole Moment Estimation

S. A. Campuzano,[1,2] F. J. Pavón-Carrasco,[3] and M. L. Osete[1,2]

Abstract—The study of the temporal evolution of the dipole moment variations is a forefront research topic in Earth sciences. It constrains geodynamo simulations and is used to correct cosmogenic isotope production, which is evidence of past solar activity, and it is used to study possible correlations between the geomagnetic field and the climate. In this work, we have analysed the main error sources in the geomagnetic dipole moment computation from palaeomagnetic data: the influence of the non-dipole terms in the average approach, the inhomogeneous distribution of the current palaeomagnetic database, and the averaging procedure used to obtain the evolution of the dipole moment. To evaluate and quantify these effects, we have used synthetic data from a global model based on instrumental and satellite data, the International Geomagnetic Reference Field: 11th generation. Results indicate that the non-dipole terms contribute on a global scale of <6 % in the averaged dipole moment, whereas the regional non-dipole contribution can show deviations of up to 35 % in some regions such as Oceania, and different temporal trends with respect to the global dipole moment evolution in other ones, such as Europe and Asia. A regional weighting scheme seems the best option to mitigate these effects in the dipole moment average approach. But when directional and intensity palaeomagnetic information is available on a global scale, and in spite of the inhomogeneity of the database, global modelling presents more reliable values of the geomagnetic dipole moment.

Key words: Geomagnetism, palaeomagnetism, archaeomagnetism, geomagnetic dipole moment, geomagnetic field modelling.

Electronic supplementary material The online version of this article (doi:10.1007/s00024-014-0919-3) contains supplementary material, which is available to authorized users.

[1] Dpto. de Física de la Tierra, Astronomía y Astrofísica I, Universidad Complutense de Madrid (UCM), Avd. Complutense s/n, 28040 Madrid, Spain. E-mail: sacampuzano@ucm.es; mlosete@ucm.es

[2] Instituto de Geociencias (IGEO) CSIC, UCM, Ciudad Universitaria, 28040 Madrid, Spain.

[3] Istituto Nazionale di Geofisica e Vulcanologia (INGV), Via Vigna Murata, 605, 00143 Rome, Italy. E-mail: javier.pavon@ingv.it

1. Introduction

The dipole field is the major contributor to the geomagnetic field at the earth's surface. Its time evolution plays a significant role in constraining geodynamo models (e.g. GLATZMAIER and ROBERTS, 1995; CHRISTENSEN *et al.*, 2010). In addition, accurate determinations of the past dipole moment are needed for appropriately correcting the production rate of cosmogenic isotopes (^{14}C, ^{10}Be) used for reconstructing scenarios of past solar activity (e.g. MUSCHELER *et al.*, 2007; VIEIRA *et al.*, 2011; ROTH and JOOS, 2013). Finally, geomagnetic dipole moment evolution at decadal and centennial time scales is necessary to address debated questions as the possible link between geomagnetic field variations and earth's climate (e.g. GALLET *et al.*, 2005; USOSKIN *et al.*, 2008; GENEVEY *et al.*, 2013).

The dipole moment (DM) can be estimated when a global geomagnetic model is available. Spherical harmonic analysis (SHA) is the methodology usually used in the generation of global models of the earth's magnetic field (WHALER and GUBBINS, 1981). This technique is based on SH expansion developed by Gauss in 1838, being the potential of the internal geomagnetic field established at any point (r, θ, λ) over the earth's surface:

$$V(r,\theta,\lambda,t) = a \sum_{n=1}^{N} \sum_{m=0}^{n} \left(\frac{a}{r}\right)^{n+1} P_n^m(\cos\theta) \\ \left(g_n^m(t) \cdot \cos m\lambda + h_n^m(t) \cdot \sin m\lambda\right) \quad (1)$$

where a is the mean radius of the earth ($a = 6{,}371.2$ km), P_n^m are the associated Legendre functions with integer degree n and integer order m, $\cos m\lambda$ and $\sin m\lambda$ the Fourier functions, and the N is the maximum degree of the spatial expansion. $g_n^m(t)$ and $h_n^m(t)$ are the spherical harmonic coefficients, also denoted as Gauss coefficients.

The DM is easily calculated from the three first Gauss coefficients. These coefficients (g_1^0, g_1^1 and h_1^1) provide the contribution of an inclined geocentric dipole, and the DM can be obtained as (see JACOBS, 1991):

$$DM = \frac{4\pi}{\mu_0} a^3 \sqrt{\left(g_1^0\right)^2 + \left(g_1^1\right)^2 + \left(h_1^1\right)^2} \qquad (2)$$

where μ_0 is the magnetic permeability of the free space ($\mu_0 = 4\pi \times 10^7$ VsA^{-1}m^{-1}).

When only the axial geocentric dipole is considered, i.e. it is aligned with the earth's rotational axis, the DM is derived in the axial dipole moment (ADM):

$$ADM = \frac{4\pi}{\mu_0} a^3 \left|g_1^0\right| \qquad (3)$$

Nowadays, a dipole tilted by approximately 11° accounts for more than 98 % of the geomagnetic field observed on the earth's surface. The international geomagnetic reference field (IGRF) models describe the evolution of the field during the last century. Their last generation, the IGRF-11 (FINLAY et al., 2010), covers the time span from 1900 to 2010 and is developed using instrumental data and satellite data (for the last few decades). During the last century, both DM and ADM are decreasing with rates around 50×10^{-3} Am^2/year.

To extend the knowledge of field variations to the past, historical directional data (JONKERS et al., 2003), which came from shipboard observations for navigational purposes, have been used in global modelling. The GUFM1 model (JACKSON et al., 2000) is the model based on historical and instrumental data collected from 1590 to 1990 AD. But, due to the lack of historical intensity data before 1832 (GAUSS, 1833), when Gauss developed a method for its measurement, this model had to assume an estimation of the temporal evolution of the first Gaussian coefficient (g_1^0) prior to this epoch. JACKSON et al. (2000) extrapolated linearly the value of this coefficient in the year 1840, and they assumed a constant rate of temporal evolution of 15 nT/year, which corresponds to the average value of the time derivative of g_1^0 from 1850 to 1990.

Prior to 1590 AD, there are no direct measurements of the geomagnetic field elements (declination

D, inclination I, and intensity F), and the description of the field is based on indirect measurements of magnetized materials, such as sediments, lava flows, or heated archaeological artefacts. Each provides different types of palaeomagnetic information due to the different processes involved in its remanence acquisition.

The archaeomagnetic and lava flow data acquire their magnetization by a thermoremanence (TRM) mechanism. Archaeomagnetic data come from heated archaeological structures as pottery, tiles, or bricks. They recorded the geomagnetic field acting during their last heating–cooling process. In the case of lava flows, the magnetization was recorded during their natural cooling after eruption. If the age of these cooling events is well-controlled, these data provide spot records of the ancient geomagnetic field. For this reason, detailed reconstructions of the geomagnetic field variations generally use this kind of information (e.g. KOVACHEVA et al., 2009; GENEVEY et al., 2013).

In contrast, sediments acquire a magnetization throughout depositional and/or post-depositional remanent magnetization processes (DRM and/or pDRM, respectively). This magnetization mechanism is delayed due to the compaction time required to lock in the magnetization. Consequently, geomagnetic field variations recorded by sediments are smoothed and global models derived from this kind of information present smaller variations of the geomagnetic field elements (KORTE et al., 2009). In addition, from sedimentary data, only relative intensities can be determined (e.g. TAUXE, 1993) in contrast to archaeomagnetic and volcanic data, which provide absolute palaeointensities.

In terms of data distribution, the present spatial and temporal distribution of the archaeomagnetic and volcanic data is very inhomogeneous (Fig. 1): for the last 14,000 years, the spatial distribution presents a clear lack of data in the Southern Hemisphere and a high concentration in the European region. In time, 83 % of the data are concentrated into the last 3,000 years, whereas the remaining 17 % is distributed between 12000 and 1000 BC (Fig. 1). The sedimentary data present a slightly better distribution in both space and time (DONADINI et al., 2009), and some authors (e.g. KORTE et al., 2009; KORTE and CONSTABLE, 2011; LICHT et al., 2013) have preferred

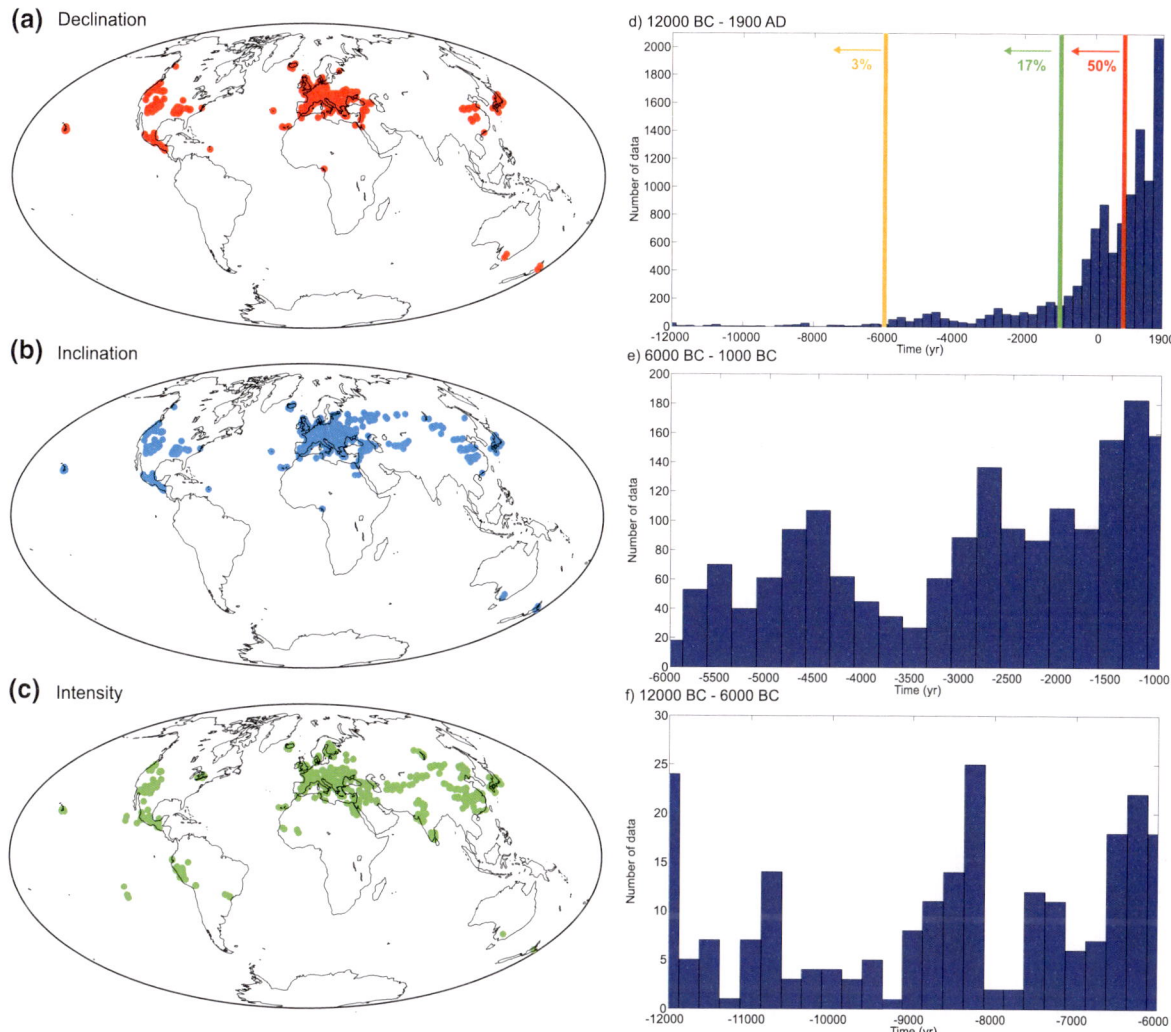

Figure 1

Spatial (**a–c**) and temporal (**d–f**) distribution of the archaeomagnetic and lava flow data for the last 14,000 years. Adapted from Pavón-Carrasco *et al.* (2014)

to include them in the geomagnetic field reconstructions.

Following the above mentioned, the time evolution of the DM given by the global models depends on the data used. On one hand, the archaeomagnetic and volcanic data provide higher temporal variability, but an overfitting of the available data could produce artificially high frequency in the temporal variability. On the other hand, the inclusion of sediment data increases the smoothness of the DM variability. If we take into account the behaviour of the geomagnetic field during the last 170 years (available time span for the GUFM1 and IGRF models), the DM presents a slow temporal evolution. This could be in agreement with the use of sediment data, but a problem arises in that the true frequency content of the DM changes is still not well-known for the last millennia. For this reason, we prefer to use only archaeomagnetic and volcanic data in our study, avoiding the different problems related to the use of sediment data.

A habitual practice to estimate the DM from palaeomagnetic data is by using the geomagnetic field elements F and I and calculating the so-called virtual dipole moment (VDM$_i$) as follows (e.g. Genevey *et al.*, 2008):

$$\mathrm{VDM}_i = \frac{4\pi F_i a^3 \sqrt{1 + 3\cos^2 I_i}}{2\mu_0} \qquad (4)$$

where the sub-index i indicates the value of the VDM for an individual point on the earth's surface. However, not all the intensity data available in the current palaeomagnetic database corresponds to inclination data. This is a common problem in archaeomagnetic studies, for example, when archaeointensities are determined from ceramic fragments where the orientation at the time of cooling is not known. In these cases, the VDM_i cannot be estimated and it is commonly substituted by the virtual axial dipole moment (VADM_i), obtained as

$$\mathrm{VADM}_i = \frac{4\pi}{\mu_0} \frac{F_i a^3}{\sqrt{1 + 3\sin^2 \phi_i}} \qquad (5)$$

where ϕ_i is the latitude of the studied site. In palaeomagnetism, the common procedure to analyse the evolution of the V(A)DM on a millennial timescale is by averaging local values of V(A)DM$_i$ to obtain a mean V(A)DM at regional or global scales (e.g. YANG et al., 2000; MACOUIN et al., 2004; GENEVEY et al., 2008). The best averaging procedure is still an open question and different authors have followed distinct approaches (see GENEVEY et al., 2008 for a deeper discussion).

The main objective of this work is to assess the reliability of the different ways proposed in the literature for magnetic DM determinations from palaeomagnetic data. First (Sect. 2), we carry out a quantitative determination of the non-dipole effect when an averaging process is considered. This effect is produced because the geomagnetic field at the earth's surface, which is recorded by a palaeomagnetic material, contains information about the whole field, not only from the dipole field. The main assumption of palaeomagnetic studies is that the non-dipole contribution of the geomagnetic field is cancelled by averaging.

The next section (Sect. 3) is focused on the impact of the sparse palaeomagnetic data distribution (see Fig. 1) on the computation of the global averages of the V(A)DM. The strong geographical bias of the intensity database toward Eurasia might likely produce erroneous estimations of the global V(A)DM. KORTE and CONSTABLE (2005b) noticed that some care

had to be taken to weight the data properly as a function of their location. GENEVEY et al. (2008) proposed a simple first-order weighting scheme. We study the reliability of this kind of averaging procedure. We denote this analysis as a regional effect.

In the last section (Sect. 4), we study the limitations of the current palaeomagnetic database to generate geomagnetic field models. During the last few decades, global geomagnetic models based on palaeomagnetic data have been developed (KORTE and CONSTABLE, 2003, 2005a, 2011; KORTE et al., 2009; KORTE and CONSTABLE, 2011; LICHT et al., 2013; PAVÓN-CARRASCO et al., 2014). These models allow a direct estimation of the (A)DM (Eqs. 2 and 3). However, they are based on strongly biased databases (see Fig. 1). The objective of this section is to evaluate the effect of the use of a sparse database as input data in the models and to determine how the Gaussian coefficients are affected indirectly for the database used. Henceforth, this effect will be known as the regional indirect effect.

2. Non-Dipole Effect

Global averages of the V(A)DM$_i$ are commonly developed in palaeomagnetism in order to determine the long-scale temporal evolution of the geomagnetic dipole moment. The main assumption is that the non-dipole contribution of the field is cancelled when these averages are calculated. To check the reliability of this assertion, we propose to work with the IGRF-11 model, which spans between 1900 and 2010.

The IGRF-11 model was generated using instrumental data collected from geomagnetic observatories and satellites (CHAMP, Ørsted, and SAC-C missions). This model is based on a spherical harmonic expansion whose maximum degree, N, is chosen so that the coefficients of the model are reliably determined given the available coverage and quality of observations. For IGRF-11, N was chosen to be 10 up to 1995; thereafter it is extended to $N = 13$ to take advantage of the accurate data provided by the Ørsted and CHAMP satellites.

To evaluate the non-dipole effect in the global computation of the V(A)DM, we used synthetic data (F and I) from the IGRF-11 model. Synthetic data computed are defined in a geocentric framework. The high accuracy and the good worldwide coverage of this model assure that our results will not be affected by the regional indirect effect. The data were synthesized in a total of 2,561 points distributed homogeneously over the earth's surface by using all the coefficients of the harmonic expansion. We have computed, every 5 years, the individual VDM_i and $VADM_i$ using Eqs. 4 and 5, respectively. Then, we have calculated the average values, i.e. the global averages, denoted as V(A)DM, which have been compared with the theoretical values DM and ADM (Eqs. 2 and 3) provided by the three first Gaussian coefficients of the IGRF-11 model. The comparison has been quantified by the difference between the V(A)DM and the (A)DM as follows:

$$\sigma_{VADM} = \frac{VADM - ADM}{ADM} \times 100\,\% \qquad (6)$$

$$\sigma_{VDM} = \frac{VDM - DM}{DM} \times 100\,\% \qquad (7)$$

The relative differences are plotted in Fig. 2a and summarized in Table 1S (Supplementary Material). Average values for the whole temporal interval gives differences of 5.4 % between the VADM estimation and the ADM, and 1.7 % between VDM and DM. This result confirms that the non-dipole terms are not completely cancelled after the averaging procedure. However, their contributions are lower (always below 6 % for the last 110 years) than the common errors on palaeointensity estimations: around 10 % (see DONADINI et al., 2009). It is also interesting to point out that all $\sigma_{V(A)DM}$ are positive, which reflects that the V(A)DM is always higher than the (A)DM in the time span from 1900 to 2010.

To investigate which are the most important non-dipole terms affecting the V(A)DM, we have computed these magnitudes varying the maximum degree N (from the dipole, $N = 1$, up to the total field, $N = 13$) of the harmonic expansion of the IGRF-11 model. In Fig. 2b, c the temporal evolution of the V(A)DM computed from the first three field contributions (dipole, $N = 1$; dipole + quadrupole, $N = 2$; and dipole + quadrupole + octupole, $N = 3$) and the total field ($N = 13$) are shown, together with the theoretical (A)DM. We can observe the well-known decrease of the dipole moment, (A)DM, during the

last century. This trend is also presented in the computed V(A)DM, with a decrease of 5.9 % for the VADM and 6.4 % for the VDM, calculated using all the harmonic contributions ($N = 13$). This decreasing tendency is observed in the entire time interval and does not depend on the degree N considered for the analysis. The first three harmonic terms (N from 1 to 3) present the highest contributions to the V(A)DM estimations. When including $N = 4$ and higher terms, no significant differences are observed (see Fig. 1S of the Supplementary Material).

3. Regional Effect

In the previous section, we analysed the influence of the non-dipole contributions to the global V(A)DM estimation with synthetic data from the IGRF-11 model and a dense grid homogeneously distributed all around the world. This kind of homogeneous database is not realistic when we are dealing with palaeomagnetic data. In this section, the objective is to study how the inhomogeneous spatial and temporal distribution of the palaeomagnetic database (Fig. 1) affects the regional averages of the $V(A)DM_i$.

First, we have calculated different regional averages of the $V(A)DM_i$ on a continental scale using a homogeneous grid for each continent (denoted as $V(A)DM_{continent}$). Secondly, we used the original locations of the intensity palaeomagnetic database of GENEVEY et al. (2008) for the last 3,000 years and computed global estimations of the $V(A)DM_i$ directly [V(A)DM] or by using a regional weighting scheme [$V(A)DM_W$]. In both cases, the data were synthesized using the IGRF-11.

We have called this procedure the regional effect (RE), to distinguish it from our previous study. However, we have to remark that the regional effect is also due to non-dipole contributions, which are highlighted by the regional average computation approach.

3.1. Regional Average of the $V(A)DM_i$ on a Continental Scale Using a Homogeneous Database

We selected six different spherical cap areas of 30° of radius, centred in the star points of Fig. 3,

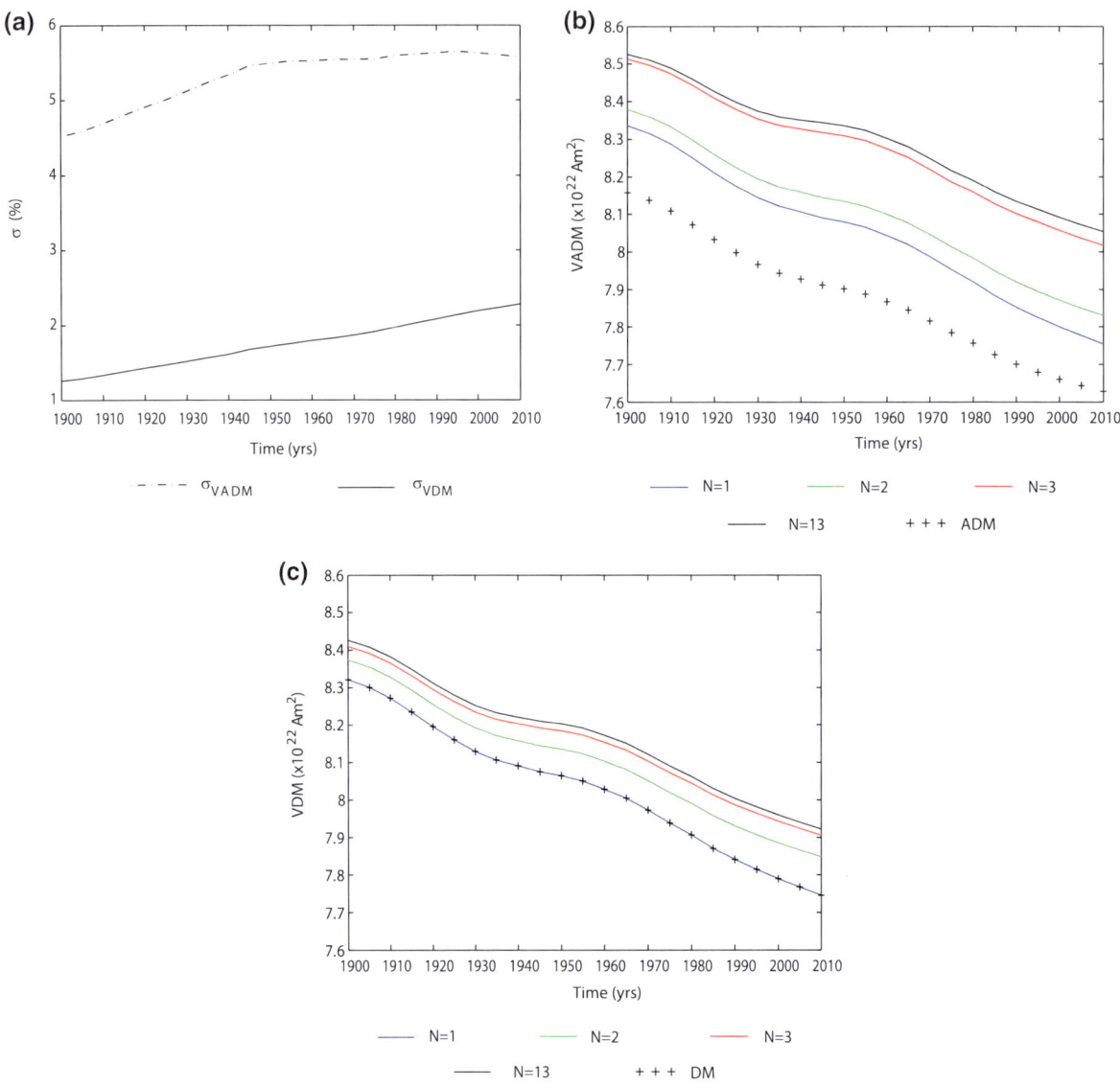

Figure 2

a Relative errors between VADM and ADM (σ_{VADM}) and VDM and DM (σ_{VDM}) calculated from Eqs. 6 and 7. Temporal evolution of the **b** VADM and **c** VDM curves obtained with synthetic data of the IGRF-11 model, for $N = 1$ (*solid blue*), $N = 2$ (*solid green*), $N = 3$ (*solid red*), and $N = 13$ (*solid black*), together with the (A)DM curves represented with *crosses*, by comparison

corresponding to the continental regions of North America, Europe and Northern Africa, Asia, South America, Central and South Africa, and Oceania. In each selected area, the synthetic data were generated considering a homogeneous distribution with a density of 173 points in each spherical cap. The quantification of the RE [Axial Regional Effect (ARE), and RE] was calculated by the relative difference between the V(A)DM$_{\text{continent}}$ for each continent (with a sub-index indicating the name of

the continent) and the theoretical values of the (A)DM as follows:

$$ARE = \frac{VADM_{\text{continent}} - ADM}{ADM} \times 100\% \quad (8)$$

$$RE = \frac{VDM_{\text{continent}} - DM}{DM} \times 100\% \quad (9)$$

To investigate the origin of the differences between the V(A)DM$_{\text{continent}}$ and (A)DM estimations we have carried out a more detailed study of the

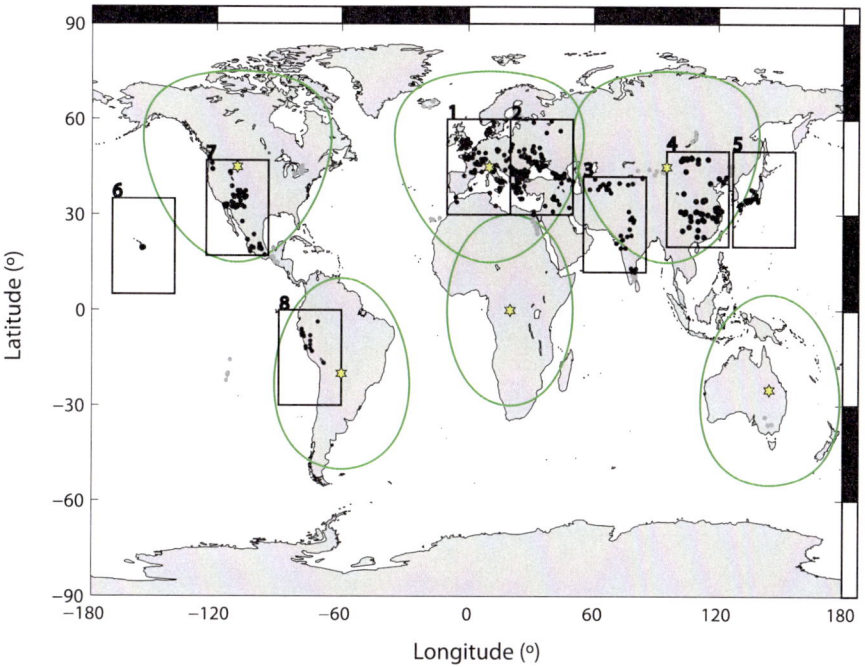

Figure 3

Map showing the considered continental areas in the study of regional effect. The spherical caps (*green circles*) are of 30° of radius, and are centred in the *yellow stars*. The geographical distribution of the ArcheoInt database (Genevey *et al.*, 2008) for the last 3,000 years are also shown (*black and grey points*). Definition of the eight regions (each 30° width, both in latitude and longitude) chosen for the next VADM and VDM computations as in Genevey *et al.* (2008): *1* Western Europe (latitudes between 30N and 60N, longitudes between 10W and 20E), *2* Central Europe and near East (latitudes between 30N and 60N, longitudes between 20E and 50E), *3* Central Asia (latitudes between 12N and 42N, longitudes between 55E and 85E), *4* Eastern Eurasia (China; latitudes between 20N and 50N, longitudes between 95E and 125E), *5* Far East (Japan; latitudes between 20N and 50N, longitudes between 127E and 157E), *6* Pacific (Hawaii; latitudes between 5N and 35N, longitudes between 190E and 220E), *7* Southwest part of North America (latitudes between 17N and 47N, longitudes between 235E and 265E), *8* Northwest part of South America (Peru; latitudes < 0, longitudes between 270E and 300E)

different multipolar contributions affecting the selected regions. Apart from the $V(A)DM_{continent}$ calculated considering the total field ($N = 13$), we have also computed the $V(A)DM_{continent}$ for the total field without the quadrupole contribution, the total field without the quadrupole and octupole contributions, and using only the dipole field ($N = 1$). The ARE/RE values for all the above mentioned contributions are given in Table 1 and along the text, and plotted in Fig. 4 along with the theoretical (A)DM.

In North America (Fig. 4a), we observe higher $V(A)DM_{North\ America}$ estimations than real (A)DM values, but they present a similar temporal trend. The difference between the $V(A)DM_{North\ America}$ calculated from $N = 1$ and $N = 13$ accounts for the importance of the higher non-dipole terms ($N > 3$) in this region. The small difference observed between the $VDM_{North\ America}$ and the DM is due to the

Table 1

Errors (rms) of $V(A)DM_{continental}$ estimations for the period 1900–2010. The (axial) regional effects, (A)RE, are computed from Eqs. 8 and 9

Regions	Axial regional effect, ARE-rms (%)	Regional effect, RE-rms (%)
North America	16.24	4.97
Europe and Northern Africa	4.16	3.74
Asia	14.53	14.36
South America	20.06	15.69
Africa	3.15	4.27
Oceania	35.29	18.96

octupole field that contributes around 3 % to the $VDM_{North\ America}$.

In Europe and Northern Africa (Fig. 4b), the main difference between regional and theoretical DM

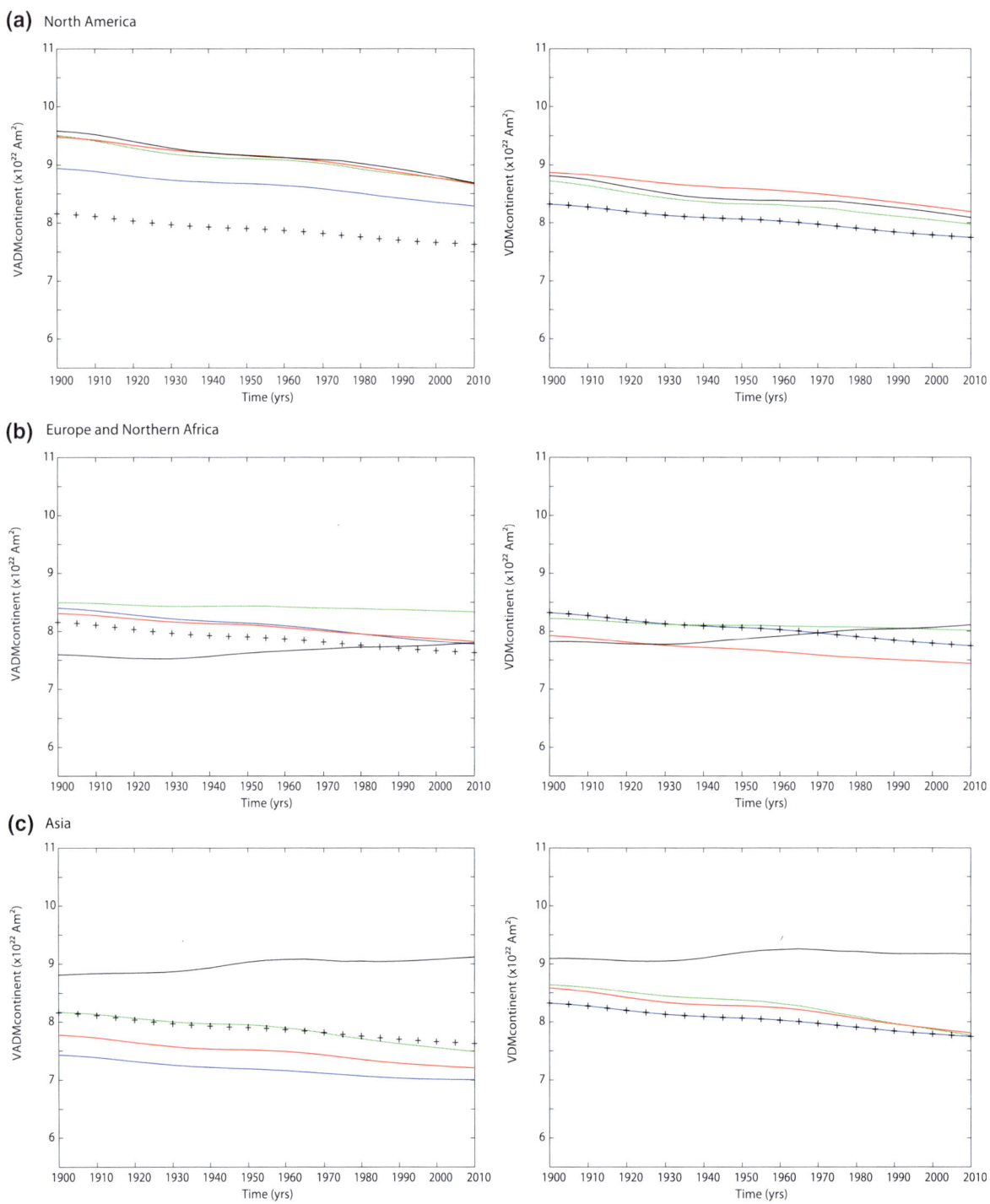

Figure 4

Regional averaged of VADM (*left column*) and VDM (*right column*) [V(A)DM$_{continent}$] curves, synthesized from IGRF-11 model to $N = 13$ (*solid black*), $N = 13$ minus quadrupole term (*solid green*), $N = 13$ minus quadrupole and octupole terms (*solid red*), and $N = 1$ (*solid blue*) in **a** North America, **b** Europe and Northern Africa, **c** Asia, **d** South America, **e** Central and South Africa, **f** Oceania. The ADM and DM curves (*crosses*) are shown for comparison

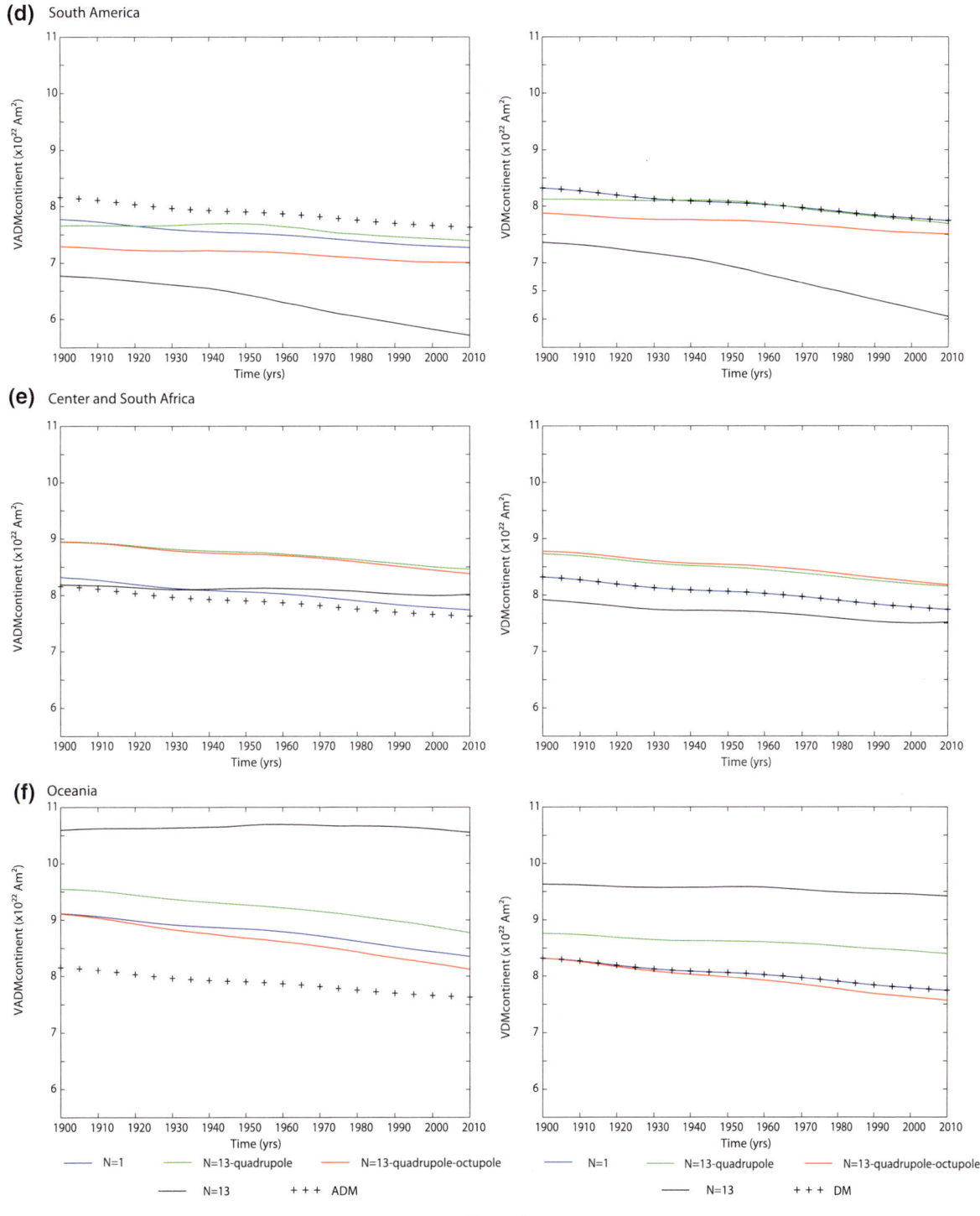

Figure 4
continued

estimations is the temporal evolution of these magnitudes. In contrast to the global decreasing trend of the (A)DM, the V(A)DM$_{Europe}$ curves present an increasing trend with a minimum around the year 1930. As can be observed in Fig. 4b, these anomalous values are related to the local effect of the quadrupole (with contributions around -9.8 % for the $VADM_{Europe}$ and -2.3 % for the VDM_{Europe}) and octupole terms (4.4 % for the $VADM_{Europe}$, 5.5 % for the VDM_{Europe}), because the increasing trend disappears when removing these contributions.

In Asia (Fig. 4c), the $V(A)DM_{Asia}$ estimations are higher than the $(A)DM$ values. A nearly constant or slightly increasing temporal trend of the VDM is suggested by the regional averages, with a small relative maximum around the year 1960. The quadrupole is the main source of differences between regional and theoretical values with a percentage of contribution of 14.2 % for the $VADM_{Asia}$ and 11.1 % for the VDM_{Asia}.

The continent with lower $V(A)DM_{continent}$ values than the $(A)DM$ is South America (Fig. 4d). Here, deviations between regional averages and $(A)DM$ estimations are greater than 19 % for the $VADM_{South\ America}$ and 15 % for the $VDM_{South\ America}$. This area is under the influence of the South Atlantic Anomaly (SAA) with intensity values lower than expected for that region. The difference between the $V(A)DM_{South\ America}$ and the $(A)DM$ is mainly due to the quadrupole term (-16.2 % for the $VADM_{South\ America}$, -14.8 % for the $VDM_{South\ America}$). The contribution of the octupole term affects around the 5.5 % to the $VADM_{South\ America}$ and 3.5 % to the $VDM_{South\ America}$). In this case, the quadrupole and octupole terms act in opposite directions. The first one decreases the value of $V(A)DM_{South\ America}$, whereas the second one increases it, being the most powerful the quadrupole term.

In Africa (Fig. 4e), the most important non-dipole term is the quadrupole, with a contribution of the -8.0 and -9.5 % for $VADM_{Africa}$ and VDM_{Africa} respectively. However, the $V(A)DM_{Africa}$ and $(A)DM$ estimations are similar, consequently the non-dipole contribution in the RE is not so strong.

Finally, Oceania is the region where the geomagnetic field is more affected by the non-dipole terms (Fig. 4f). Here the $V(A)DM_{Oceania}$ reaches the highest values (up to 10.5×10^{22} Am^2 for the $VADM_{Oceania}$ and 9.5×10^{22} Am^2 for the $VDM_{Oceania}$), associated with the quadrupole (18.2 % for the $VADM_{Oceania}$, 11.8 % and for the $VDM_{Oceania}$) and with the octupole terms (7.3 % for the $VADM_{Oceania}$, 8.2 % for the $VDM_{Oceania}$). Differences between

$V(A)DM_{Oceania}$ and $(A)DM$ estimations are about 35 % for the $VADM_{Oceania}$ (ARE) and 19 % for the $VDM_{Oceania}$ (RE).

The values contained in Table 1 show that, in general, ARE is greater than RE, and that these errors can be locally very high. The high errors and the differences observed between $VADM_{continent}$ and $VDM_{continent}$ suggests that the use of mixed VADM/VDM curves, commonly combined in palaeomagnetism due to the lack of inclination values (e.g. GENEVEY et al., 2008), introduces an additional source of errors. Then, it is not an appropriate approach.

On the other hand, the palaeomagnetic database for the last 14,000 years is clearly biased (Fig. 1): for the last 8,000 years the archaeomagnetic data are concentrated in Eurasia, while for the earlier times, from 12000 to 6000 BC, the available data are mainly lava flows from Hawaii and North America. This means that if $V(A)DM$ estimations are not adequately averaged, they might be clearly influenced by the regional effect. However, we must point out that the RE depends on the geomagnetic field structure and then it is time-dependent, i.e. our values cannot be directly extrapolated for the past, but provide a reliable idea about the order of magnitude of the regional effect.

3.2. Regional Average of the $V(A)DM_i$ Using Simulations of the Palaeointensity Data Distribution

One of most important problems in the ancient DM estimation, in both $V(A)DM$ and $(A)DM$, is the inhomogeneous palaeomagnetic database. Most of the palaeomagnetic data are concentrated in the Northern Hemisphere (around 95 % of the archaeomagnetic and lava flow data for the last 3,000 years (DONADINI et al., 2009)). This heterogeneous spatial distribution generates problems in the $V(A)DM$ (global average) such as an overestimation of the regions with more available data, as is the case of Eurasia (GENEVEY et al., 2008). In order to correct this RE in the $V(A)DM$ estimation, GENEVEY et al. (2008) proposed a simple, first-order regional weighting scheme based on the definition of eight regions (rectangle regions in Fig. 3). These regions were selected taking into account the locations of the palaeointensity data compiled in the database Archeo-Int (GENEVEY et al., 2008) for the last 3,000 years. They considered that each selected region contains enough palaeointensity information. Here, in order to check the reliability of the regional weighting scheme of GENEVEY et al. (2008), we simulate their procedure but using synthetic data from the IGRF-11 model.

The data were synthesized at the locations of the ArcheoInt database (Fig. 3). Since the database contains palaeomagnetic data for the last 3,000 years, whereas the IGRF-11 model only spans from 1900 to 2010, we had to adapt linearly the time interval covered by the database to the last century. That is to say, we simulated a synthetic database with field information given by the IGRF-11 model at the locations of the ArcheoInt database (sites represented in Fig. 3) and we attributed to each data point a fictitious age (linearly adapted) within the 1900–2010 time interval. That is to say, the assigned age has been estimated as follows: $t_2 = m \times (1,000 + t_1) + 1900$, where $m = 110/2,900$, t_2 the time adapted in the new synthetic database, and t_1 the time given by the ArcheoInt database.

Two important points to remark: (1) we used all the locations of the ArcheoInt database. That is, we did not introduce the selection criteria used by GENEVEY et al. (2008) to consider only high quality palaeointensity data. (2) We have synthesized both inclination and intensity data at all locations. However, some of the data of ArcheoInt provide only intensity values without inclination data (58 % of the intensity data) and, therefore, the VDM_i could not be always calculated. This is the reason why the authors used mixed VADM/VDM curves. Consequently, we are considering the best case scenario (i.e. lower errors are expected) for the regional averaging procedure proposed by GENEVEY et al. (2008).

The regional weighting scheme of GENEVEY et al. (2008), consists of calculating eight regional VADM and VADM/VDM curves for each selected region, by using the classical sliding overlapping windows technique, and then computing the averaged global VADM and VADM/VDM curves (assuming equal weight for each region).

To estimate the temporal evolution of the $V(A)DM_{regional}$ we have transformed the original 500-year window shifted by 250 years and a 200-year

window shifted by 100 years, into a 20-year window shifted by 10 years and 10-year window shifted by 5 years, respectively. We calculated the regional average $V(A)DM_{regional}$ from each region and time window, and an estimation of the global weighted averaged V(A)DM, denoted as $V(A)DM_W$, was obtained. The different $V(A)DM_{regional}$ for each region are plotted in the supplementary Fig. 2S and the global $V(A)DM_W$ is plotted in the Fig. 5. For comparison, we have also added the global V(A)DM directly calculated from all data, without the regional weighting procedure. The theoretical (A)DM curves are represented as well.

In order to provide a more realistic result, we have perturbed our synthetic database using a set of 500 random perturbations obtained from Gaussian distributions with mean values equal to zero and standard deviations equal to the standard deviation of the archaeomagnetic data for the last 3,000 years (4.2° for inclinations and 8.6 µT for intensities, DONADINI et al., 2009). We have repeated the previous process using the new datasets of perturbed data. The results provide the bands at ~ 65 % confidence level (dashed lines in Fig. 5) for the $V(A)DM_W$ and for the global V(A)DM (without the regional weighting scheme).

Our results indicate that the variability reported is related to the spatial and temporal distribution of the data. The data distribution is more different among shorter windows, leading to differences in influence of regional bias from one window to the next. The higher variability is an artefact of the RE varying with the data distribution.

The VADM presents an increasing temporal trend, with a maximum value around the year 1970 which is a clear artefact. An increasing trend was also observed in the $VADM_{Europe}$ curve for the European continent in our previous study (see Fig. 4b). This means that when VADM is obtained from global averaging, the European zone is overestimated because it is the region with more available data (up to 55 %, regions 1 and 2 in Fig. 3). So, when the regional weighting scheme is applied, the influence of European data is weakened and the VADM evolution is more similar to the ADM trend. Although still higher $VADM_W$ values than ADM are obtained, which means that the RE has not been completely cancelled. Deviations between the global $V(A)DM_W/$

(a) 20-yr windows shifted 10 yr.

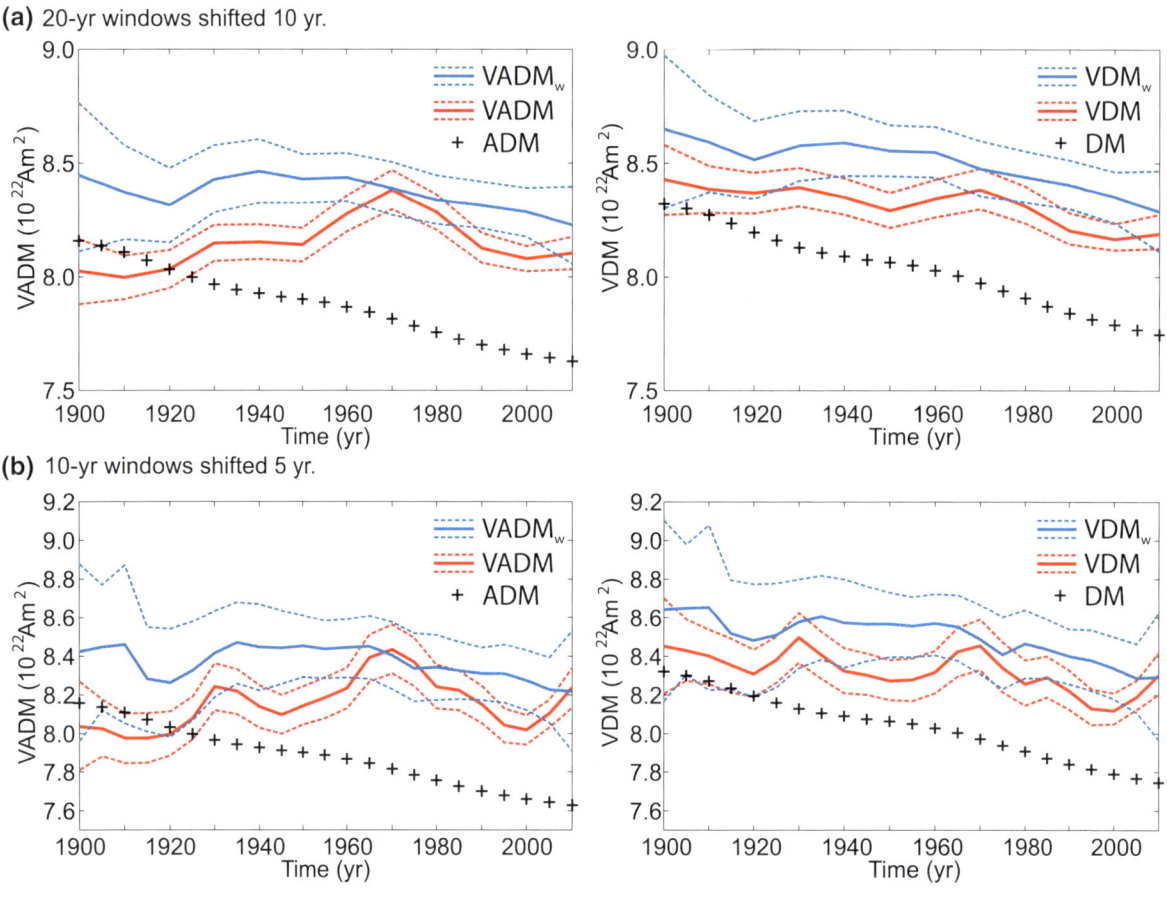

(b) 10-yr windows shifted 5 yr.

Figure 5
Effect of the geographic bias in the distribution of the synthetic data on the estimates of the (*left*) global VADM and (*right*) global VDM variation curves. Computations are performed using the selected data (see Fig. 3) smoothed over overlapping sliding windows of **a** 20 years shifted by 10 years, **b** 10 years shifted by 5 years. *Solid blue* V(A)DM$_W$ computed with the weighting scheme of GENEVEY et al., (2008), *solid red* V(A)DM calculated from all data without the regional weighting scheme, *dashed lines* show the error band with a level of confidence of 65 %, computed as the involving of 500 perturbed databases (see text for more details), *crosses* (A)DM calculated from the first three Gaussian coefficients

Table 2

Errors (rms) of regionally weighting averaged V(A)DM [V(A)DM$_W$] estimations for the period 1900–2010 and the V(A)DM without regional weighting scheme, together with the error band (confidence level of ∼65 %). The comparisons are developed as a function of (A)DM for the same temporal interval

rms (%)	Sliding windows	
	20 years shifted by 10 years	10 years shifted by 5 years
VADM$_W$	6.6 ± 1.6	6.6 ± 2.4
VADM	4.6 ± 0.7	4.8 ± 1.0
VDM$_W$	6.0 ± 1.7	6.0 ± 2.5
VDM	3.9 ± 0.9	4.0 ± 1.3

V(A)DM and the (A)DM are outlined in Table 2. Lower relative errors between V(A)DM and (A)DM than between V(A)DM$_W$ and (A)DM are obtained. However, this result does not mean that the use of the regional weighting scheme is inappropriate. As we discussed previously, the results are more consistent with the theoretical trend when a regional weighting is considered.

The lower differences between VDM$_W$ and VDM trends and the lower errors in relation with DM values (see Table 2), are related to the use of more field information: the inclination in addition to the intensity. With this additional information changes in the tilt of the dipole are also considered and, therefore, a more accurate description of the DM is expected. It is important to note that the original ArcheoInt database contains inclination information of around 48 % of the sites. Consequently the errors that we have obtained are the lowest that could be reached.

Finally, we would like to point out that the error bands of the V(A)DM are narrower than those of the V(A)DM$_W$. The reason is the average procedure: the V(A)DM is obtained with all the data and this high number of data gives lower standard deviations. This is not the case for the V(A)DM$_W$, where the lower number of regions to be averaged (eight regions) increases the standard deviations.

4. Regional Indirect Effect

In this section, we want to analyse the influence of a sparse database in the models generated from palaeomagnetic/archaeomagnetic data (e.g. KORTE et al., 2009, 2011; KORTE and CONSTABLE, 2011; LICHT et al., 2013; PAVÓN-CARRASCO et al., 2014) and especially, its effects on the (A)DM estimation.

We have developed a geomagnetic global model by using the same synthetic database of the previous Sect. 3b, including a new set of synthetic data for the declination, which is necessary to develop the global model. The global model, called IGRF-11$_S$, was obtained by using the classical approach from palaeomagnetic data (KORTE and CONSTABLE, 2003): the SHA technique in space and the penalized cubic B-splines (DE BOOR, 2001) in time. In terms of the SHA, the potential of the internal geomagnetic field can be established at any point (r, θ, λ) over the earth's surface as (1). The usual time-dependent Gaussian coefficients $[g_n^m(t)$ and $h_n^m(t)]$ may be developed using penalized cubic B-splines defined by the matrix $B_q(t)$, as follows:

$$g_n^m(t) = \sum_{q=1}^{Q} g_{n,q}^m B_q(t)$$
$$h_n^m(t) = \sum_{q=1}^{Q} h_{n,q}^m B_q(t)$$
(10)

where Q is the maximum degree of the temporal expansion and $g_{n,q}^m(t)$ and $h_{n,q}^m(t)$ are the time-dependent spherical harmonic coefficients.

In palaeomagnetic studies, the measures of the geomagnetic field are D, I, and F. These components cannot be expressed as a linear combination of the Gaussian coefficients. For this reason, any scalar element of the geomagnetic field d (declination, inclination, or intensity) must be given as a non-linear function f, related to Eq.1 and depending on the time-dependent Gaussian coefficients:

$$d = f(\vec{m}) + \varepsilon$$
(11)

where $\vec{m}$ contains all the Gaussian coefficients and ε is the error. To find the optimal set of time-dependent Gaussian coefficients, we chose the regularized least-squares inversion applying the Newton–Raphson iterative approach (GUBBINS and BLOXHAM, 1985):

$$\vec{m}_{i+1} = \vec{m}_i + \left(\hat{A}_i' \times \hat{A}_i + \alpha \times \hat{S} + \tau \times \hat{T}\right)^{-1}$$
$$\times \left(\hat{A}_i' \times \vec{\gamma}_i - \alpha \times \hat{S} \times \vec{m}_i - \tau \times \hat{T} \times \vec{m}_i\right)$$
(12)

where $\hat{A}$ is the matrix of parameters which depends on the SH functions in space and time (the so-called Frechet matrix) and $\hat{A}'$ is the transpose of $\hat{A}$. $\vec{\gamma}$ is the vector of differences between the input data and modelled data for the ith iteration. The $\hat{S}$ and $\hat{T}$ matrices are the spatial and temporal regularization matrices, respectively, with damping parameters α and τ. The index i indicates the number of the iteration, which requires a first initial solution $\vec{m}_0$. To create the B-spline base we have selected knot points between 1899 and 2011 every 4 years.

The spatial regularization minimizes the Ohmic dissipation at the core-mantle boundary (CMB) (GUBBINS, 1975), which can be written as:

$$\hat{S} = \frac{4\pi}{t_e - t_s} \int_{t_s}^{t_e} \sum_{n=1}^{N} \frac{(n+1)(2n+1)(2n+3)}{n} \left(\frac{a}{c}\right)^{2n+3}$$
$$\times \sum_{m=0}^{n} \left[\left(g_n^m(t)\right)^2 + \left(h_n^m(t)\right)^2\right] dt$$
(13)

where t_s and t_e are the initial and final epoch respectively and c is the mean radius of the CMB. The temporal regularization minimizes the second time derivative of the radial field at the CMB (BLOXHAM AND JACKSON, 1992), as follows:

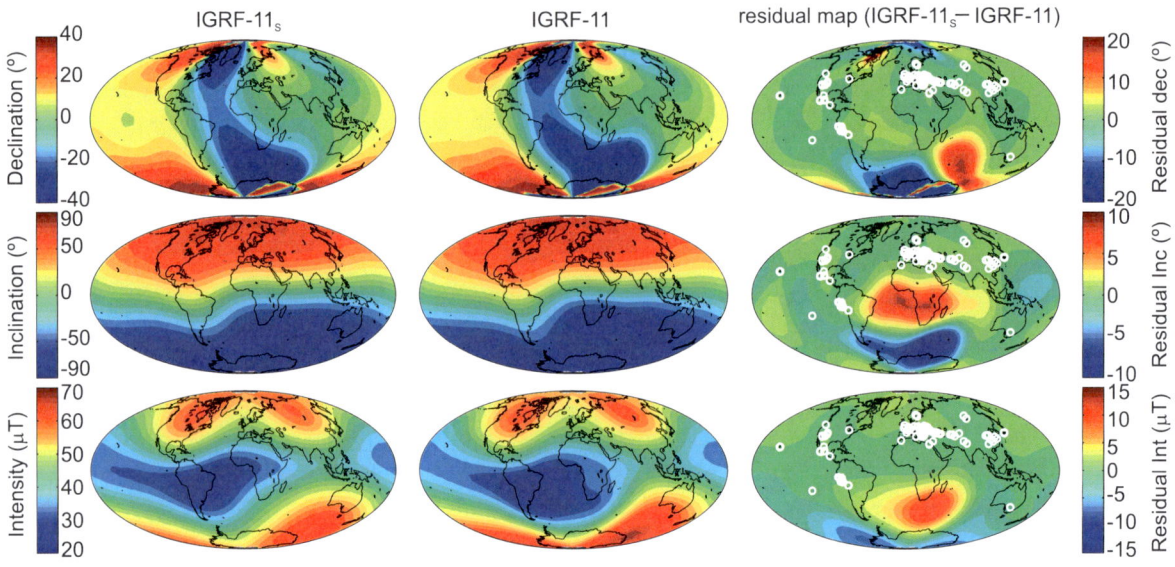

Figure 6

Declination, *D*, inclination, *I*, and intensity, *F*, maps at 1955 for (*left column*) global model generated using synthetic data from ArcheoInt IGRF-11$_S$. (*center column*) IGRF-11 model. Maps for different years are given as Supplementary Material, (*right column*). Residual between both models is also shown

$$\hat{T} = \frac{1}{t_e - t_s} \int\limits_{t_s}^{t_e} \oint\limits_{\Omega} \left(\partial_t^2 B_r\right)^2_{r=c} \mathrm{d}\Omega \mathrm{d}t \qquad (14)$$

where dΩ is the differential solid angle over the sphere Ω. The choice of the best regularization is applied to obtain a model with the minimal complexity and a reasonable fit to the data. After carrying out several tests with different values of damping parameters (e.g. LICHT *et al.*, 2013; PAVÓN-CARRASCO *et al.*, 2014), we have chosen α and τ equal to $5 \times 10^{-9} \ \mu T^{-2}$ and $10^{-3} \ \mu T^{-2} \ year^4$, respectively. Again, in order to provide a more realistic result, we have also used the 500 perturbations from the dataset of the previous section. In this case, the new element (the declination) was perturbed by a Gaussian distribution with mean 0° and standard deviation equal to 6.1° (from DONADINI *et al.*, 2009). A total of 500 models were developed providing the error modelling as the standard deviation of the Gaussian coefficients.

The IGRF-11$_S$ model is compared with the original IGRF-11. In Fig. 6 the maps of *D*, *I*, *F* at 1955 (central epoch of the considered time interval: 1900–2010) from the IGRF-11$_S$ and IGRF-11 models are represented. The differences between these geomagnetic elements are also represented, together with the locations of the data in the time span from 1950 to

1960 (the three knot points considered for 1955). These data represent less than 10 % of the total data. We can observe that the IGRF-11$_S$ reproduces very well the main characteristics of the geomagnetic field (see also the maps provided in the Supplementary Material). The good representation of the Southern Atlantic Anomaly by the IGRF-11$_S$ model is highlighted, in spite of a lack information available from this region (and from the Southern Hemisphere in general).

The major differences between both models are located in the regions with absence of data: Africa and Antarctica. The highest differences in declination are found in Antarctica and the Southern Indian Ocean. Discrepancies in inclination are low, with the exception of a small dipole in central Southern Africa that produces inclination differences up to 8° and −10°. The major disagreement in intensity is observed in Southern Africa and the Southern Atlantic Ocean, with higher intensities than the IGRF model (around 12 µT). These artefacts are due to the absence of information to reproduce adequately the SAA, i.e. the main differences between the IGRF-11$_S$ and the IGRF-11 are located in the region affected by the SAA from where there is not enough available palaeomagnetic information.

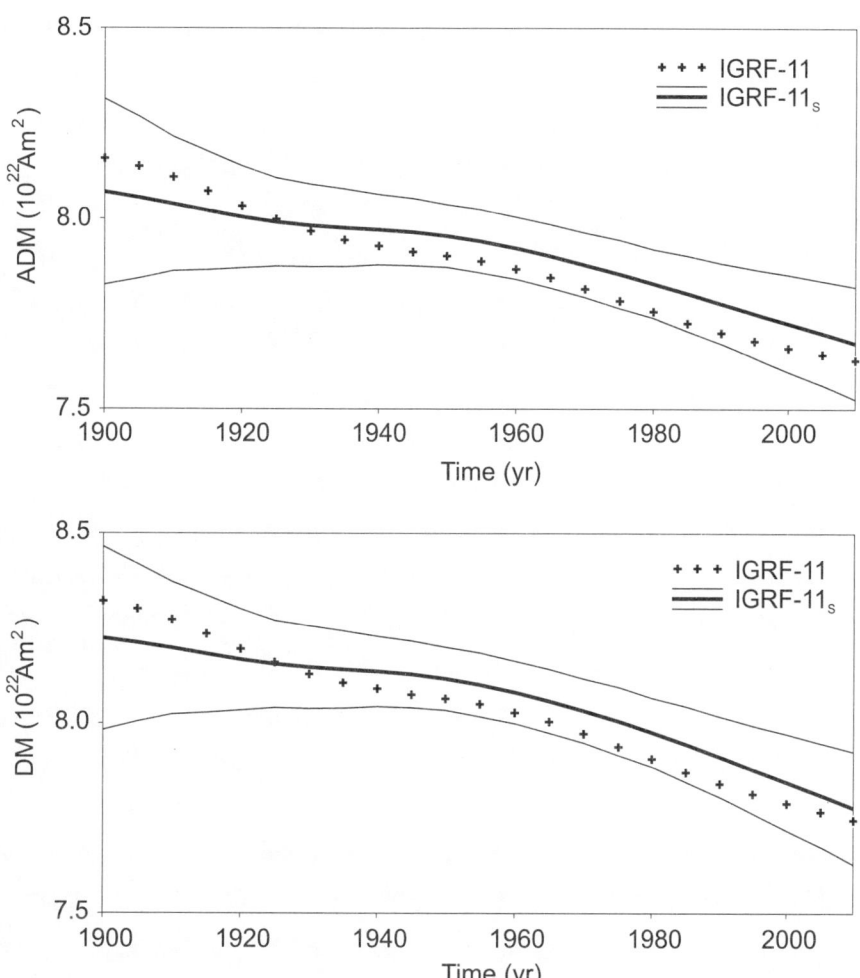

Figure 7

For IGRF-11$_S$ **a** ADM$_S$ and **b** DM$_S$ curves together with the error band at ∼65 % confidence level are shown. The theoretical (A)DM is also plotted for comparison

Table 3

Errors (rms) for the deviation between ADM and DM of the geomagnetic field calculated with the Gauss coefficients of IGRF-11$_S$ and IGRF-11 models from 1900 to 2010, every 5 years, together with the error band (confidence level of ∼65 %)

	rms (%)
ADM$_S$	0.7 ± 0.2
DM$_S$	0.8 ± 0.1

Finally, and using the new set of Gaussian coefficients provided by the IGRF-11$_S$ model, we have calculated the (A)DM curves, denoted as (A)DM$_S$. The coefficients' error is used to obtain the error bands at ∼65 % confidence level. Figure 7 shows the (A)DM$_S$ curves together with the (A)DM curves of the original IGRF-11 model, and Table 3 summarizes the relative errors between them. We can observe a similar temporal trend between all (A)DM, with lower values at the beginning of the time interval, and higher values at the end, likely due to the inhomogeneous data distribution. In spite of this, we can observe that when we consider the error band, the theoretical (A)DM curves lie in the error band.

In contrast to the regional weighting scheme where artificial variations of the DM were obtained for small sliding windows, dipole variations obtained by modelling reproduces much better the theoretical dipole moment. But we have to take into account that we have used synthetic data, i.e. we have considered the best situation. And more importantly, in this last case we have increased the amount of palaeomagnetic information (including declinations) with respect to the previous section. From this analysis, we could conclude that when directional palaeomagnetic information is available, the best method to compute the geomagnetic dipole moment evolution is from global modelling.

5. Conclusions

In this work, we have evaluated and quantified different sources of error introduced in the geomagnetic dipole moment estimation. The principal errors considered in this study come from (1) the averaged procedure of the non-dipole contribution, because palaeomagnetic data record all the contributions of the geomagnetic field and not only the dipole field (non-dipole effect); and (2) the effect of the current palaeomagnetic database distribution in the averaged procedure (regional effect) and in the generation of the global geomagnetic field models (regional indirect effect). To evaluate these errors we have used the IGRF-11 model.

Firstly, we have estimated the non-dipole effect in the global and regional averages for the last century. Although the main assumption is that the non-dipole terms are cancelled in the averaged procedure, we have reported that this contribution is not cancelled completely. In the global averages, it can reach 5.8 % for the VADM and 2.3 % for the VDM. The most important terms are the quadrupole and octupole. For the last 110 years, the non-dipole effect is small (never greater than 6 %) if we compare with the values of the palaeointensity errors (around 10 %).

In the regional averages, the non-dipole effect can give rise to deviations between the V(A)DM and (A)DM higher than 35 % in some continental regions, such as Oceania. Again, the quadrupole and octupole terms are the most important non-dipolarity sources, the quadrupole effect associated with the Southern Atlantic Anomaly being especially important. This term produces a decrease of about 15–16 % in the $V(A)DM_{continent}$ over the Southern American region. Another interesting artefact is the anomalous evolution trend of $V(A)DM_{continent}$ observed in Europe and Asia. The $V(A)DM_{Europe}$ and $V(A)DM_{Asia}$

are increasing whereas (A)DM is decreasing for the time span from 1900 to 2010.

Because of the sparse palaeomagnetic database, which is clearly biased towards the European region (with more available data), the regional evolution of the $V(A)DM_{Europe}$ could affect the global averages. One of the methods proposed to avoid this overestimation is the first-order regional weighting scheme proposed by GENEVEY et al. (2008). Our results confirm the improvement of the V(A)DM when the regional weighting scheme is considered, with respect to the global average on the total database. However, mathematical artefacts are created with this procedure depending on the size of the temporal sliding windows used and the number of data available. The smaller the temporal sliding window is, the more artefacts appear. This varying regional bias effect might also affect the power distribution between dipole and non-dipole contributions in palaeomagnetic SHA models.

Finally, we have analyzed the effect of the current data distribution for the last 3,000 years on the generation of the global geomagnetic field models. We have generated a synthetic model (IGRF-11$_S$) with the D, I, F synthesized in the locations of the ArcheoInt database with the temporal interval adapted linearly to the last century. The results confirm that the main differences between the model created and the IGRF-11 are located in regions with lack of data (e.g. Africa). Moreover, we can observe a good agreement between the (A)DM calculated from the IGRF-11$_S$ and the IGRF-11 models, lying in the error band with a confidence level of ~65 %.

From this analysis, we might conclude that when directional and intensity palaeomagnetic information is available, in spite of the inhomogeneity of the database, the best method to compute the geomagnetic dipole moment evolution is from global modelling. The $(A)DM_{model}$ seems to be the most appropriate parameter to correct the cosmogenic isotopes production or to study the possible correlations between the geomagnetic field and the climate.

Acknowledgments

The authors are grateful to the Spanish research project CGL2011-24790 of the Spanish Ministerio de

Economía y Competitividad and the FPI grant BES-2012-052991, which has allowed to the author S.A. Campuzano a stay for 3 months at INGV in Rome. All algorithms have been developed in Matlab® codec (Matlab 7.11.0, R2010b) along with the figures. The authors also thank two anonymous reviewers for the constructive comments and suggestions which have helped to improve substantially this manuscript.

References

Bloxham, J., Jackson, A. (1992), *Time-dependent mapping of the magnetic field at the core-mantle boundary*. J. Geophys. Res. *97*, 19537–19563.

Christensen, U.R., Aubert, J., Hulot, G. (2010), *Conditions for Earth-like geodynamo models*. Earth Planet. Sci. Lett. *296*(3–4), 487–496.

De Boor, C., A Practical Guide to Splines (Springer, New York 2001).

Donadini, F., Korte, M., Constable, C. G. (2009), *Geomagnetic field for 0–3 ka: 1. New data sets for global modeling*. Geochem. Geophys. Geosyst. *10*(6), Q06007.

Finlay, C. C., (2010), *International Geomagnetic Reference Field: the eleventh generation*. Geophys. J. Int. *183*, 1216–1230.

Gallet, Y., Genevet, A., Fluteau, F. (2005), *Are there connections between the Earth's magnetic field and climate?* Earth Planet. Sci. Lett. *236*, 339–347.

Gauss, C.F., Intensitas vis Magneticae Terrestris ad Mensuram Absolutam Revocata (Dieterich, Göttingen 1833).

Genevey, A., Gallet, Y., Constable, C., Korte, M., Hulot, G. (2008), *ArcheoInt: An upgraded compilation of geomagnetic field intensity data for the past ten millennia and its application to the recovery of the past dipole moment*. Geochem. Geophys. Geosyst. *9*(4), Q04038.

Genevey, A., Gallet, Y., Thébault, E., Jasset, S., Le Goff, M. (2013), *Geomagnetic field intensity variations in Western Europe over the past 1100 years*. Geochem. Geophys. Geosyst. *14*(8), 2858–2872.

Glatzmaier, G.A., Roberts, P.H. (1995). *A three-dimensional self-consistent computer simulation of a geomagnetic field reversal*. Nature. *377*, 203–209.

Gubbins, D. (1975), *Can the Earth's magnetic field be sustained by core oscillations?* Geophys. Res. Lett. *2*, 409–412.

Gubbins, D., and Bloxham, J. (1985), *Geomagnetic field analysis. III. Magnetic fields on the core–mantle boundary*. Geophys. J. R. Astron. Soc. *80*, 695–713.

Jackson, A., Jonkers, A.R.T., Walker, M.R. (2000), *Four centuries of geomagnetic secular variation from historical records*. Philos. Trans. R. Soc. Lond. A *358*(1768), 957–990.

Jacobs, J. A., Geomagnetism (Vol. *1*) (Academic Press 1991).

Jonkers, A. R. T., Jackson, A., Murray, A. (2003), *Four centuries of geomagnetic data from historical records*. Rev. Geophys. *41*(2), 1006.

Korte, M., and Constable, C.G. (2003), *Continuous geomagnetic field models for the past 3000 years*. Phys. Earth Planet. Interiors. *140*, 73–89.

Korte, M., and Constable, C.G. (2005a), *Continuous geomagnetic field models for the past 7 millennia: 2 CALS7 K*. Geochem. Geophys. Geosyst. *6*, Q02H16.

Korte, M., and Constable, C.G. (2005b), *The geomagnetic dipole moment over the last 7000 years—new results from a global model*. Earth Planet. Sci. Lett. *236*, 348–358.

Korte, M., and Constable, C.G. (2011), *Improving geomagnetic field reconstructions for 0–3 ka*. Phys. Earth Planet. Interiors. *188*, 247–259.

Korte, M., Donadini, C., Constable, C.G. (2009), *The geomagnetic field for 0–3 ka, part II: a new series of time-varying global models*. Geochem. Geophys. Geosyst. *10*, Q06008.

Kovacheva, M., Boyadziev, Y., Kostadinova, M., Jordanova, N., Donadini, F. (2009), *Updated archeomagnetic data set of the past 8 millennia from the Sofia laboratory, Bulgaria*, Geochem. Geophys. Geosyst., *10*, Q05002, doi:10.1029/2008GC002347.

Licht, A., Hulot G., Gallet Y., Thébault E. (2013), *Ensembles of low degree archeomagnetic field models for the past three millennia*. Phys. Earth Planet. Interiors. *224*, 38–67.

Macouin, M., Valet, J.P., Besse, J. (2004), *Long-term evolution of the geomagnetic dipole moment*, Phys. Earth Planet. Interiors. *147*, 239–246.

Muscheler, R., Joos, F., Beer, J., Muller, S.A., Vonmoos, M., Snowball, I. (2007), *Solar activity during the last 1000 yr inferred from radionuclide records*. Quatern. Sci. Rev. *26*, 82–97.

Pavón-Carrasco, F.J., Osete, M.L., Torta, J.M., De Santis, A. (2014), *A geomagnetic field model for the Holocene based on archaeomagnetic and lava flow data*. Earth Planet. Sci. Lett. *388*, 98–109.

Roth, R., and Joos, F. (2013), *A reconstruction of radiocarbon production and total solar irradiance from the Holocene 14C and CO2 records: implications of data and model uncertainties*, Clim. Past Discuss. *9*, 1165–1235.

Tauxe, L. (1993), *Sedimentary records of relative paleointensity of the geomagnetic field: theory and practice*. Rev. Geophys. *31*, 319–354.

Usoskin, I., Korte, M., Kovaltsov, G.A. (2008), *Role of centennial geomagnetic changes in local atmospheric ionization*. Geophys. Res. Lett. *35*, L05811.

Vieira, L.E.A., Solanki, S.K., Krivova, N.A., Usoskin, I.G. (2011), *Evolution of the solar irradiance during the Holocene*. Astron. Astrophys. *531*, A6.

Whaler, K. A. & Gubbins, D. (1981), *Spherical harmonic analysis of the geomagnetic field: an example of a linear inverse problem*, Geophys. J. R. astr. SOC., *65*, 645–693.

Yang S., Odah, H., Shaw, J. (2000), *Variations in the geomagnetic dipole moment over the last 12,000 years*. Geophys. J. Int. *140* (1), 158–162.

(Received April 10, 2014, revised July 9, 2014, accepted July 28, 2014, Published online August 13, 2014)

107

Pure Appl. Geophys. 172 (2015), 109–120
© 2014 Springer Basel
DOI 10.1007/s00024-014-0913-9

Wind Forecasting Based on the HARMONIE Model and Adaptive Finite Elements

ALBERT OLIVER,[1] EDUARDO RODRÍGUEZ,[1] JOSÉ MARÍA ESCOBAR,[1] GUSTAVO MONTERO,[1] MARIANO HORTAL,[2]
JAVIER CALVO,[2] JOSÉ MANUEL CASCÓN,[3] and RAFAEL MONTENEGRO[1]

Abstract—In this paper, we introduce a new method for wind field forecasting over complex terrain. The main idea is to use the predictions of the HARMONIE meso-scale model as the input data for an adaptive finite element mass-consistent wind model. The HARMONIE results (obtained with a maximum resolution of about 1 km) are refined in a local scale (about a few metres). An interface between both models is implemented in such a way that the initial wind field is obtained by a suitable interpolation of the HARMO-NIE results. Genetic algorithms are used to calibrate some parameters of the local wind field model in accordance to the HARMONIE data. In addition, measured data are considered to improve the reliability of the simulations. An automatic tetrahedral mesh generator, based on the meccano method, is applied to adapt the discretization to complex terrains. The main characteristic of the framework is a minimal user intervention. The final goal is to validate our model in several realistic applications on Gran Canaria island, Spain, with some experimental data obtained by the AEMET in their meteorological stations. The source code of the mass-consistent wind model is available online at http://www.dca.iusiani.ulpgc.es/Wind3D/.

Key words: Adaptive finite element method, HARMONIE model, Local wind field forecasting, Mass-consistent model, Complex topography, Genetic algorithms.

1. Introduction

Over the last few years, the use of wind power to produce electric power has augmented considerably. Wind models are tools that allow the study of several problems related to the atmosphere, such as the effect of wind on structures, pollutant transport (OLIVER et al. 2012, 2013), fire spreading (FERRAGUT et al. 2007), and wind farm location (RODRÍGUEZ 2004).

In this paper, we propose a method for wind forecasting by coupling the HARMONIE meso-scale model with a local mass-consistent wind model specially suited for complex terrain (RODRÍGUEZ et al. 2012); similar coupling methods have been proposed by GASSET et al. (2012) and CARVALHO et al. (2013). HARMONIE is used experimentally at AEMET with promising results (NAVASCUÉS et al. 2013). Despite the high-resolution of the HARMONIE mesoscale model, the minimum horizontal resolution is about 1 km, which is a drawback when the microscale (about 1 m) is considered. For this reason, the results of the HARMONIE mesoscale model are coupled with the local wind field model. An initial wind field is required: it is obtained by a vertical extrapolation and a horizontal interpolation. The vertical extrapolation is based on a log-linear wind profile (LALAS and RATTO 1996). Both the mass-consistent model and the interpolation are defined by a set of parameters. Some of these parameters are known, but others have to be estimated. In order to calibrate these parameters, genetic algorithms are used (MONTERO et al. 2005). Algorithm 1 synthesises the main steps of the model.

---

[1] University Institute for Intelligent Systems and Numerical Applications in Engineering (SIANI), University of Las Palmas de Gran Canaria (ULPGC), Campus de Tafira, 35017 Las Palmas de Gran Canaria, Spain. E-mail: albert.oliver@upc.edu; eduardo.rodriguez@ulpgc.es; jescobar@dsc.ulpgc.es; gustavo@dma.ulpgc.es; rafa@dma.ulpgc.es

[2] Agencia Estatal de Meteorología (AEMET), Leonardo Prieto Castro, 8, 28040 Madrid, Spain. E-mail: mhortalr@aemet.es; j.calvo@aemet.es

[3] Department of Economics and Economic History, Faculty of Economics and Business, University of Salamanca, 37007 Salamanca, Spain. E-mail: casbar@usal.es

---

Algorithm 1 Overall algorithm

---

1. Mesh generation with the Meccano method

2. Assimilation of HARMONIE weather meso-scale model data for its use in the local wind field model

3. Calibration of the wind field model parameters using genetic algorithms

---

This paper is organised as follows. Sections 2, 3, and 4 explain in detail the main parts of this work: the meccano mesh generation (Sect. 2), the HARMONIE mesoscale model (Sect. 3), and the local wind field model (Sect. 4). Section 5 discusses the genetic algorithms and the parameters to be estimated. Section 6 shows an experiment of this method applied to Gran Canaria island.

2. Meccano Mesh Generation

The main steps of the meccano tetrahedral mesh generation algorithm are summarized in this section. This method was previously introduced in MONTENE-GRO et al. (2009, 2010) and CASCÓN et al. (2013). The input data of the algorithm is the definition of the solid boundary (for example, a surface triangulation or CAD description) and a given precision (corresponding to the approximation of the solid boundary). Algorithm 2 describes our mesh generation approach.

from a topological point of view, i.e., their surfaces must have the same genus.

Once the meccano is assembled, we have to define an admissible one-to-one mapping between the boundary faces of the meccano and the boundary of the solid. If the solid is genus-zero and its boundary is given by a triangulation, we propose in MONTENEGRO et al. (2010) an automatic method to construct a parametrization of the solid surface triangulation to a cube boundary. For this purpose, we first divide the solid surface triangulation into six patches with the same topological connection as the cube faces. Then, a discrete mapping from each surface patch to the corresponding cube face is built using the mean value parametrization proposed in FLOATER (2003).

At the moment, if the genus of the surface of the solid is greater than zero, the meccano should be defined by the user. An automatic construction of the meccano could be difficult when the topology of the solid is complex. We also remark that a non-optimal meccano can introduce large distortion in mesh

---

Algorithm 2 Meccano tetrahedral mesh generation

---

1. Meccano: Construct a meccano, $\mathcal{M}$, approximation of the solid, Ω, formed by polyhedral pieces.

2. Mapping: Define an admissible mapping, Π, between the meccano boundary faces, $\partial\mathcal{M}$, and the solid boundary, $\partial\Omega$, i.e., $\Pi : \partial\mathcal{M} \to \partial\Omega$.

3. Coarse Mesh: Construct a coarse tetrahedral mesh, $\mathcal{T}_0(\mathcal{M})$, of the meccano.

4. Refined Mesh: Generate a local refined tetrahedral mesh, $\mathcal{T}(\mathcal{M})$, from $\mathcal{T}_0(\mathcal{M})$, such that the surface triangulation, $\tau(\Omega)$, obtained after Π-mapping of $\mathcal{T}(\mathcal{M})$ boundary nodes, approximates the solid boundary $\partial\Omega$ for a given precision, ε.

5. External Node Mapping: Move the boundary nodes of $\mathcal{T}(\mathcal{M})$ to the solid surface according to Π.

6. Relocation and Optimization: Relocate the inner nodes of $\mathcal{T}(\mathcal{M})$ and optimize the resulting tetrahedral mesh by applying the simultaneous untangling and smoothing procedure to obtain the final tetrahedral mesh, $\mathcal{T}(\Omega)$, that approximates the solid.

The first step of the procedure is to construct a meccano approximation by connecting polyhedral pieces. The meccano and the solid must be equivalent

generation. To avoid this issue, an optimization of the boundary parametrization could be included (WAN et al. 2011).

In step 3, the meccano is decomposed into a coarse tetrahedral mesh $\mathcal{T}_0(\mathcal{M})$ by an appropriate subdivision of its initial polyhedral pieces. Although any tetrahedralization algorithm could be used, we propose a partition of meccano compatible with the Kossaczký refinement algorithm (KOSSACZKÝ 1994).

This mesh is locally refined in step 4 to obtain an approximation of the solid boundary within a given precision. To be more precise, we have to introduce some notations. Given a tetrahedral mesh of the meccano $\mathcal{T}(\mathcal{M})$, we denote as $\tau(\mathcal{M})$ its boundary triangulation and $\tau(\Omega)$ the surface triangulation obtained after Π-mapping of $\tau(\mathcal{M})$ nodes. Note that $\tau_o(\Omega)$ is a coarse approximation of $\partial\Omega$. In order to improve this approximation we build a refined mesh $\mathcal{T}(\mathcal{M})$ of $\mathcal{T}_0(\mathcal{M})$ such that the distance between $\tau(\Omega)$ and $\partial\Omega$ is less than a prescribed tolerance ε. The concept of distance between surfaces can be defined and implemented in several ways. In our case it is as follows: Let $T = \langle a, b, c \rangle$ be a triangle of $\mathcal{T}(\mathcal{M})$, where a, b and c are their vertices, and let $p_k \in \{p_i\}_{i=1}^{N_q}$ be a Gauss quadrature point of T. We define the distance, $d(t)$, between the triangle $\langle \Pi(a), \Pi(b), \Pi(c) \rangle \in \tau(\Omega)$ and $\partial\Omega$ as the maximum of the volumes of the tetrahedra formed by $\Pi(a), \Pi(b), \Pi(c)$ and $\Pi(p_k)$. Then, the distance between $\tau(\Omega)$ and $\partial\Omega$, $d(\tau(\Omega)), \partial\Omega$, is the maximum of all $d(T)$, that is

$$d(\tau(\Omega), \partial\Omega) = \max_{T \in \tau(\mathcal{M})} d(T) \tag{1}$$

We recall that local refinement stops when $d(\tau(\Omega), \partial\Omega) < \varepsilon$. Note that this is an approximation of the maximum missed (or overestimated) volume per face of $\tau(\Omega)$. A more accurate approach of distance based on Hausdorff envelope can be found in (BOROUCHAKI and FREY 2005).

Then, we construct a mesh of the solid $\mathcal{T}(\Omega)$ by mapping the boundary nodes of $\mathcal{T}(\mathcal{M})$ and by relocating the inner nodes at a reasonable position. After these two steps, the resulting mesh is generally tangled. Therefore, a simultaneous untangling and smoothing procedure (ESCOBAR et al. 2003, 2010) is applied and a valid adaptive tetrahedral mesh of the solid is obtained. In short, this last procedure finds the new positions of the inner nodes of $\mathcal{T}(\Omega)$ optimizing an objective function. Such a function is based on a certain measurement of the quality of the local sub-mesh $N(q)$, formed by the set of tetrahedra connected to the free node q. In fact, we use a suitable modification of the objective function such that it is regular over all $\mathbb{R}^3$, (ESCOBAR et al. 2003).

An example of the different steps of this method is shown in Fig. 1.

3. HARMONIE Mesoscale Weather Model

HIRLAM-ALADIN Research on Mesoscale Operational NWP in Europe (HARMONIE) is a weather prediction model design for operational use at convective scale resolutions. The system was mainly developed by Meteo-France and ALADIN Consortium in collaboration with ECMWF and HIRLAM Consortium.

This model uses a 3D-Var data assimilation (FISCHER et al. 2005) which shares most of the code with the ECMWF and ARPEGE models. For surface variables, a statistical interpolation algorithm is used. The non-hydrostatic dynamics is based on (BUBNOVÁ et al. 1995), and the physics is adapted from Meso-NH research model.

AEMET has been running HARMONIE with AROME configuration at 2.5 km horizontal resolution since October 2011. This configuration is close to the one used operationally at Météo–France (SEITY et al. 2011). Local and extreme forecasts are improved significantly with the HARMONIE 2.5 km model compared to coarser grid models like HIRLAM or ECMWF (NAVASCUÉS et al. 2013). The model is run four times per day over two domains (Iberian Peninsula and Canary Islands) with a forecast length of 48 h.

4. Local Wind Field Simulation

Once the tetrahedral mesh is constructed, we consider a mass-consistent model (MONTERO et al. 1998, 2005; FERRAGUT et al. 2010) to compute a wind field $\mathbf{u}$ in the three-dimensional domain Ω, with a boundary $\Gamma = \Gamma_a \cup \Gamma_b$, that verifies the continuity equation and the impermeability condition on the terrain Γ_a,

(a)   **(b)**   **(c)**

(d)   **(e)**   **(f)**

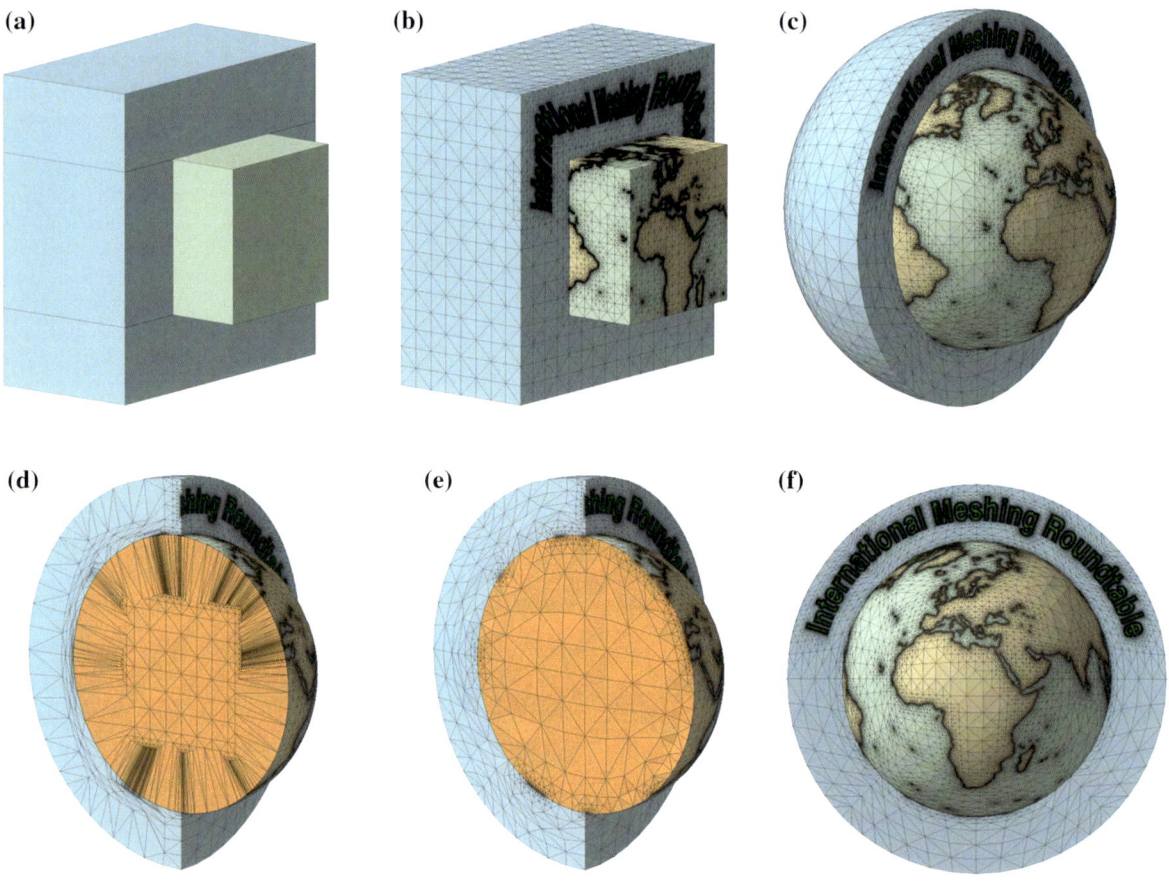

Figure 1

The different meccano steps. **a** Parametric space, **b** triangular projection of the surface in the parametric space, **c** resulting surface mesh, **d** tangled interior nodes, **e** optimized mesh, **f** frontal view

$$\nabla \cdot \mathbf{u} = 0 \quad \text{in} \quad \Omega$$
$$\mathbf{n} \cdot \mathbf{u} = 0 \quad \text{on} \quad \Gamma_a \tag{2}$$

where $\mathbf{n}$ is the outward-pointing normal unit vector, being Γ_b the boundary where the impermeability condition is not imposed.

The model formulates a least-squares problem in the domain Ω to find a wind field $\mathbf{u} = (u, v, w)$, such that it is adjusted as much as possible to an interpolated wind field $\mathbf{u}_0 = (u_0, v_0, w_0)$. The adjusting functional for a field $\mathbf{v} = (\tilde{u}, \tilde{v}, \tilde{w})$ is defined as

$$e(\mathbf{v}) = \frac{1}{2} \int_\Omega (\mathbf{v} - \mathbf{u}_0)^t \mathbf{P}(\mathbf{v} - \mathbf{u}_0) \mathrm{d}\Omega \tag{3}$$

where $\mathbf{P}$ is a 3×3 diagonal matrix with $p_{1,1} = p_{2,2} = 2\alpha_1^2$ and $p_{3,3} = 2\alpha_2^2$. The Lagrange multiplier technique is used to minimise the functional (3), with the

restrictions (2). Considering the Lagrange multiplier λ, the Lagrangian is defined as

$$l(\mathbf{v}, \lambda) = e(\mathbf{v}) + \int_\Omega \lambda \nabla \cdot \mathbf{v} \, \mathrm{d}\Omega \tag{4}$$

and the solution $\mathbf{u}$ is obtained by finding the saddle point $(\mathbf{u}, \phi)$ of the Lagrangian (4). This resulting wind field verifies the Euler–Lagrange equation,

$$\mathbf{u} = \mathbf{u}_0 + \mathbf{P}^{-1} \nabla \phi \tag{5}$$

where ϕ is the Lagrange multiplier. As α_1 and α_2 are constant in Ω, the variational approach results in an elliptic problem in ϕ, by substituting (5) in (2), that is solved by using the finite element method.

$$-\nabla \cdot \left(\mathbf{P}^{-1} \nabla \phi \right) = \nabla \cdot \mathbf{u}_0 \quad \text{in} \quad \Omega \tag{6}$$

$$-\mathbf{n} \cdot \mathbf{P}^{-1} \nabla \phi = \mathbf{n} \cdot \mathbf{u_0} \quad \text{on} \quad \Gamma_a \qquad (7)$$

$$\phi = 0 \quad \text{on} \quad \Gamma_b \qquad (8)$$

4.1. Construction of the Initial Field

The interpolated wind field $\mathbf{u_0}$ is constructed from the HARMONIE data, specifically, the values of the 10 m wind, $\mathbf{u_n^h}$, given at point n of the HARMONIE grid, the geostrophic wind $\mathbf{u_g}$. Therefore, we consider a horizontal interpolation and a vertical extrapolation of the available measurements to construct $\mathbf{u_0}$ in the whole computational domain.

4.1.1 Horizontal Interpolation

A common technique of interpolation at a given point, placed at a height z_m over the terrain, is formulated as a function of the inverse of the squared distance between that point and the measurement stations, and the inverse of their height differences (MONTERO et al. 1998)

$$\mathbf{u_0}(z_m) = \xi \frac{\sum_{n=1}^{N} \frac{\mathbf{u_n^h}}{d_n^2}}{\sum_{n=1}^{N} \frac{1}{d_n^2}} + (1 - \xi) \frac{\sum_{n=1}^{N} \frac{\mathbf{u_n^h}}{|\delta h_n|}}{\sum_{n=1}^{N} \frac{1}{|\delta h_n|}} \qquad (9)$$

where the value of $\mathbf{u_n^h}$ is the velocity measured at HARMONIE point n, N is the number of HARMONIE points considered in the interpolation, d_n is the horizontal distance from the station n to the point of the domain where we are computing the wind velocity, $|\delta h_n|$ is the height difference between station n and the studied point, and ξ is a weighting parameter ($0 \leq \xi \leq 1$), that allows us to give more importance to one of these interpolation criteria.

4.1.2 Vertical Extrapolation

In this work, a log-linear wind profile is considered (LALAS and RATTO 1996) in the surface layer, which takes into account the horizontal interpolation (MONTERO and SANÍN 2001) and the effect of roughness on the wind intensity and the direction. These values also depend on the air stability (neutral, stable, or unstable atmosphere) according to the Pasquill stability class. Above the surface layer, a linear

interpolation is carried out using the geostrophic wind. The logarithmic profile is given by,

$$\mathbf{u_0}(z) = \frac{\mathbf{u}^*}{k} \left(\log \frac{z}{z_0} - \phi_m \right) \quad z_0 < z \leq z_{sl} \qquad (10)$$

where $\mathbf{u}^*$ is the friction velocity, k is von Karman's constant, z_0 is the roughness length (MCRAE et al. 1982) and z_{sl} is the height of the surface layer. The values of ϕ_m depend on the air stability. For neutral atmosphere its value is $\phi_m = 0$, for stable atmosphere $\phi_m = -5\frac{z}{l}$, and for unstable atmosphere

$$\phi_m = \log \left[\left(\frac{\theta^2 + 1}{2} \right) \left(\frac{\theta + 1}{2} \right)^2 \right] - 2 \arctan \theta + \frac{\pi}{2} \qquad (11)$$

where, $\theta = (1 - 16\frac{z}{l})^{1/4}$ and $\frac{1}{l} = az_0^b$, with a, b, depending on the Pasquill stability class (ZANNETTI 1990). The friction velocity is obtained from (10) at any point (x, y) by using the horizontal interpolated velocity $\mathbf{u_0}(z_m)$

$$\mathbf{u}^* = \frac{k \, \mathbf{u_0}(z_m)}{\log \frac{z_m}{z_0} - \phi_m} \qquad (12)$$

The height of the boundary layer z_{pbl} above the ground is chosen such that the wind intensity and direction are constant at that height,

$$z_{pbl} = \frac{\gamma |\mathbf{u}^*|}{f} \qquad (13)$$

where $f = 2\omega \sin \phi$ is the Coriolis parameter (ω is Earth's rotation and ϕ the latitude), and γ is a parameter depending on the atmospheric stability, and is between 0.15 and 0.3 (PANOFSKY and DUTTON 1984). The height of the mixed layer h is considered to be equal to z_{pbl} in neutral and unstable conditions. In stable conditions, it is approximated by

$$h = \gamma' \sqrt{\frac{|\mathbf{u}^*| \, l}{f}} \qquad (14)$$

where $\gamma' = 0.4$ (GARRATT 1982). The height of surface layer is $z_{sl} = \frac{h}{10}$. From z_{sl} to z_{pbl}, a linear interpolation with geostrophic wind $\mathbf{u_g}$ is carried out,

$$\mathbf{u_0}(z) = \rho(z) \mathbf{u_0}(z_{sl}) + [1 - \rho(z)]\mathbf{u_g} \quad z_{sl} < z \leq z_{pbl} \qquad (15)$$

$$\rho(z) = 1 - \left(\frac{z - z_{sl}}{z_{pbl} - z_{sl}}\right)^2 \left(3 - 2\frac{z - z_{sl}}{z_{pbl} - z_{sl}}\right) \quad (16)$$

Finally, this model assumes $\mathbf{u}_0(z) = \mathbf{u}_g$ if $z > z_{pbl}$ and $\mathbf{u}_0(z) = 0$ if $z \leq z_0$.

5. Genetic Algorithms

Genetic algorithms are optimisation tools based on the natural evolution mechanism. They produce successive trials that have an increasing probability to obtain a global optimum.

In this work, we apply the model developed by LEVINE (1994). The most important aspects of genetic algorithms are the construction of an initial population, the evaluation of each individual in the fitness function, the selection of the parents of the next generation, the crossover of those parents to create the children, and the mutation to increase diversity. The initial population has been randomly generated, and we use iteration limit exceeded as stopping criterion. The fitness function plays the role of the environment. It evaluates each string of a population. This is a measure, relative to the rest of the population, of how well that string satisfies a problem-specific metric. The values are mapped to a non-negative and monotonically increasing fitness value. Two population replacements are commonly used. The first, the generational replacement, replaces the entire population each generation (HOLLAND 1992). The second, used in this work, is known as steady-state and only replaces a few individuals in each generation (WHITLEY 1989). We have chosen the stochastic universal selection and the uniform crossover operator (SPEARS 1991). The mutation operator is better used after crossover (DAVIS 1991). It allows us to reach individuals in the search space that could not be evaluated otherwise. The mutation operator used in this work replaces the gene value with a random one within the initialisation range.

5.1. Parameters to Calibrate

In the numerical experiments with this wind model, the parameters to be estimated are α, ξ and γ. For this purpose, the fitting function to be minimised

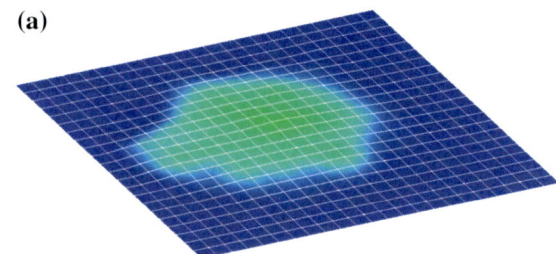

(a)

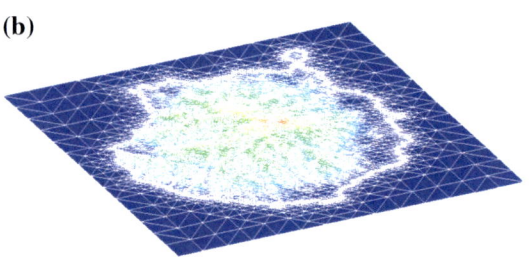

(b)

Figure 2
Terrain discretization. **a** HARMONIE grid, **b** Meccano mesh

is the root mean square error (RMS) of the wind velocities given by the model with respect to the measures at the HARMONIE points.

$$E(\alpha, \xi, \gamma) = \sqrt{\frac{\sum_{n=1}^{N}\left[\left(u_n^\star - u_n\right)^2 + \left(v_n^\star - v_n\right)^2\right]}{2N}}$$

(17)

where $u_n^\star$ and $v_n^\star$ are the X and Y components of the HARMONIE wind field in the point n used for the calibration, u_n and v_n are the X and Y components of the resulting wind field of the mass-consistent model (5), and N is the number of points where the calibration is performed.

The first parameter to be estimated is the stability parameter (α), which is defined as

$$\alpha = \frac{\alpha_1}{\alpha_2} \quad (18)$$

where α_1 and α_2 are the components of the matrix $\mathbf{P}$ defined in (3). The parameter α defines the predominant component of the flow adjustment, being the vertical component when $\alpha > 1$, and the horizontal component when $\alpha < 1$.

The second parameter is the weighting coefficient ξ in the Eq. (9). Note that $\xi = 1$ implies the inverse of

(a)

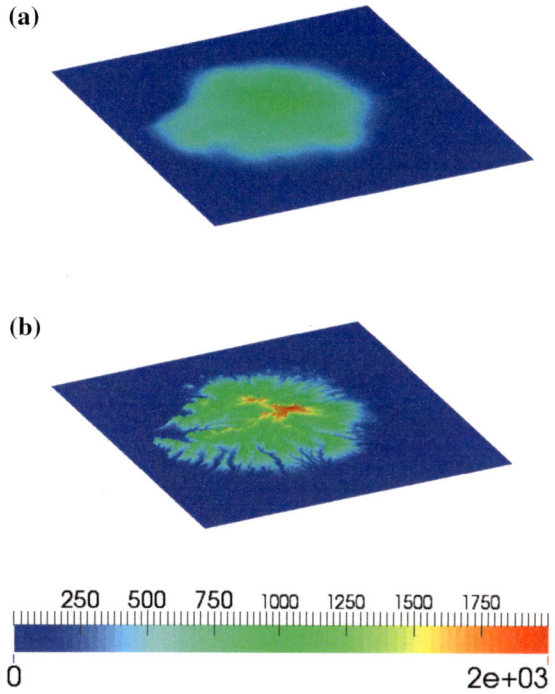

(b)

(a)

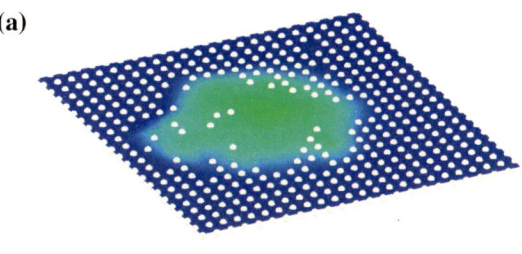

(b)

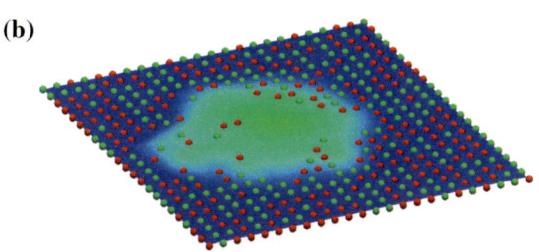

Figure 4
HARMONIE points used in simulation. **a** Point fulfilling the tolerance, **b** points used as a stations (*green*), and points used as control points for calibration (*red*)

250 500 750 1000 1250 1500 1750

0   2e+03

Figure 3
Terrain heights (m). **a** HARMONIE grid, **b** Meccano mesh

the squared distance interpolation, while $\xi = 0$ stands for a height difference interpolation.

The parameter γ is related to the height of the boundary layer z_{pbl}, and depends on the atmospheric stability. As stated in Sect. 4.1.2, its value varies in a range between 0.15 and 0.3.

6. Results

In this section, an example is presented using the methodology described in this paper. The example is located in Gran Canaria island, using the results from the HARMONIE model. Finally, a validation of the method is performed using measurement stations data.

6.1. Mesh Generation

The first step in the forecasting of the wind field is the generation of the air volumetric mesh. The mesh is created with the meccano method from a digital

terrain model of the Gran Canaria island. The height of the domain is set to 10,000 m. The resulting mesh has 251,808 nodes and 1,090,366 tetrahedra.

Figure 2 shows the terrain of the resulting meccano mesh and the terrain of the HARMONIE grid. The figure shows the difference in the discretization between the two models. This is the main motivation in coupling both models.

6.2. HARMONIE Data for the Mass-Consistent Model

The HARMONIE data assimilation has been done using the velocity at 10 m, and the geostrophic wind.

The most straightforward method is to use the whole data at 10 m, selecting a subset of that data as the measured data in the genetic process, and another subset as control points.

Figure 3 shows the terrain height in the meccano mesh and the HARMONIE grid. The great difference in heights indicates that probably not all values of the HARMONIE velocity at 10 m are appropriate. For this reason, we propose to use only those HARMONIE points whose heights differ from the meccano height in less than a certain tolerance.

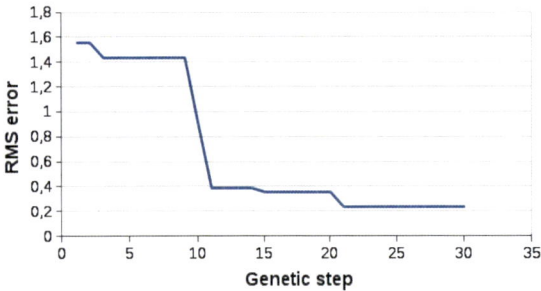

Figure 5
Error (ms^{-1}) at each genetic step

(a)

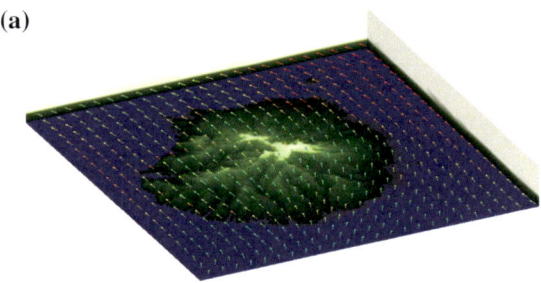

(b)

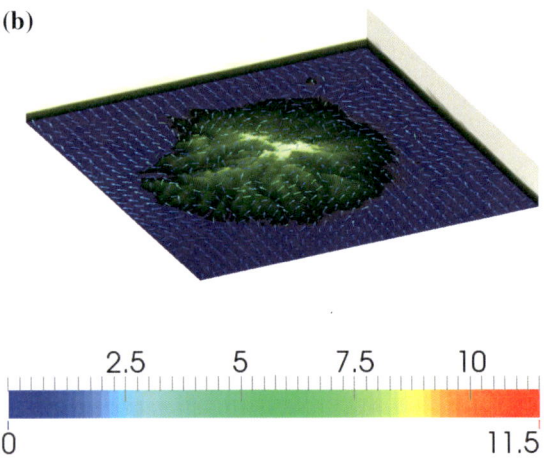

Figure 6
Wind field at 10 m (ms^{-1}). **a** HARMONIE wind field, **b** resulting wind field

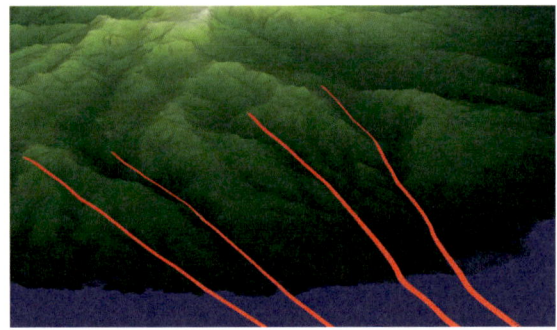

Figure 7
Streamlines of the resulting wind field

Table 1

Measurement stations UTM coordinates

Station	X	Y
C619X	429,982.2	3,108,577.3
C629Q	429,966.6	3,073,034.7
C635B	443,504.2	3,088,472.9
C639X	455,377.2	3,076,514.6
C639Y	443,283.4	3,070,534.1
C649R	462,851.1	3,095,782.8

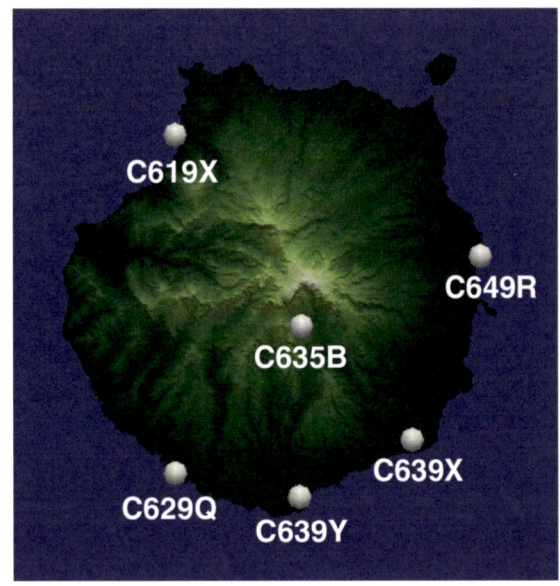

Figure 8
Location of the measurement stations

In Fig. 4, the points that will be used in the simulation are represented. The tolerance in this case has been set to 50 m. As noted before, the points fulfilling the tolerance are divided in two different subsets, one used as stations (the green ones), and the other used as control points (the red ones).

6.3. Model Calibration

Once the stations and the control points are fixed, the calibration of the parameters can be done.

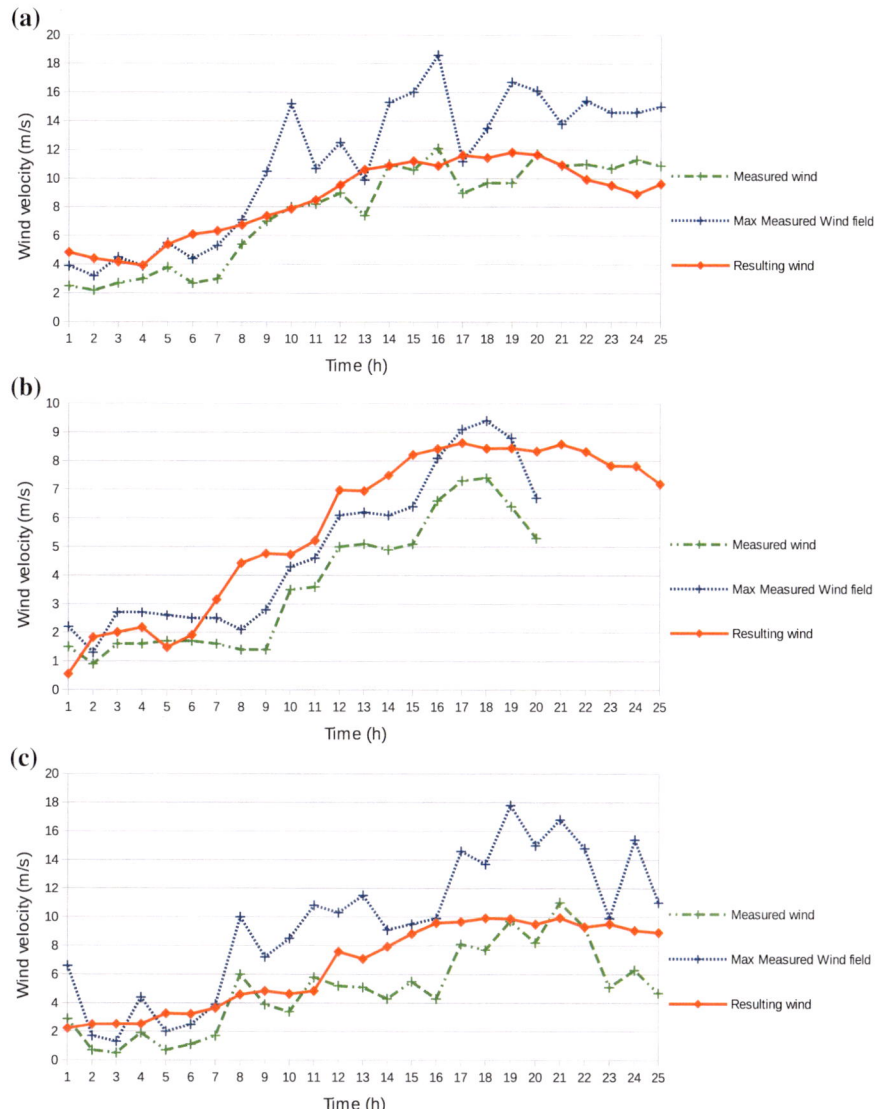

Figure 9

Comparison of the wind field with the instant measured data and the maximum measured data. **a** C619X measurement station, **b** C629Q measurement station, **c** C635B measurement station, **d** C639X measurement station, **e** C639Y measurement station, **f** C649R measurement station

Using the genetic algorithm described above, a simulation of 30 genetic steps has been computed. Figure 5 shows the diminution of the fitness function (17) in the subsequent genetic steps.

The run time for an episode is about 90 min with a population size of 60 individuals using six cores in parallel.

6.4. Resulting Wind Field

With the parameters calibrated by the genetic algorithm, we can finally compute the resulting wind field.

Figure 6 shows the wind field at 10 m for the HARMONIE mesoscale model (a), and the resulting wind field (b). It can be observed that the

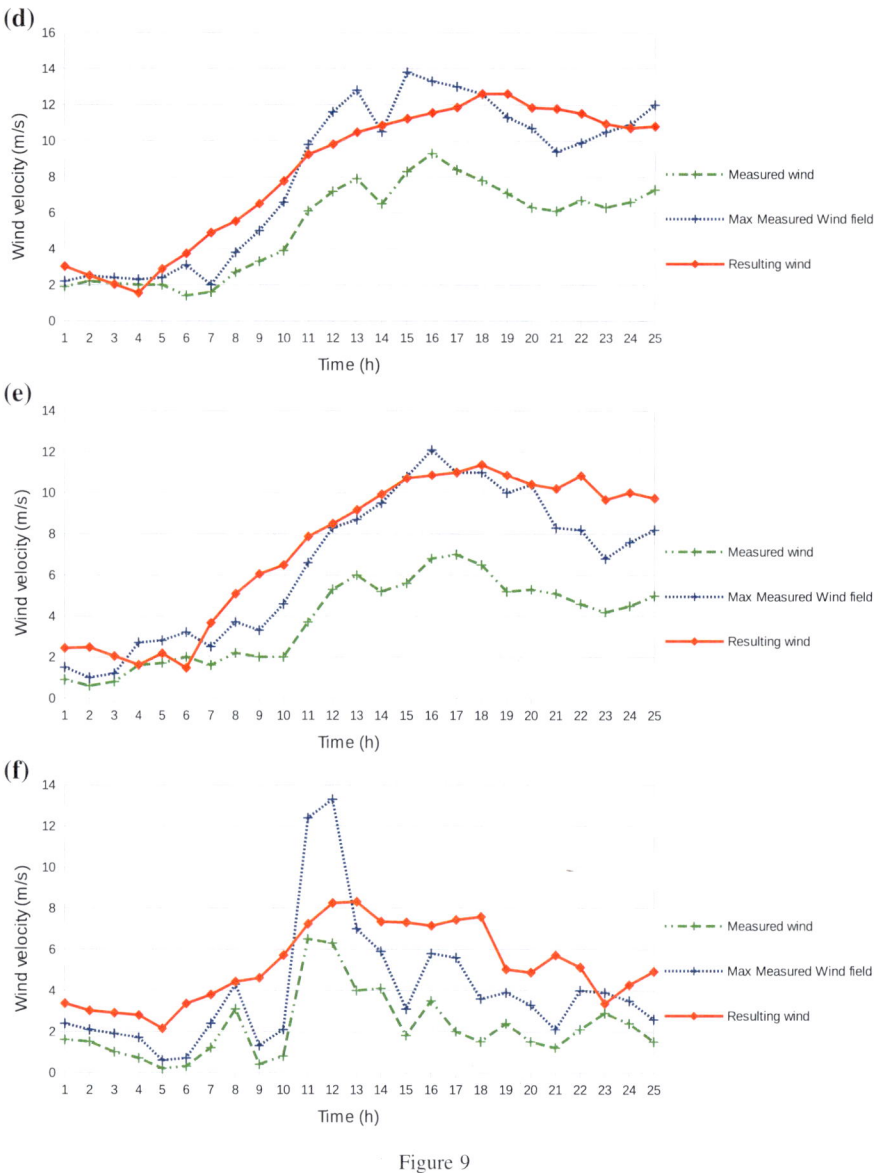

Figure 9
continued

HARMONIE wind field crosses the island almost ignoring the orography, while the resulting wind field changes its direction when it reaches the island.

In Fig. 7 we can see a detail of the streamlines of the resulting wind field in several points. It has to be remarked that the wind field tries to align with the valleys of the island.

6.5. Validation against Measurement Station Data

To validate the method presented in this work, a comparison between the resulting wind field and real wind measurements has been performed.

Six measurement stations have been considered. Their UTM coordinates are summarised in Table 1, and their location is represented in Fig. 8. Most stations are near the sea level, except for C635B with a height of 960 m.

In order to conduct the comparison, a whole day forecast has been executed. The HARMONIE data used in this forecast is a 24 h execution starting at midnight on 2010/02/20. Measurement data are available for the whole day in all the stations, except for C629Q with only 20 measurements.

Figure 9 shows the comparisons for the stations. For each station, the instant wind velocity and the maximum wind velocity have been represented. It has to be noted that, in general, the resulting wind velocity is reasonably similar to the measured wind velocity, being in some cases closer to the maximum (for example, in C629Q, C639Y); in other cases, it remains within the range (C635B and C619X). The comparison at station C649R is the only one where the method has predicted, in the whole episode, a higher wind velocity than the measured one.

Examining the comparisons it can be noted that this method smooths the wind velocity. Looking at the measured data, there are abrupt changes among time steps that are not captured by the method.

A special mention has to be made of the C635B station. In the location of this station, the difference between the HARMONIE grid and the meccano mesh was greater than the tolerance, so no data from HARMONIE were used nearby. Nevertheless, the resulting wind is a good forecast, proving the feasibility of this method.

7. Conclusions

This paper presents a strategy to simulate the wind field forecast in complex orography locations. It is based on the coupling of the HARMONIE mesoscale model and a mass-consistent model. The results show the importance of the terrain in the resulting wind field. Genetic algorithms have proved to be useful in this kind of problems, allowing us to calibrate the unknown parameter to the HARMONIE model wind field. The numerical experiment shows a reasonable behaviour of the proposed method.

Acknowledgments

This work has been supported by the Spanish Government, "Ministerio de Ciencia e Innovación", Grant Contracts: CGL2011-29396-C03-01 and CGL2011-29396-C03-02, and by "Junta de Castilla León", "Consejería de Educación", Grant Contract SA266A12-2.

REFERENCES

BOROUCHAKI H, FREY P (2005) *Simplification of surface mesh using hausdorff envelope.* Computer Methods in Applied Mechanics and Engineering *194*(48–49), pp. 4864–4884, DOI:10.1016/j.cma.2004.11.016.

BUBNOVÁ R, HELLO G, BÉNARD P, GELEYN J (1995) *Integration of the fully elastic equations cast in the hydrostatic pressure terrain-following coordinate in the framework of the arpege/aladin nwp system.* Mon Wea Rev *123*, pp. 515–535, DOI:10.1175/1520-0493(1995)123<0515:IOTFEE>2.0.CO;2.

CARVALHO D, ROCHA A, SANTOS CS, PEREIRA R (2013) *Wind resource modelling in complex terrain using different mesoscale microscale coupling techniques.* Applied Energy *108*(0), pp. 493–504, DOI:10.1016/j.apenergy.2013.03.074.

CASCÓN JM, RODRÍGUEZ E, ESCOBAR JM, MONTENEGRO R (2013) *Comparison of the meccano method with standard mesh generation techniques.* Engineering with Computers pp 1–14, DOI:10.1007/s00366-013-0338-6.

DAVIS L (1991) Handbook of genetic algorithms. Van Nostrand Reinhold, New York, NY, USA.

ESCOBAR JM, RODRÍGUEZ E, MONTENEGRO R, MONTERO G, GONZÁLEZ-YUSTE JM (2003) *Simultaneous untangling and smoothing of tetrahedral meshes.* Computer Methods in Applied Mechanics and Engineering *192*(25), pp. 2775–2787, DOI:10.1016/S0045-7825(03)00299-8.

ESCOBAR JM, RODRÍGUEZ E, MONTENEGRO R, MONTERO G, sc González-Yuste JM (2010) SUS code: simultaneous mesh untangling and smoothing code. http://www.dca.iusiani.ulpgc.es/SUScode

FERRAGUT L, ASENSIO M, MONEDERO S (2007) *A numerical method for solving convection-reaction-diffusion multivalued equations in fire spread modelling.* Advances in Engineering Software *38*(6), pp. 366–371, DOI:10.1016/j.advengsoft.2006.09.007.

FERRAGUT L, MONTENEGRO R, MONTERO G, RODRÍGUEZ E, ASENSIO M, ESCOBAR JM (2010) *Comparison between 2.5-D and 3-D realistic models for wind field adjustment.* Journal of Wind Engineering and Industrial Aerodynamics *98*(10–11), pp. 548–558, DOI:10.1016/j.jweia.2010.04.004.

FISCHER C, MONTMERLE T, BERRE L, AUGER L, ŞTEFĂNESCU SE (2005) *An overview of the variational assimilation in the aladin/france numerical weather-prediction system.* Quarterly Journal of the Royal Meteorological Society *131*(613), pp. 3477–3492, DOI:10.1256/qj.05.115.

FLOATER MS (2003) *Mean value coordinates.* Computer Aided Geometric Design *20*(1), pp. 19–27, DOI:10.1016/S0167-8396(03)00002-5.

GARRATT J (1982) *Observations in the nocturnal boundary layer.* Boundary-Layer Meteorology 22(1), pp. 21–48, DOI:10.1007/BF00128054.

GASSET N, LANDRY M, GAGNON Y (2012) *A comparison of wind flow models for wind resource assessment in wind energy applications.* Energies 5(11), pp. 4288–4322, DOI:10.3390/en5114288.

HOLLAND JH (1992) Adaptation in Natural and Artificial Systems. MIT Press, Cambridge, MA, USA.

KOSSACZKÝ I (1994) *A recursive approach to local mesh refinement in two and three dimensions.* Journal of Computational and Applied Mathematics 55(3), pp. 275–288, DOI:10.1016/0377-0427(94)90034-5.

LALAS D, RATTO C (1996) Modelling of Atmospheric Flow Fields. World Scientific Publishing, Singapore.

LEVINE D (1994) A parallel genetic algorithm for the set partitioning problem.

MCRAE GJ, GOODIN WR, SEINFELD JH (1982) *Development of a second-generation mathematical model for urban air pollution-I. Model formulation.* Atmospheric Environment (1967) 16(4):679–696, DOI:10.1016/0004-6981(82)90386-9.

MONTENEGRO R, CASCÓN JM, ESCOBAR JM, RODRÍGUEZ E, MONTERO G (2009) *An automatic strategy for adaptive tetrahedral mesh generation.* Applied Numerical Mathematics 59(9), pp. 2203–2217, DOI:10.1016/j.apnum.2008.12.010.

MONTENEGRO R, CASCÓN JM, RODRÍGUEZ E, ESCOBAR JM, MONTERO G (2010) The meccano method for automatic three-dimensional triangulation and volume parametrization of complex solids. In: Topping B, Adam J, Pallarés F, Bru R, Romero M (eds) Developments and Applications in Engineering Computational Technology, seventh edn, Saxe-Coburg Publications, Stirlingshire, chap 2, pp 19–48, DOI:10.4203/csets.26.2.

MONTERO G, SANÍN N (2001) *3-D modelling of wind field adjustment using finite differences in a terrain conformal coordinate system.* Journal of Wind Engineering and Industrial Aerodynamics 89(5), pp. 471–488, DOI:10.1016/S0167-6105(00)00075-1.

MONTERO G, MONTENEGRO R, ESCOBAR JM (1998) *A 3-D diagnostic model for wind field adjustment.* Journal of Wind Engineering and Industrial Aerodynamics 74–76(0), pp. 249–261, DOI:10.1016/S0167-6105(98)00022-1.

MONTERO G, RODRÍGUEZ E, MONTENEGRO R, ESCOBAR JM, GONZÁLEZ-YUSTE JM (2005) *Genetic algorithms for an improved parameter estimation with local refinement of tetrahedral meshes in a wind model.* Advances in Engineering Software 36(1), pp. 3–10, DOI:10.1016/j.advengsoft.2004.03.011.

NAVASCUÉS B, CALVO J, MORALES G, SANTOS C, CALLADO A, CANSADO A, CUXART J, DÍEZ M, DEL RÍO P, ESCRIBÀ P, GARCÍA-COLOMBO O, GARCÍA-MOYA J, GEIJO C, GUTIÉRREZ E, HORTAL M, MARTÍNEZ I, ORFILA B, PARODI J, RODRÍGUEZ E, SÁNCHEZ-ARRIOLA J, SANTOS-ATIENZA I, SIMARRO J (2013) *Long-term verification of HIRLAM and ECMWF forecasts over southern europe: History and perspectives of numerical weather prediction at AEMET.* Atmospheric Research 125–126(0), pp. 20–33, DOI:10.1016/j.atmosres.2013.01.010.

OLIVER A, MONTERO G, MONTENEGRO R, RODRÍGUEZ E, ESCOBAR JM, PÉREZ-FOGUET A (2012) *Finite element simulation of a local scale air quality model over complex terrain.* Advances in Science and Research 8, pp. 105–113, DOI:10.5194/asr-8-105-2012.

OLIVER A, MONTERO G, MONTENEGRO R, RODRÍGUEZ E, ESCOBAR JM, PÉREZ-FOGUET A (2013) *Adaptive finite element simulation of stack pollutant emissions over complex terrains.* Energy 49(0), pp. 47–60, DOI:10.1016/j.energy.2012.10.051.

PANOFSKY H, DUTTON J (1984) Atmospheric turbulence. Models and methods for engineering applications. Wiley, New York.

RODRÍGUEZ E (2004) Modelización numérica de campos de viento mediante elementos finitos en 3-D. PhD thesis, Instituto Universitario de Sistemas Inteligentes y Aplicaiones Numéricas en Ingeniería (IUSIANI), Universidad de Las Palmas de Gran Canaria (ULPGC).

RODRÍGUEZ E, MONTERO G, ESCOBAR JM, MONTENEGRO R, OLIVER A (2012) Wind3D. http://www.dca.iusiani.ulpgc.es/Wind3D

SEITY Y, BROUSSEAU P, MALARDEL S, HELLO G, BÉNARD P, BOUTTIER F, LAC C, MASSON V (2011) *The arome-france convective-scale operational model.* Mon Wea Rev 139, pp. 976–991, DOI:10.1175/2010MWR3425.1.

SPEARS W (1991) On the virtues of parameterized uniform crossover.

WAN S, YIN Z, ZHANG K, ZHANG H, LI X (2011) *A topology-preserving optimization algorithm for polycube mapping.* Computers & Graphics 35(3), pp. 639–649, DOI:10.1016/j.cag.2011.03.018.

WHITLEY D (1989) The genitor algorithm and selection pressure: Why rank-based allocation of reproductive trials is best. In: Proceedings of the Third International Conference on Genetic Algorithms, Morgan Kaufmann, pp 116–121.

ZANNETTI P (1990) Air Pollution Modeling. Computational Mechanics Publications, Boston.

(Received March 31, 2014, revised July 12, 2014, accepted July 20, 2014, Published online August 11, 2014)

Pure Appl. Geophys. 172 (2015), 121–139
© 2014 Springer Basel
DOI 10.1007/s00024-014-0893-9

A Wildland Fire Physical Model Well Suited to Data Assimilation

L. Ferragut,[1,3] M. I. Asensio,[1,3] J. M. Cascón,[1,2] and D. Prieto[3]

Abstract—In this article, we focus on a simplified two-dimensional fire model with some three-dimensional effects. The model takes into account the moisture content and the energy lost in the vertical direction and to radiation from the flames. We couple this model with a local wind model, well adapted to fire modelling. The topography, fuel type, mass fraction of the fuel and the meteorological data required by the model (temperature, humidity and wind) are provided by geographic information systems. We incorporate data assimilation techniques to our fire model in order to improve the approximations obtained with the model. The data assimilated are the temperature of the solid fuel (which is related to the position of the fire front) and the mass fraction of fuel at certain points in the domain. The numerical examples show that this procedure is able to correct the approximations obtained by the model simulations, providing more realistic predictions. The process is implemented using parallel computing.

Key words: Forest fire modelling, data assimilation.

1. Introduction

In recent years, advances in the computational power and increases in the capabilities of spatial information technologies [remote sensing and geographic information systems (GIS)] have offered better potential for the effective modelling of wildland fire behaviour than was available previously. This has intensified the interest in fire modelling, as can be appreciated in the reviews on this that have appeared recently (Pastor *et al.* 2003; Sullivan 2009a, b, c).

Owing to such advances in computational power, apart from wildfire models based on semi-empirical fire spread equations as in Rothermel (1972), physics-based wildfire models are now considered a serious alternative. Following the models in Cox (1995) and Margerit and Séro-Guillaume (2002), and previous models reported by the authors in Asensio and Ferragut (2002) and Ferragut *et al.* (2007a, b), herein we consider a simplified two-dimensional model with some three-dimensional effects. This model takes the following into account: radiation from the flames, the wind, the slope of the orography, the moisture content and the energy lost due to convection.

Data assimilation is a technique used to incorporate data into a running model using sequential statistical estimation. Data assimilation periodically adjusts the model state, incorporating new data, by using statistical methods. Because fire is highly nonlinear and irreversible, data assimilation for fire models poses special challenges, and standard methods such as the ensemble Kalman filter (EnKF) do not always work satisfactorily (Kalman 1960; Evensen 2009; Mandel *et al.* 2008).

This paper is organized as follows: in Sect. 2 we discuss the fire model and present a mathematical description of the model, as well as the numerical approximation. In Sect. 3 we apply the model to the simulation of a real fire, comparing the simulation results with the real fire data. Finally, in Sect. 4 we consider the problem of data assimilation, showing by means of a numerical experiment that data assimilation using the deterministic ensemble Kalman filter (DEnKF) (Sakov and Oke 2008) can be useful to improve the numerical simulation results,

---

[1] Instituto Universitario de Física Fundamental y Matemáticas, University of Salamanca, C. Parque s.n., 37008 Salamanca, Spain. E-mail: ferragut@usal.es; mas@usal.es

[2] Departamento de Economía e Historia Económica, University of Salamanca, Edificio FES, Campus Miguel de Unamuno, 37007 Salamanca, Spain. E-mail: casbar@usal.es

[3] Departamento de Matemática Aplicada, University of Salamanca, C. Parque s.n., 37008 Salamanca, Spain. E-mail: dpriher@usal.es

particularly when the precise position of the ignition point is not known. We end the paper with some conclusions in Sect. 5.

2. Fire Model

The three generally accepted forms of heat transfer in wildland fires are conduction, convection and radiation. The most important physical processes driving heat transfer in a wildland fire are convection and radiation. In low wind conditions, the dominating mechanism is radiation (WEBER 1986), but in conditions where wind is not insignificant, it is convection that dominates (WEBER 1991). The relative importance of radiation and convection varies from fire to fire and estimation of their exact combination is not simple, although diffusion can be considered negligible. In this model, we focus on radiation and hence the model is applicable to low wind conditions.

Although this is a two-dimensional model, it takes into account some important three-dimensional factors. First, the energy lost by natural convection in the vertical direction is represented by a zero-order term in the partial differential equation of the temperature. The second is the radiation from the flames above the surface where the fire takes place, taking into account the effect of the surface slope through the flame tilt in the computation of the radiation. This means that the topography of the surface is included in the model. The effect of wind on the slope of the flame is another three-dimensional effect that is also taken into account.

An important simplification is that we only consider the solid phase of the combustion process: the mass fraction of solid fuel c is a non-dimensional variable between 0 and 1, and the maximum value of non-dimensional solid fuel temperature u is u_p, the non-dimensional pyrolysis temperature. The gaseous phase is parameterized through the flame temperature T_f and the flame height F in the radiation term.

The effect of the moisture content of the solid fuel is included through the multivalued operator relating enthalpy and temperature.

2.1. Mathematical Description of the Model

Let $d = [0, l_x] \times [0, l_y] \subset \mathbb{R}^2$ be a rectangle representing the projection of the surface S where the fire takes place, defined by the mapping:

$$S : d \longmapsto \mathbb{R}^3$$
$$(x, y) \longmapsto (x, y, h(x, y)),$$

where $h(x, y)$ is a known function representing the orography of the surface S.

We shall assume that vegetation can be represented by a given fuel load M, (kg m^{-2}) together with a moisture content M_v, (kg of water/kg of dry fuel). M and M_v are scalar functions defined on d. Moreover, we shall assume that the height of flames F is bounded by δ.

In order to take into account some three-dimensional effects, specifically the radiation from the flames above the surface, S, we shall consider the following three-dimensional domain,

$$D = \{(x, y, z) : x, y \in d, h(x, y) < z < h(x, y) + \delta\}.$$

The equations governing the fire model are based on the energy and mass conservation equation of the solid fuel on the surface, S, and the radiation equation in D.

The non-dimensional simplified equations for the fire model are:

$$\partial_\tau e + \alpha u = r \quad \text{in } S \quad \tau \in (0, \tau_{\max}), \quad (1)$$

$$e \in G(u) \quad \text{in } S \quad \tau \in (0, \tau_{\max}), \quad (2)$$

$$\partial_\tau c = -g(u)c \quad \text{in } S \quad \tau \in (0, \tau_{\max}). \quad (3)$$

We complete the problem with homogeneous Dirichlet boundary conditions and the following initial conditions:

$$u(x, y, 0) = u_0(x, y) \quad \text{in } S, \quad (4)$$

$$c(x, y, 0) = c_0(x, y) \quad \text{in } S. \quad (5)$$

The unknowns $e = \frac{E}{MCT_\infty}$, non-dimensional enthalpy, $u = \frac{T - T_\infty}{T_\infty}$, the non-dimensional temperature of the solid fuel, and $c = \frac{M}{M_0}$, the mass fraction of solid fuel, are bidimensional variables defined in $S \times (0, \tau_{\max})$, where $\tau_{\max}$ is the time τ of study. The physical quantities E, T and M are enthalpy, the temperature of solid fuel, and the fuel load, respectively; C is the heat capacity of solid fuel; T_∞ a reference temperature, and M_0 is the initial fuel load.

The non-dimensional enthalpy e is an element of a multivalued maximal monotone operator G (BRÉZIS 1973), given by:

$$G(u) = \begin{cases} u & \text{if} \quad u < u_v, \\ [u_v , u_v + \lambda_v] & \text{if} \quad u = u_v, \\ u + \lambda_v & \text{if} \quad u_v < u < u_p, \\ [u_p + \lambda_v , \infty] & \text{if} \quad u = u_p, \end{cases}$$

where u_v and u_p are the non-dimensional evaporation temperature of the water and the non-dimensional pyrolysis temperature of the solid fuel, respectively. The quantity λ_v is the non-dimensional evaporation heat related to the evaporation latent heat Λ_v

$$\lambda_v = \frac{M_v \Lambda_v}{C T_\infty}.$$

It should be noted that in the burnt zone, the multivalued operator does not exactly represent the physical phenomena because the water vapor is no longer in the porous medium. This drawback can be avoided by setting $\lambda_v = 0$ in the burnt area. For further details about the multivalued operators, see BRÉZIS (1973).

The term αu represents the energy losses by natural convection in the vertical direction. The parameter α is related to physical quantities by $\alpha = \frac{B[\tau]}{MC}$, where B is the natural convection coefficient and $[\tau]$ is the time scale.

The right hand side of Eq. (3) represents the loss of solid fuel due to combustion. Thus, $g(u) = 0$ when $u < u_p$, and $g(u)$ is constant when $u = u_p$, where the constant is inversely proportional to the half-life time of combustion of each type of fuel.

2.1.1 Radiation

The right hand side of Eq. (1) describes the thermal radiation reaching the surface S from the flame above the layer,

$$r = \frac{[\tau]}{MCT_\infty} R.$$

R represents the incident energy at a point $\mathbf{x} = (\mathbf{x}, \mathbf{y}, \mathbf{h}(\mathbf{x}, \mathbf{y}))$ of the surface, S, due to radiation from

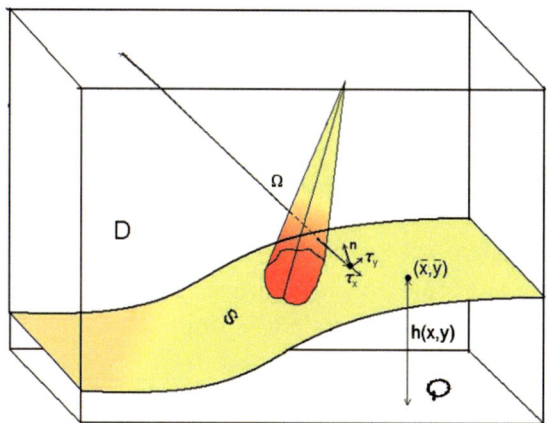

Figure 1
Scheme of the domain, the flame and the incident ray

the flame above the surface per unit time and per unit area, obtained by summing up the contribution of all directions $\mathbf{\Omega}$, i.e.,

$$R(\mathbf{x}) = \int_{\omega=0}^{2\pi} \mathbf{I}(\mathbf{x}, \mathbf{\Omega}) \mathbf{\Omega} \cdot \mathbf{N} \, d\omega, \qquad (6)$$

where ω is the solid angle and $\mathbf{N}$ is the unit normal vector to the surface. We consider only the hemisphere above the fuel layer, and each contribution depends on the flame height F (see Fig. 1). I is the total radiation intensity, i.e., the integral overall wavelength of the radiation energy passing through an area per unit time, per unit of projected area, and per unit of solid angle.

2.2. Wind Model

The model takes into account the effect of wind and slope through the radiation term tilting the flame. We used the high definition wind model developed by the authors in ASENSIO *et al.* (2005) and FERRAGUT *et al.* (2011). The starting point of our wind model is a vertical diffusion wind-field model. If the significant phenomena that we wish to simulate occur in a zone where the horizontal dimensions are much greater than the vertical one, then an asymptotic approximation of the primitive Navier–Stokes equations can be derived. The most salient feature of this asymptotic approach is that it provides a three-dimensional velocity wind field (which satisfies the incompressibility condition in the air layer) governed by a two-dimensional linear equation such that it can be coupled with the temperature surface distribution in order to take into account thermal effects such as the high temperatures that occur during a fire, although in this work the influence of fire in the wind has been neglected. The validity of this model has the following constraints: the nonlinear terms are neglected, and it is assumed that the air temperature decreases linearly with height. In addition, air compressibility is neglected. However, the model does take into account buoyancy forces, slope effects, and mass conservation. The three-dimensional domain is now

$$\overline{D} = \{(x, y, z) : (x, y) \in d, h(x, y) < z < \overline{\delta}\},$$

where $\overline{\delta} \geq h(x, y) + \delta$. The ground surface S is again the lower boundary of $\overline{D}$, and $A = \{(x, y, z) : (x, y) \in d, z = \overline{\delta}\}$ is now the new air upper boundary of $\overline{D}$.

Using the fact that the thickness, $\overline{\delta}$, of the air layer considered is small in comparison with its width, and assuming that the wind is not too strong, the asymptotic wind equations in $\overline{D}$ are:

$$-\partial_{zz}^2 \mathbf{V} + \nabla_{xy} P = 0, \tag{7}$$

$$\partial_z P = \lambda T, \tag{8}$$

$$\nabla_{xy} \cdot \mathbf{V} + \partial_z W = 0, \tag{9}$$

where $\mathbf{V} = (\mathbf{V_1}, \mathbf{V_2})$ are the horizontal components of the three-dimensional wind, W is the vertical component, P is the pressure, T is the temperature and λ is related to the Grashof number.

The boundary conditions of the asymptotic model are:

$$\partial_z \mathbf{V} = \zeta \mathbf{V}, \quad (\mathbf{V}, W) \cdot \mathbf{N} = 0, \quad \textbf{on S}, \tag{10}$$

$$\partial_z \mathbf{V} = 0, \quad W = 0, \quad \textbf{on A}, \tag{11}$$

$$\overline{\mathbf{V}} \cdot \boldsymbol{\eta} = (\overline{\delta} - \mathbf{h}) \mathbf{v_m} \cdot \boldsymbol{\eta}, \quad \textbf{on } \partial d, \tag{12}$$

where $\mathbf{N}$ is the inner unit normal vector field to $\partial \overline{D}$, $\boldsymbol{\eta}$ is the inner unit normal vector field to ∂d, ζ is the friction coefficient, $\mathbf{v_m}$ is the horizontal component of the meteorological wind, not depending on z and with a null total flux through the lateral boundary. $\overline{\mathbf{V}}$ is the

horizontal flux at a point $(x, y) \in d$ and time τ, and is defined by

$$\overline{\mathbf{V}}(\tau, \mathbf{x}, \mathbf{y}) = \int_{\mathbf{h}(\mathbf{x}, \mathbf{y})}^{\overline{\delta}} \mathbf{V}(\tau, \mathbf{x}, \mathbf{y}, \mathbf{z}) \, d\mathbf{z}. \tag{13}$$

If we assume that the air temperature decreases linearly with height and vanishes on the upper boundary, by analytical integration of Eq. (8), the asymptotic problem can be reduced to solving the following bidimensional problem:

$$-\nabla_{xy} \cdot (a \nabla_{xy} p) = \nabla_{xy} \cdot (b \nabla_{xy} t) \quad \text{in } d, \tag{14}$$

$$a \nabla_{xy} p \cdot \boldsymbol{\eta} = -b \nabla_{xy} t \cdot \boldsymbol{\eta} + v \quad \text{on } \partial d, \tag{15}$$

where the potential p depends only on the two first spatial variables (x, y) and on time τ; t is a rescaled bidimensional temperature depending on ground surface temperature and

$$v = v(\tau, x, y) = (\overline{\delta} - h(x, y)) \mathbf{v_m}(\tau, \mathbf{x}, \mathbf{y}) \cdot \boldsymbol{\eta}$$

is the horizontal flux on ∂d.

Thus, the horizontal wind is given by,

$$\mathbf{V}(\mathbf{x}, \mathbf{y}, \mathbf{z}) = \mathbf{m}(\mathbf{x}, \mathbf{y}, \mathbf{z}) \nabla_{xy} \mathbf{p}(\tau, \mathbf{x}, \mathbf{y}) + \mathbf{n}(\mathbf{x}, \mathbf{y}, \mathbf{z}) \nabla_{xy} \mathbf{t}(\tau, \mathbf{x}, \mathbf{y}),$$
$$\tag{16}$$

and $\mathbf{v}(\mathbf{x}, \mathbf{y}) = \mathbf{V}(\mathbf{x}, \mathbf{y}, \mathbf{0})$ is the value used in the convective term of the fire model in Eq. (1).

Functions a, b, m and n depend on $\overline{\delta}$, h and the inverse of the friction coefficient. For more details of this wind model see ASENSIO et al. (2005) and FERRAGUT et al. (2011).

This model requires knowledge of the horizontal wind flux v through ∂d, which is currently unknown. Usually, the meteorological wind at a finite number of points over the surface S is known. In FERRAGUT et al. (2011), we propose a way to solve the bidimensional potential problem given by Eqs. (14) and (15) as an optimal control problem: given n experimental measurements of the wind velocity $\mathbf{V_i}$, $\mathbf{i} = \mathbf{1}, \ldots, \mathbf{n}$, at n given points $P_i = (x_i, y_i, z_i)$, $i = 1, \ldots, n$, we search for the value of v, the solution of Eqs. (14) and (15) such that the $\mathbf{V}(\mathbf{x_i}, \mathbf{y_i}, \mathbf{z_i})$ given by Eq. (16) are as close as possible to the experimental values $\mathbf{V_i}$.

2.3. Numerical Method

In practice, we solve a discretized version of Eqs. (1), (2) and (3). To be useful in decision support, model simulation must be achieved faster than real time. In order to reduce the computational time, we only solve the equations of the model surrounding the fire front, defining the set of active nodes for each time step. We define a uniform and fine mesh at the beginning of the numerical process, and we solve the corresponding equations only in the set of active nodes formed by the nodes located inside the neighborhood of the fire front. This reduces the computational time, since we do not have to solve the equations of the model where the solution does not change at all.

2.3.1 Time Integration

Let $\Delta\tau = \tau^{n+1} - \tau^n$ be a time step and let c^n, e^n and u^n denote approximations at time step τ^n to the exact solution, c, e and u, respectively.

At each time step, we solve:

$$\frac{e^{n+1} - e^n}{\Delta\tau} + \alpha u^{n+1} = r^n, \qquad (17)$$

$$e^{n+1} \in G(u^{n+1}), \qquad (18)$$

$$\frac{c^{n+1} - c^n}{\Delta\tau} = -g(u^{n+1})c^{n+1}. \qquad (19)$$

The basic idea is to treat the positive terms implicitly. The non-local radiation term r depends strongly on the temperature u and on the fuel mass c, and therefore it will be evaluated explicitly at time τ^n. Once the radiation r^n is known, the problem given by Eqs. (17), (18) and (19) continues to be non-linear due to the multivalued operator G. However, the solution of this problem can be reduced to explicit calculations.

2.3.2 Numerical Solution of the Multivalued Operator

The multivalued operator in Eq. (18) is maximal monotone, and hence its resolvent $J_\mu = (Id + \mu G)^{-1}$ for any $\mu > 0$ is a well-defined univalued operator. Moreover, the Yosida approximation of G, $G_\mu = \frac{Id - J_\mu}{\mu}$ is a Lipschitz operator, and the inclusion in Eq. (18) is equivalent (BERMÚDEZ AND MORENO 1981) for all $\mu > 0$ to the equation

$$e^{n+1} = G_\mu(u^{n+1} + \mu e^{n+1}), \qquad (20)$$

or

$$u^{n+1} = J_\mu(u^{n+1} + \mu e^{n+1}). \qquad (21)$$

Rearranging Eq. (17), we have

$$u^{n+1} + \frac{1}{\alpha\Delta\tau}e^{n+1} = \frac{1}{\alpha\Delta\tau}e^n + \frac{1}{\alpha}r^n. \qquad (22)$$

Taking $\mu = 1/(\alpha\Delta\tau)$ in Eq. (21), we obtain

$$u^{n+1} = J_{1/\alpha\Delta\tau}\left(\frac{1}{\alpha\Delta\tau}e^n + \frac{1}{\alpha}r^n\right). \qquad (23)$$

Once u^{n+1} has been obtained by solving Eq. (23), we calculate e^{n+1} and c^{n+1} explicitly

$$e^{n+1} = e^n - \alpha\Delta\tau u^{n+1} + \Delta\tau r^n, \qquad (24)$$

$$c^{n+1} = \frac{c^n}{1 + \Delta\tau g(u^{n+1})}. \qquad (25)$$

It remains to explain how to calculate u^{n+1} in Eq. (23). For a given $b = \frac{1}{\alpha\Delta\tau}e^n + \frac{1}{\alpha}r^n$, to compute $\bar{u} = J_{1/\alpha\Delta\tau}(b)$ is equivalent to solving

$$(\alpha\Delta\tau\, Id + G)\bar{u} \ni \bar{b} = \alpha\Delta\tau\, b. \qquad (26)$$

Thus, $\bar{u}$ is given by

$$
\begin{aligned}
&\text{if} \quad \bar{b} < (1 + \alpha\Delta\tau)u_v && \text{then} \quad \bar{u} = \frac{\bar{b}}{1 + \alpha\Delta\tau}, \\
&\text{if} \quad (1 + \alpha\Delta\tau)u_v < \bar{b} < (1 + \alpha\Delta\tau)u_v + \mu_v && \text{then} \quad \bar{u} = u_v, \\
&\text{if} \quad (1 + \alpha\Delta\tau)u_v + \mu_v < \bar{b} < (1 + \alpha\Delta\tau)u_p + \mu_v && \text{then} \quad \bar{u} = \frac{\bar{b} - \lambda_v}{1 + \alpha\Delta\tau}, \\
&\text{if} \quad (1 + \alpha\Delta\tau)u_p + \mu_v < \bar{b} < \infty && \text{then} \quad \bar{u} = u_p.
\end{aligned}
$$

Note that Eqs. (23), (24) and (25) can be solved simultaneously in all active nodes, and hence parallel computation can be used to improve the computational time. Indeed, calculation of the variables u^{n+1}, e^{n+1} and c^{n+1} has been achieved by means of a parallel calculation using the API OpenMP (CHAPMAN et al. 2007), i.e., the loop over all active nodes is parallelized.

2.3.3 Computation of the Radiation Term

For the basic definitions of radiation theory, see SIEGEL and HOWELL (1972).

The differential equation describing the total radiation intensity I at any position along a given path s in a gray medium may be written neglecting scattering

$$\frac{dI}{ds} + a(s)I(s) = a(s)I_b(s), \qquad (27)$$

where I_b is the total black body radiation intensity and is given by Boltzman's law, corresponding to the integral over all wavelengths of the emissive power of a black body

$$I_b = \frac{\sigma}{\pi}T^4, \qquad (28)$$

where $\sigma = 5.6699 \times 10^{-8}\,\mathrm{W\,m^{-2}\,K^{-4}}$ is the Boltzman constant. a is the radiation absorption coefficient inside the flame.

Now defining the optical thickness as

$$\kappa = \int_0^s a(s^*)ds^*, \qquad (29)$$

Equation (27) can be integrated giving

$$I(\kappa) = I(0)e^{-\kappa} + \int_0^\kappa e^{-(\kappa-\kappa^*)}I_b(\kappa^*)d\kappa^*. \qquad (30)$$

2.3.4 Flame Model

Since radiation essentially comes from the flames, we shall consider a physical model in which the gases produced by pyrolysis burn above the fuel layer, producing a flame over that layer. The flame emits radiation that reaches the points ahead of the flame,

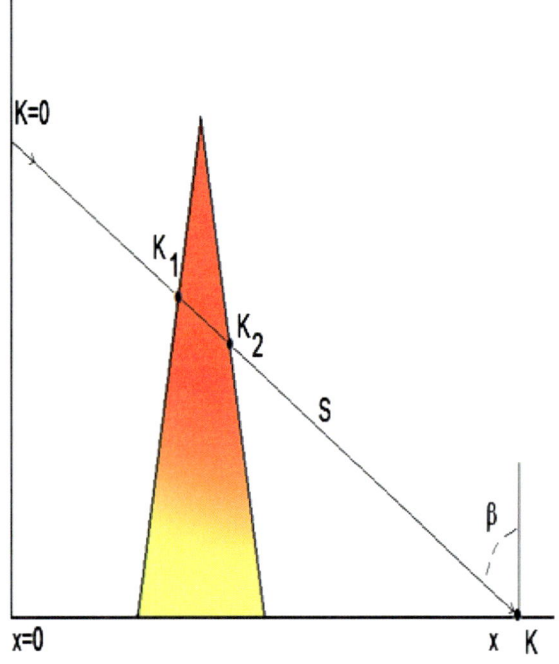

Figure 2
Flame model

heating the surrounding non-burned fuel and thus allowing the fire to propagate.

For a general flame such as the one shown in Fig. 2, we wish to calculate the heat due to radiation from the flame. At the point with coordinates x (or optical thickness κ), we have Eq. (27) with the boundary condition $I(0) = 0$ if we assume that the radiation produced outside the flame is negligible.

Inside the flame, $I_b(\kappa) = \frac{\sigma T_f^4}{\pi}$ between κ_1 and κ_2, where T_f is the temperature inside the flame. We thus have

$$I(\kappa) = \frac{\sigma T_f^4}{\pi}e^{-\kappa}(e^{\kappa_2} - e^{\kappa_1}).$$

Assuming that the absorption coefficient a is constant inside the flame and null outside, that is, $\kappa - \kappa_1 = a(s_2 - s_1)$ and $\kappa - \kappa_2 = 0$, we have

$$I(\kappa) = \frac{\sigma T_f^4}{\pi}\left(1 - e^{-a(s_2-s_1)}\right),$$

and for small values of $a(s_2 - s_1)$, (thin flames), we finally obtain

$$I(\kappa) \approx \frac{a\sigma T_f^4}{\pi}(s_2 - s_1).$$

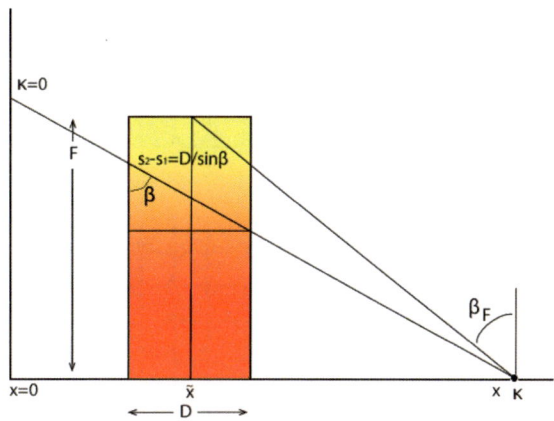

Figure 3
Rectangular flame

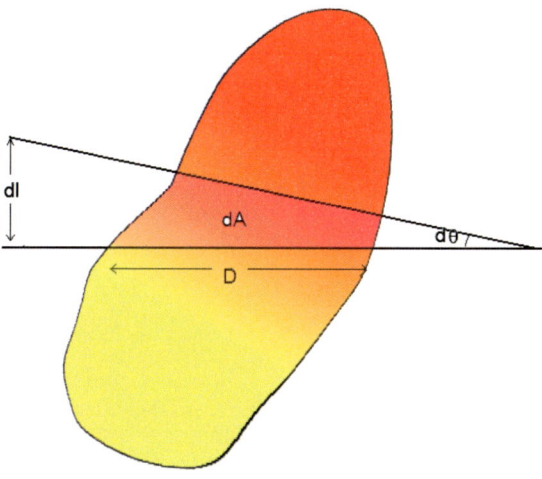

Figure 4
Flame base

The incident energy per unit time and per unit area at a point x due to a beam in the direction (β, θ), and in an interval of solid angle $d\omega$ centered in the direction (β, θ) is

$$\frac{a\sigma T_f^4}{\pi}(s_2 - s_1)\cos\beta d\omega = \frac{a\sigma T_f^4}{\pi}(s_2 - s_1)\cos\beta\sin\beta d\beta d\theta.$$

2.3.5 Vertical Flame

We now consider a rectangular and vertical flame like the one shown in Fig. 3. The horizontal axis in the figure represents the tangent plane on the point $\tilde{x}$ located at the bottom of the flame.

Following Fig. 3, we have $\sin\beta = \frac{D}{s}$. Summing up the heat flow for all β, we have

$$\frac{a\sigma T_f^4}{\pi}D\left(\int_{\beta=\beta_F}^{\beta=\frac{\pi}{2}}\cos\beta d\beta\right)d\theta = \frac{a\sigma T_f^4}{\pi}D(1 - \sin\beta_F)d\theta.$$

We now relate this integration to a surface integral,

$$\sin\beta_F = \frac{\tan\beta_F}{\sqrt{1 + \tan^2\beta_F}} = \frac{\frac{\|x - \tilde{x}\|}{F}}{\sqrt{1 + \frac{\|x - \tilde{x}\|^2}{F^2}}},$$

where $\tilde{x}$ is the point at the center of the bottom of the flame.

If D is the width of the flame and F is the height, we have (see Fig. 4) $d\theta = \frac{dl}{\|x - \tilde{x}\|}$ and $Ddl = d\tilde{A}$. Thus,

$$dR = \frac{a\sigma T_f^4}{\pi}\frac{\sqrt{F^2 + \|x - \tilde{x}\|^2} - \|x - \tilde{x}\|}{\|x - \tilde{x}\|\sqrt{F^2 + \|x - \tilde{x}\|^2}}d\tilde{A},$$

which is the energy per unit area received per unit time at the point x due to the flames with base $d\tilde{A}$. Summing up the total contributions of the flames, the energy per unit area at the point x will be given by

$$R(x) = \frac{a\sigma T_f^4}{\pi}\int_{\Omega_f}\frac{\sqrt{F^2 + \|x - \tilde{x}\|^2} - \|x - \tilde{x}\|}{\|x - \tilde{x}\|\sqrt{F^2 + \|x - \tilde{x}\|^2}}d\tilde{A}, \tag{31}$$

where Ω_f is the burning surface.

2.3.6 Tilted Flame

For a more general tilted flame we consider the flame split into several layers, as shown in Fig. 5, and we sum up the contribution of each layer. Thus, for a flame with N layers the energy per unit area in x due to layer $l = 0, \ldots, N-1$ is

$$dR_l = \frac{a\sigma T_f^4}{\pi}(\sin\beta_l - \sin\beta_{l+1})Dd\theta. \tag{32}$$

Finally, taking into account

$$\sin\beta_l = \frac{N\|x - \tilde{x}\|}{\sqrt{l^2 F^2 + N^2\|x - \tilde{x}\|^2}},$$

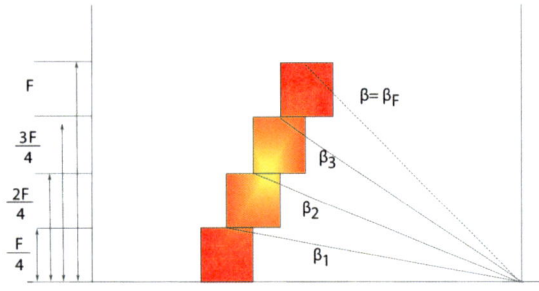

Figure 5
Tilted flame

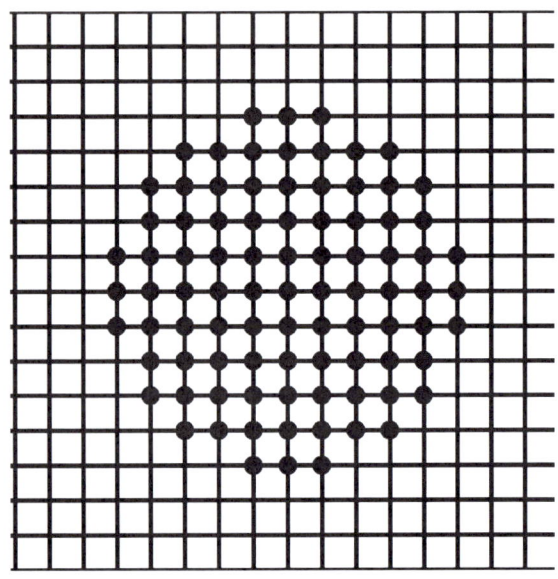

Figure 6
Radiation molecule

and integrating twice, we have

$$R_l(x) = \frac{a\sigma T_f^4}{\pi}$$

$$\times \int_{\Omega_l} \frac{\sqrt{(l+1)^2 F^2 + N^2 ||x-\tilde{x}||^2} - \sqrt{l^2 F^2 + N^2 ||x-\tilde{x}||^2}}{\sqrt{l^2 F^2 + N^2 ||x-\tilde{x}||^2}\sqrt{(l+1)^2 F^2 + N^2 ||x-\tilde{x}||^2}} d\tilde{A}.$$

(33)

2.3.7 Discretization of the Radiation Term

As noted above, for a rectangular flame (resp. for each layer of a tilted flame), the radiation term R (resp. R_l) has the form

$$R(x) = \frac{a\sigma T_f^4}{\pi} \int_\Omega \chi(\tilde{x}) f(x - \tilde{x}) d\tilde{A}, \qquad (34)$$

where χ is the characterisitic function of the flames; i.e., $\chi(x) = 1$ if $T(x) = T_p$ and $\chi(x) = 0$ if $T(x) < T_p$, where T_p is the pyrolysis temperature.

We shall represent the function f with the help of a finite element basis. Let $\{\phi_j\} j = 1, \ldots, D$ be a finite element basis and let us set

$$f(x - \tilde{x}) = \sum_j f^j(x)\phi_j(\tilde{x}) \quad \text{where} \quad f^j(x) = f(x - x_j).$$

Thus, R can be calculated easily by

$$R(x) = \frac{a\sigma T_f^4}{\pi} \sum_j f^j(x) A_j \chi(x_j),$$

where $A_j = \int_\Omega \phi_j(\tilde{x}) d\tilde{A}$ is the area associated with the basis function ϕ_j. Finally, for each node x_i of the mesh, the radiation term can be written as

$$R(x_i) = \frac{a\sigma T_f^4}{\pi} \sum_j R_{ij} A_j \chi(x_j), \qquad (35)$$

where $R = (R_{ij} = f^j(x_i))$ is the radiation matrix. In practice, only the terms in the neighborhood of the flame are calculated, and this calculation is performed using parallel computing with OpenMP. Thus, we consider a radiation molecule like the one shown in Fig. 6

When wind is present, a radiation matrix R_l corresponding to each layer of the tilted flame is considered. The radiation matrix is computed once (outside the time loop), and the loop over all active nodes is parallelized.

2.4. Summary of the Steps in the Algorithm

1. The wind is computed with the wind model whenever new meteorological data become available (each hour in the example).

2. At each time step, with knowledge of the values of the solid fuel temperature u^n, the enthalpy e^n and the mass fraction of solid fuel c^n, the following are computed in parallel in each of the active nodes:

(a) The heat at each point due to radiation r^n from the expression of R (resp. R_l) given by Eq. (35). In the computation of Eq. (35) the following are taken into account: the area where the flame is as the area where $u = u_p$, and the influence of the wind and the surface slope, on the flame tilt. In the expression of R_l through the term $\|x - \tilde{x}\|$ in Eq. (33);

(b) The solid fuel temperature u^{n+1} in the new time step τ^{n+1}, by Eq. (23).

(c) The enthalpy e^{n+1} in the new time step τ^{n+1}, by Eq. (24).

(d) The mass fraction of solif fuel c^{n+1} in the new time step τ^{n+1}, by Eq. (25).

3. Application to a Real Fire

3.1. Fire Description

In order to test the model, we simulated a real fire that occurred at Serradilla del Llano (a village located in the south of Salamanca) in September 2012. The fire began on the 14 September at 16:20h and was under the control of firefighters at 21:30h. It reappeared the following day, and was under control on the 15 September at 20:30h, and was completely extinguished on the 16 September at 20:00h. The fire burned 93.98 Ha of grass and 128.19 Ha of shrubs. Weather variables related to forest fires—wind (direction and velocity), temperature, relative humidity and precipitation—were provided for the fire area by MeteoLogica S.A. In the simulation, the fire breaks or counterfires made by the firefighters were taken into account (green lines in Fig. 7), modifying the fuel data. The wind was calculated with the wind model described in Sect. 2.2 from the meteorological wind data. The simulation area was a rectangle of 3.5×3.3 km; the minimum height was 712 m, and the maximum was 955 m. The mesh size was 700×652 nodes. The topographic relief was obtained from the digital elevation model (DEM) provided by the Instituto Tecnológico Agrario of the Junta de Castilla y León (http://www.itacyl.es/opencmswf/opencms/informacioalciudadano/wms/). The topographic data file used for the simulation has a resolution of 5×5 m, obtained by resampling the corresponding data from the DEM, which have an original resolution of 10×10 m. The fuel data were provided by the Spanish Forest Map 1:50.000 (MFE50) (http://www.magrama.gob.es/es/biodiversidad/temas/ecosistemas-y-conectividad/mapa-forestal-deespana/mfe50.aspx).

3.2. Parameters

Using global sensitivity analysis techniques (SALTELLI et al. 2008, 2004), the model parameters that are most influential on the rate of spread of the fire can be determined. The two methods of global sensitivity analysis more widely used are the extended Fourier amplitude sensitivity test (extended FAST) and Sobol's method, both included in the Simlab program (version 2.2) SIMLAB (2009). This analysis allowed us to conclude that in low wind conditions, when radiation is the dominating mechanism of heat transfer, the most influential parameter is by far the radiation absorption coefficient a, followed by the flame temperature T_f, and—to a lesser extent—the water content M_v and the flame height F. The natural convection coefficient B had little influence on the rate of spread.

Following the conclusions of the global sensitivity analysis and the information available in the literature, we fixed the values of the model parameters. First, we chose the values of the characteristic parameters of the fuel stratum. Here we used the following roughly estimated values: flame temperature $T_f = 1{,}300$ K, pyrolysis temperature $T_p = 500$ K, pyrolysis rate $g = 0.02\,\mathrm{s}^{-1}$, surface density $M = 0.9\,\mathrm{kg/m}^2$ for grass and $M = 1\,\mathrm{kg/m}^2$ for shrubs, fuel calorific capacity $C = 1{,}300\,\mathrm{J\,kg}^{-1}\,\mathrm{K}^{-1}$ and water content $M_v = 0.3$ kg water/kg dry fuel. Second, in order to fit the results with the observed data, and taking into account the sensitivity anlaysis, we adjusted the radiation absorption coefficient a (27) empirically, which determines the heat source due to radiation, since this is the most influential parameter. The value $a = 0.065\,\mathrm{m}^{-1}$ achieved good agreement between the observed data and the numerical results. In the case of laboratory experimental fires (MENDES-LOPES et al. 2003), when the rate of spread is well known, to adjust the most relevant parameter, a least squares Fibonacci was used. This is not the case of the real fire considered here. In a future work, we will consider the adjustment of relevant parameters for real cases as part of the data assimilation procedure. Another parameter involved with radiation that appears as a parameter in this model is flame height F. Flame height F could be obtained from a simplified gas phase model, as in BALBI et al.

Table 1

Meteorological wind data

Time (h)	Velocity (m/s)	Direction
0	5	210
1	6	220
2	7	240
3	8	240

(2009), or fitted from a priori estimation using the characteristics of the vegetation stratum or even by direct observation. The values chosen in this work were $F = 2.1$ m for grass and $F = 2.5$ m for shrubs, taking into account the height of the fuel. We took twice the height of the fuel as the flame height F. The convection coefficient B, which determines the adimensional coefficient $\alpha = \frac{B|\tau|}{MC}$ in (1), has little influence on the rate of spread, and taking $B = 10\ \mathrm{J\,m^{-2}\,s^{-1}\,K^{-1}}$ we obtain a good approximation.

A high-definition wind field was computed each hour with the model described in Sect. 2.2 from the meteorological data of Table 1, where the direction is measured clockwise from the North and the data are assumed to have been taken at a point in the center of the domain at a height of 10 m.

3.3. Results

In Fig. 7, we show the initial ignition point, the final perimeter of the fire on the first day and the results of the simulation. The green line shows the area of the firefighter's activities, which are included in the simulation modifying the fuel data. The different textures represent the simulated burned surface at the end of each hour in a 4-h period.

The simulation is in good agreement with the final observed front. The Sørensen similarity coefficient s (FILIPPI *et al.* 2014) was calculated in order to compare the burned area predicted by the model with the real burned area, where s is defined by:

$$s = \frac{2|A^0(\tau) \cap A(\tau)|}{|A^0(\tau) + A(\tau)|}. \qquad (36)$$

Here $|A(\tau)|$ is the area of the burned surface A at time τ, and the superscript ^{0} represents the real (observed) data. The result is a value between 0 and 1; 1 means

perfect agreement between observation and simulation, and 0 means that there is no agreement. The Sørensen similarity coefficient obtained after four hours of simulation was $s = 0.9424$, pointing to very good agreement.

The simulation time was in the order of 194 s per 4 h of real fire, performed on a laptop (1.8 GHz Intel Core i7 processor with two cores and 4 GB RAM memory). The simulation time included the hourly wind computations. Each wind computation took around 9 s. These simulation times are comparable to or even lower than the times used by codes based on empirical models.

4. Data Assimilation

4.1. State Estimation

In this section, we focus on the problem of how to combine a modelled prediction of a state with a set of measurements available at a particular time. It is assumed that the error statistics of the predictions as well as the measurements are known and characterized by their respective error covariances.

We deal with model states represented by a real vector $x \in \mathbb{R}^n$. Let us assume that some observed data are available and are represented by a vector $d \in \mathbb{R}^d$. These data are related to the model state by a known matrix, $H \in \mathbb{R}^{d \times n}$ (a function in a general case). The state and data vectors will be assumed to be random vectors

$$x^f = x + \xi, \qquad (37)$$

$$d = Hx + v, \qquad (38)$$

where x stands for the true state and x^f for the forecast state. ξ denotes the unknown error in the forecast and v is the unknown measurement error. We make the usual statistical assumptions

$$E[\xi] = 0 \quad E[\xi\xi^t] = C \in \mathbb{R}^{n \times n}, \qquad (39)$$

$$E[v] = 0 \quad E[vv^t] = R \in \mathbb{R}^{d \times d}, \qquad (40)$$

$$E[\xi v^t] = 0, \qquad (41)$$

where $E[.]$ denotes the mathematical expectation.

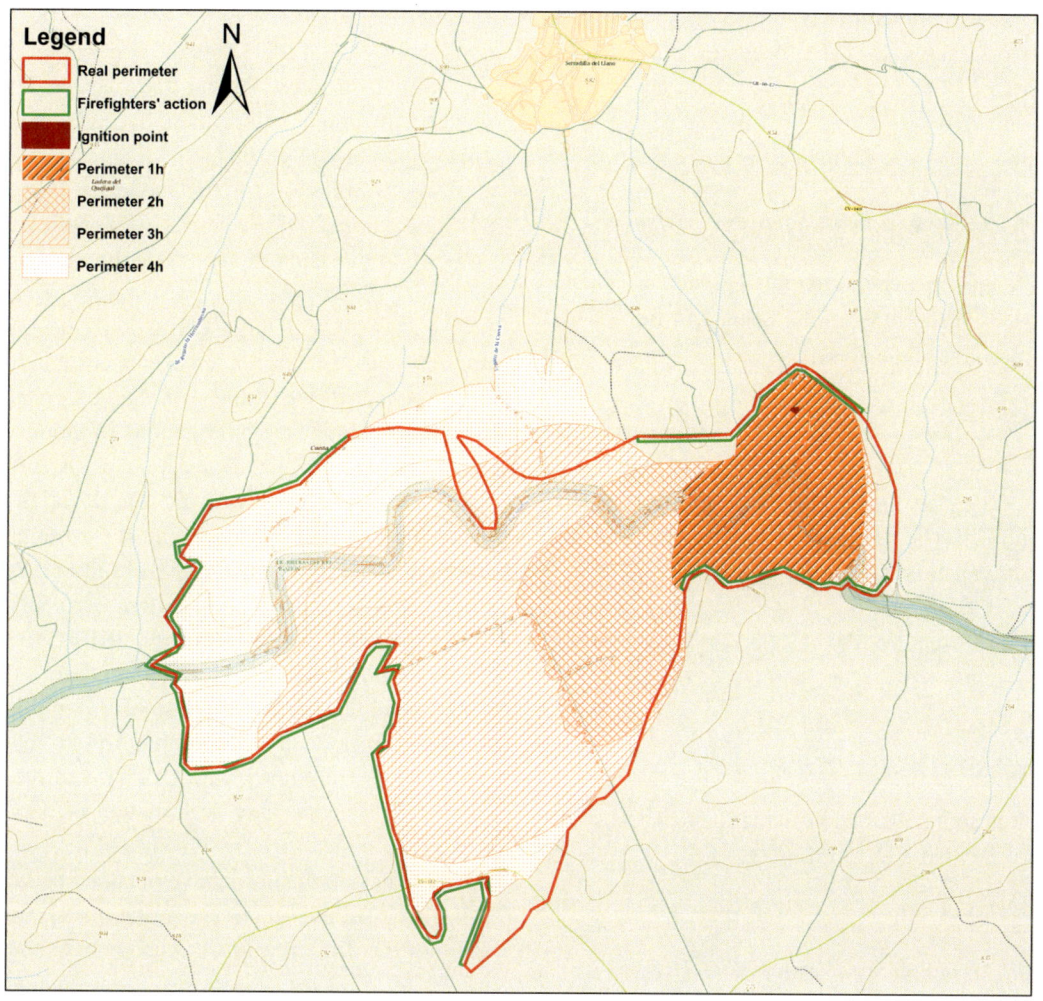

Figure 7
Perimeter of the Serradilla del Llano fire and simulation results

4.2. Best Linear Unbiased Estimator

A linear estimator is defined by the following interpolation equation

$$x^a = x^f + K(d - Hx^f). \qquad (42)$$

Now we search for the best linear unbiased estimator (BLUE) of x, i.e., the value of K such that minimizes the variance of the analysed error $x_a - x$.

The covariance matrix of the analysed error $x^a - x$ is

$$C^a = E[(x^a - x)(x^a - x)^t]. $$

From Eq. (42), for any K we have:

$$C^a = (I - KH)C(I - KH)^t + KRK^t, \qquad (43)$$

where we have taken into account Eq. (41).

We shall now search for the value of K that minimizes the trace of C^a, $\mathrm{tr}(C^a)$. Since C and R are symmetric matrices we have

$$\mathrm{tr}(C^a) = tr(C) + \mathrm{tr}(KHCH^tK^t) - 2\mathrm{tr}(CH^tK^t) + \mathrm{tr}(KRK^t).$$

The trace is a scalar continuous differentiable function of the terms of K. The condition for the minimum gives:

$$K = CH^t(HCH^t + R)^{-1}. \qquad (44)$$

Finally, by inserting the optimal value of K given by

131

Eq. (44) into expression (43), the covariance matrix of the errors analysed, $x^a - x$, becomes

$$C^a = (I - KH)C. \qquad (45)$$

K is often called the Kalman gain KALMAN (1960).

4.2.1 The BLUE as a Least Squares Problem

It is well known that the BLUE analysis given by Eqs. (42), (44) and (45) can be obtained from a least squares problem. Indeed, set

$$J(x) = (C^{-1}(x - x^f), x - x^f) + (R^{-1}(d - Hx), d - Hx), \qquad (46)$$

where $(.,.)$ denotes the usual Euclidean inner product in $\mathbb{R}^n$ and $\mathbb{R}^d$. The value x^a where $J(.)$ reaches the minimum value is given by solving the equation $\nabla J(x^a) = 0$. Since

$$\nabla J(x) = C^{-1}(x - x^f) - H^t R^{-1}(d - Hx),$$

we have

$$x^a = (H^t R^{-1} H + C^{-1})^{-1}(H^t R^{-1} d + C^{-1} x^f), \quad (47)$$

which after some algebraic calculations can be written in the recursive form Eq. (42), where K is given by Eq. (44).

4.3. Data Assimilation for Non-Linear Highly Dimensional Problems

When the true state represented by a vector x evolves according to a linear dynamical model and the errors ξ and v are Gaussian, the Kalman algorithm for dynamical linear systems can be applied (KALMAN 1960). The Kalman filtering process is designed to estimate a state given by a linear model. Here we deal with a fire propagator model that is clearly non-linear. Another drawback is the high dimensionality, which makes manipulation of the system error covariance impractical.

The ensemble Kalman filter (EnKF) introduced by EVENSEN (2009) makes it possible to apply the Kalman filter to non-linear high-dimensional problems. Although the EnKF still relies on the Gaussian assumption, in practice it is used for non-linear problems in which the Gaussian assumption is not satisfied.

The main characteristic of the EnKF is the application of the Kalman correction step of Eq. (42) to an ensemble of forecasts and the fact that the covariance matrix is replaced by the covariance matrix of the ensemble.

Data assimilation for fire models was studied in MANDEL et al. (2008), where it was applied to a fire convection-diffusion model. As pointed out in MANDEL et al. (2008), the standard ensemble EnKF does not work well and large corrections may result in non-physical states. In order to sort out this problem, in JOHNS and MANDEL (2008), a regularized technique is proposed. The regularization technique has a stabilizing effect, but the location of the fire in the data cannot be too far from the location in the ensemble. In order to circumvent these limitations in BEEZLEY and JAN MANDEL (2008), a morphing EnKF provides both additive and position correction.

In this work, a variant of the EnKF is considered, the deterministic ensemble Kalman filter (DEnKF) (SAKOV and OKE 2008), because of its simplicity and the good results that, according to our numerical experience, it provides. The DEnKF has the advantage of not needing any perturbation of data, which makes it easier to implement. Morever, in our numerical experience, we have obtained better results using the DEnKF than using the EnKF, avoiding some of the difficulties mentioned above, at least for moderate distances of data from the location of the fire front in the ensemble.

4.3.1 Ensemble Kalman Filter

The EnKF is based on the correction step in Eq. (42), with K given by Eq. (44). In the EnKF an ensemble of model states $X = [X_1, \ldots, X_m]$ is considered and we take the mean value as an a priori estimation of the state,

$$x^f = \frac{1}{m} \sum_{i=1}^{m} X_i. \qquad (48)$$

Each ensemble member is corrected using

$$X_i^a = X_i^f + K(d + \varepsilon_i - HX_i^f), \ i = 1, \ldots, m, \quad (49)$$

where ε_i is a vector of perturbations with zero average needed to prevent the so called ensemble collapse (BURGERS et al. 1998). The covariance matrix C is replaced by the sample covariance, using the expression

$$C^f = \frac{1}{m-1}\sum_{i=1}^{m}(X_i - x^f)(X_i - x^f)^t. \qquad (50)$$

It can be seen that at the limit of an infinite ensemble size, the EnKF will give the same results in the computation analysis as the Kalman filter when applied to dynamic linear systems.

4.3.2 Deterministic Ensemble Kalman Filter

Without perturbations of the observations, the adapted error covariance matrix would be that of SAKOV and OKE (2008),

$$C^a = (I - KH)C^f(I - KH)^t, \qquad (51)$$

which is less than the value given by the Kalman filter and results in a excessive reduction of the ensemble spread.

The deterministic ensemble Kalman filter (DEn-KF) proposed in SAKOV and OKE (2008) is based on the following observation. Because

$$C^f H^t K^t = C^f H^t (HC^f H^t + R)^{-1}HC^f = KHC^f,$$

the value of C^a in the case without perturbations, Eq. (51), can be written

$$C^a = C^f - KHC^f - C^f H^t K^t + KHC^f H^t K^t$$
$$= C^f - 2KHC^f + KHC^f H^t K^t.$$

If KH is small, the quadratic term $KHC^f H^t K^t$ is much smaller than the linear term $2KHC^f$; by neglecting the quadratic term and dividing the Kalman gain by two, we can match the theoretical covariance given by Eq. (45). This argument forms the theoretical basis for the DEnKF. The proposed algorithm in SAKOV and OKE (2008) is:

4.3.3 Deterministic Ensemble Kalman Algorithm

Let M_k represent the model operator giving the state at time t_{k+1} from the state at time t_k and let $X_0^a = [X_{1,0}^a, \ldots, X_{m,0}^a]$ be a set of m initial states and let us calculate its mean value using Eq. (48) and its covariance matrix given by Eq. (50)

From $k = 1, 2, \ldots$ we perform the following steps:

1. Prediction:

$$X_{i,k}^f = M_{k-1}X_{i,k-1}^a \quad i = 1, \ldots, m,$$
$$x_k^f = \frac{1}{m}\sum_{i=1}^{m}X_{i,k}^f,$$
$$A_{i,k}^f = X_{i,k}^f - x_k^f.$$

2. Calculation of the prior error covariance matrix: First we write

$$A_k^f = [A_{1,k}^f, \ldots, A_{m,k}^f]$$

and calculate

$$C_k^f = \frac{1}{m-1}A_k^f(A_k^f)^t.$$

3. Calculation of the Kalman gain

$$K_k = C_k^f H_k^t (H_k C_k^f H_k^t + R_k)^{-1}.$$

4. Calculation of the posterior analysed value for the mean and the analysed deviations by

$$x_k^a = x_k^f + K_k(d_k - H_k x_k^f),$$
$$A_k^a = A_k^f - \frac{1}{2}K_k H_k A_k^f.$$

5. Setting of the new values of the states

$$X_k^a = A_k^a + [x_k^a, \ldots, x_k^a].$$

4.4. Example with Data Assimilation

In this subsection, we apply the DEnKF to the fire described in Sect. 3 and to the model described in Sect. 2. The whole process is implemented using parallel computing in order to achieve competitive computational times. We use the MPICH3 library with a master-slave scheme in which the master process controls the flow of the algorithm, sends the job to the slaves, receives outcomes and carries out the assimilation step. The master process uses the API OpenMP and the ATLAS parallel algebra library (ATLAS 2007) to further improve the computational time. Each slave process is performed simultaneously to the other processes. They are responsible for executing the simulation corresponding to each of the states in the ensemble.

Figure 8
Spatial distribution of the ignition points of the ensemble (*grey points*) and real ignition point (*black point*)

We consider a reference solution, generated from an a posteriori, well-known initial ignition point from which we extract the data sought. Since in practical situations we do not know the location of the ignition point with precision, we start the modelling by generating a random ensemble of initial ignition points.

The number of elements in the ensemble is 50. The ignition points of the elements of the ensemble were generated taking samples from a Gaussian distribution centered on the location of the ignition point of the reference solution with a standard deviation equal to 100 m. Figure 8 shows the distribution of the location of the ignition points of the ensemble.

The state of the system is $x = [u, c]$, where u is the solid temperature and c is the mass fraction of solid fuel. The data are extracted from the perturbed reference solution by taking values every 50 m. New data are taken each hour and an assimilation step is performed.

The covariance matrix of the measurement error in Eq. (40) is the diagonal matrix $R = \text{diag}\{0.25\}$, i.e., the standard deviation for the observed solid temperature and for the mass fraction of fuel is equal to 0.5. Taking into account that the a-dimensional temperature range is between 0 and 0.7 and that the mass fraction of the fuel range is between 0 and 1, we assume implicitly that the observation of the mass fraction of fuel is more reliable than the solid temperature on the fire front. We also consider some variability of the wind, introducing a standard deviation equal to 2 km/h for the module and a

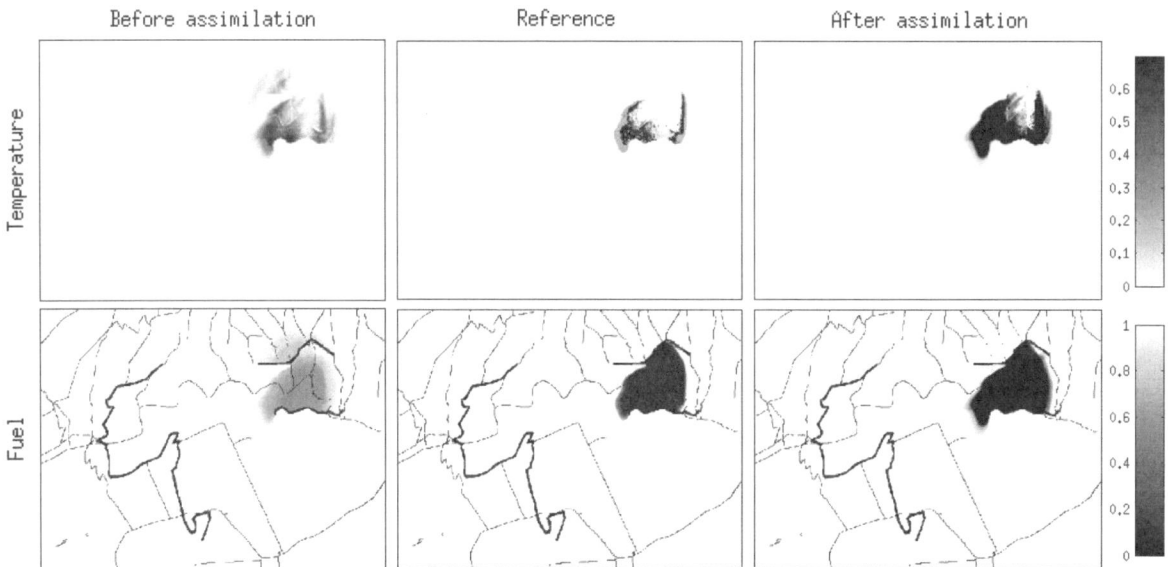

Figure 9
First hour. *First row* temperature. *Second row* fuel mass fraction. *Left column* mean ensemble results before assimilation. *Central column* reference solution. *Right column* mean ensemble results after assimilation

standard deviation equal to 3.16° for the wind direction. After the data assimilation step, the state values out of range are projected onto the range.

In Figs. 9, 10, 11 and 12, we show the contours of the temperature (first row) and of the mass fraction of fuel (second row) of the reference solution (central column) against the contours of the average of the ensemble of the 50 solutions computed before assimilation (left column) and after assimilation (right column). The goal of the assimilation analysis

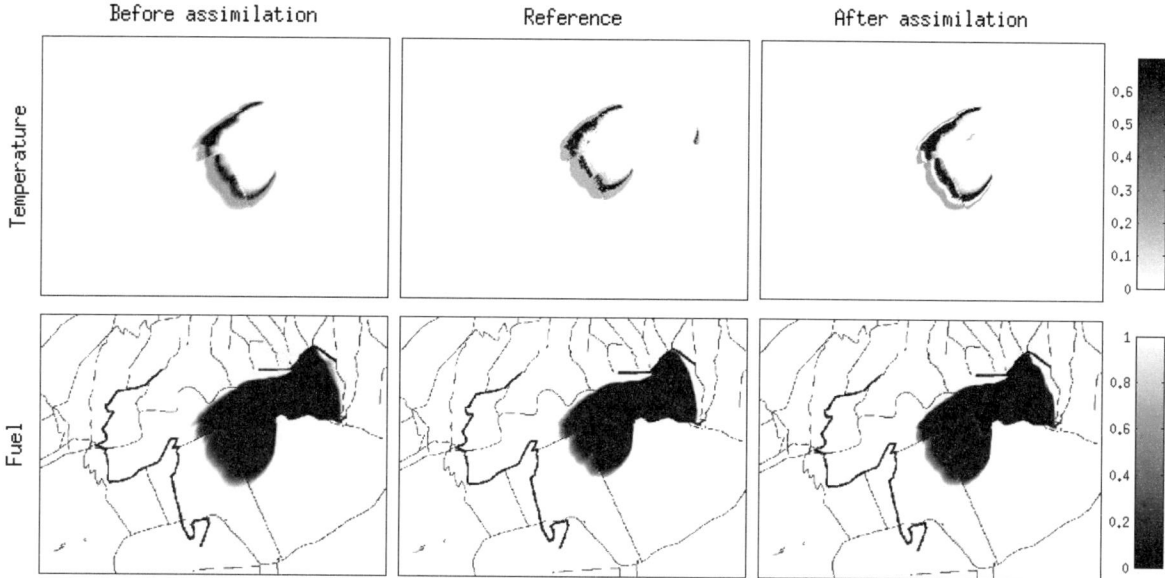

Figure 10
Second hour. *First row* temperature. *Second row* fuel mass fraction. *Left column* mean ensemble results before assimilation. *Central column* reference solution. *Right column* mean ensemble results after assimilation

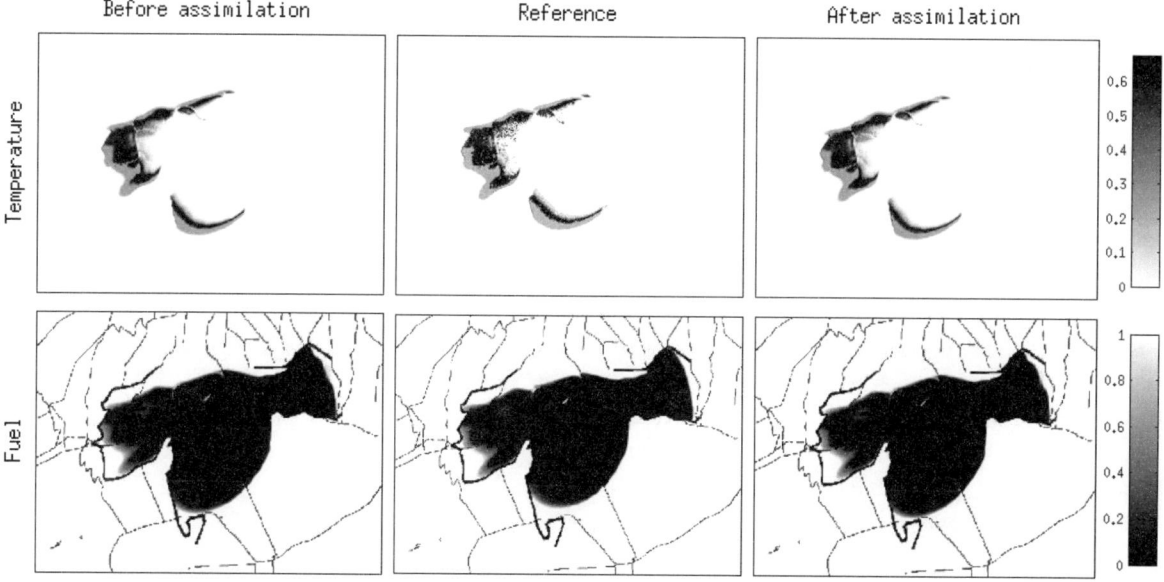

Figure 11
Third hour. *First row* temperature. *Second row* fuel mass fraction. *Left column* mean ensemble results before assimilation. *Central column* reference solution. *Right column* mean ensemble results after assimilation

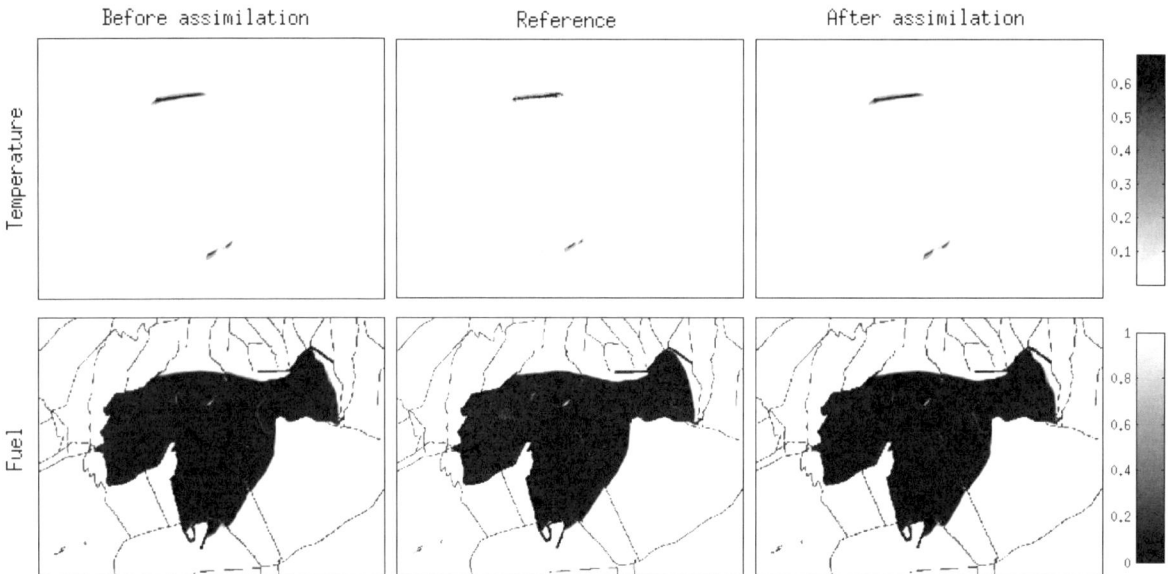

Figure 12
Fourth hour. *First row* temperature. *Second row* fuel mass fraction. *Left column* mean ensemble results before assimilation. *Central column* reference solution. *Right column* mean ensemble results after assimilation

is to correct the average solution of the ensemble of solutions with the help of observations from real data. Figure 9-left stands for the forecast solution after 1 h of fire, i.e., the average of the ensemble of simulations. Owing to the dispersion of the position of the different ignition points in the ensemble, the average of the solid temperature and the average of the mass fraction of solid fuel are far from the extreme values. This is the reason why the average is shaded in the figure. Figure 9-right is the solution analysed after the assimilation, i.e., the average of the ensemble of simulations after assimilation. Clearly, the main part of the error due to the lack of knowledge of the initial ignition point position is eliminated. Figures 10, 11 and 12 are the corresponding forecast solution, reference solution and analysed solution at 2, 3 and 4 h after the ignition time.

In order to quantify the error, i.e., the difference between the computed solution and the reference solution, before and after assimilation we use the total variance (trace of the covariance matrix) of the state of the system $x = [u, c]$. In Table 2, we report the variances before and after assimilation. Most of the dispersion diminishes after assimilation after the first hour. At each hour there is dispersion, due to the computations and to the error of the measurements, that diminishes after each assimilation.

Another error quantification method is the Sørensen similarity coefficient (FILIPPI *et al.* 2014). This coefficient measures the error only on one of the variables of the state, the mass fraction of solid fuel, since it compares the burned area predicted by the model with the real burned area. In Table 3, we summarize the Sørensen similarity coefficient before and after assimilation each hour. The greatest improvement of the Sørensen similarity coefficient is obtained after the assimilation after the first hour. The assimilation improves the Sørensen similarity coefficient each hour.

Table 2

Total variance before and after assimilation

Variance	1 h	2 h	3 h	4 h
Before assimilation	2,927.77	283.278	250.681	171.275
After assimilation	256.809	67.7551	227.259	129.713

Table 3

Sørensen similarity coefficient before and after assimilation

Sørensen	1 h	2 h	3 h	4 h
Before assimilation	0.510952	0.916155	0.956606	0.968328
After assimilation	0.9387	0.93943	0.957507	0.968536

In Fig. 13, we show the spatial distribution of the error for the mass fraction of solid fuel, before (left column) and after (right column) the assimilation for the 4 h. After the first hour, the assimilation reduces significantly the error. The assimilation reduces the error each hour as can be seen after the fourth hour where the shadow on the left out of the fire perimeter (fourth row, left column, in Fig. 13) disappears after the assimilation (fourth row, right column, in Fig. 13).

The overall computation was run on a Dell Precision T7610 workstation equipped with two Intel Xeon E5-2670 v2 processors (10 cores, each working at 2.5 GHz) and 64 GB RAM. On this workstation, for an ensemble of 50 elements, the simulation of one hour of real fire takes 1.5 min of computation. The assimilation step takes 10.5 min. Owing to the use of MPI, the simulation can be carried out on a cluster of several computers, which is particularly helpful when we increase the number of elements in the ensemble or when we use less powerful computers.

We ran the same example, changing only the number of elements in the ensemble. With 100

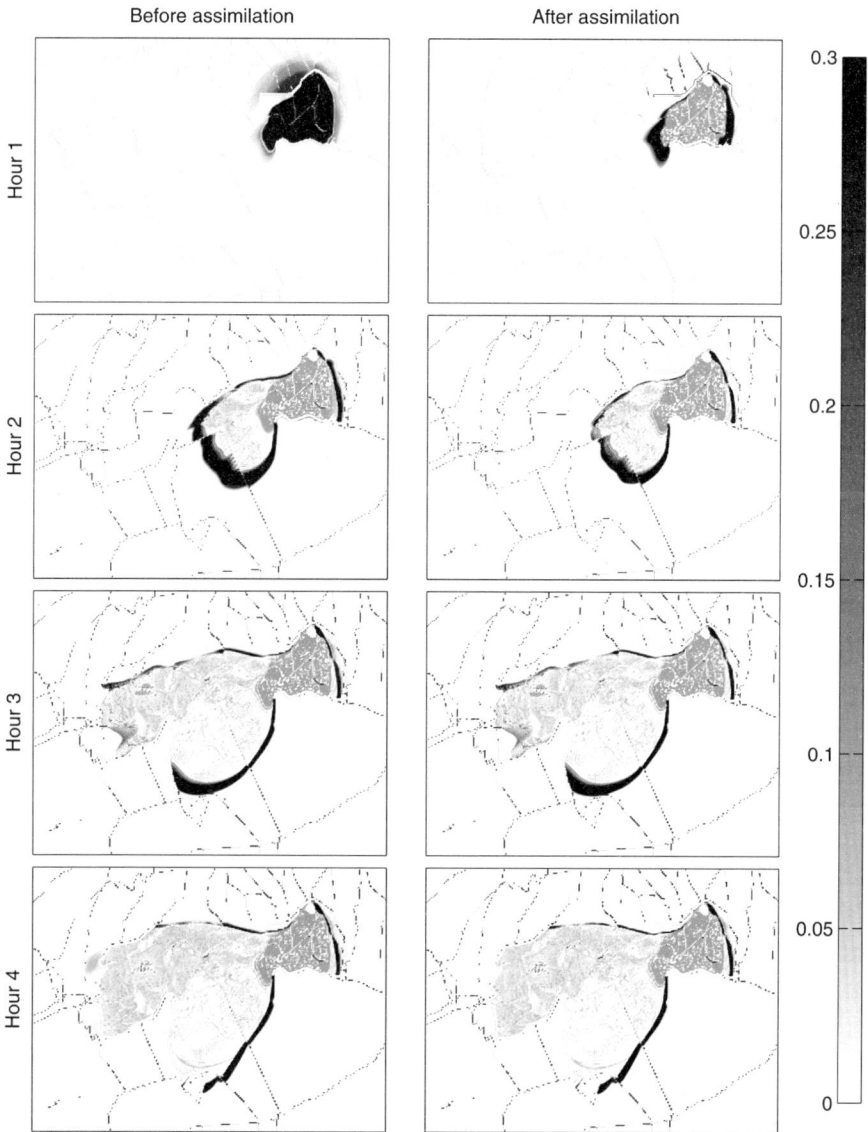

Figure 13

Mass fraction of solid fuel error. *Left column* before assimilation. *Right column* after assimilation

elements the results were similar to the case with 50 elements. With only 25 elements in the ensemble, the assimilation step is not always able to satisfactorily correct the errors in the forecast.

5. Conclusions

We present a simplified physical fire model taking into account the topography of the surface, the fuel load and type, temperature, humidity, wind and radiation. The model has been applied to a real case simulating 4 h of a real fire in a few minutes and allowing firefighting actions to be incorporated by modifying the fuel data.

The model is simple enough to require the adjustment of only a few parameters, allowing data assimilation, but sufficiently precise to reflect important phenomena for the evolution of a fire.

The simplicity of the model and the numerical techniques proposed allow real-time simulations to be achieved in very competitive computational times.

The model includes its own wind model, which allows a high-definition wind field to be computed over the simulation domain that takes into account topography and meteorological wind data at a few points, in contrast to other models that consider winds to be constant in space, i.e., with no orographic effects on winds.

The model is being integrated into a GIS as a module that allows the necessary data to be obtained and provides friendly results for the GIS.

Regarding data assimilation, we have shown that the DEnKF can be used to improve the solution if enough experimental data are available (position of the front, burned surface), which can be related to the state variables, i.e., temperature of the solid and mass fraction of fuel. Primarily, data assimilation is effective with respect to the removal of the initial uncertainty about the position of the ignition point.

Finally, although the simulation time of the model can be considered competitive with respect to other empirical models, the computation time of the assimilation step should be improved. This could be achieved by using supercomputing and improving the parallelization of the assimilation algorithm.

Acknowledgments

This work has been partially supported by the *Secretaría de Estado de Investigación, Desarrollo e Innovación* and *Centro para el Desarrollo Tecnológico Industrial* of the *Ministerio de Economía y Competitividad* of the Spanish Government, Grant Contract: CGL2011-29396-C03-02 and CEN-20101010 (associated contract Art83LOU with Tecnosylva S.L.) and by the *Conserjería de Educación* of the *Junta de Castilla y León*, Grant Contract: SA266A12-2. The authors are also grateful to Ignacio Juárez Relaño, chief of the *Sección de Protección de la Naturaleza* of the *Servicio Territorial de Medio Ambiente* of Salamanca, for his technical support, providing all the necessary information about the *Serradilla del Llano* fire.

References

Asensio, M.I., and Ferragut, L., (2002), *On a Widland fire model with radiation*, Int. J. Num. Methods Eng. *54*, 137–157.

Asensio, M.I., Ferragut, L., and Simon, J. (2005), *A convection model for fire spread simulation*, Applied Mathematical Letters, *18*, 673–677.

ATLAS (2007), Automatically Tuned Algebra Software. Available at http://math-atlas.sourceforge.net/ (Verified 15 July 2014).

Balbi, J.H., Morandini, F., Silvani, X., Filippi, J.B., and Rinieri F. (2009) *A physical model for wildland fires*, Combustion and Flame, *156*, pp. 2217–2230.

Beezley, J.D., and Jan Mandel, J. (2008) Morphing ensemble Kalman filters. Tellus, 60A, 131–140.

Bermúdez, A., and Moreno, C. (1981), *Duality methods for solving variational inequalities*, Comp. and Math. Appl., *7*, 43–58.

Brézis H., Operateurs maximaux monotones et semi-groupes de contractions dans les espaces de Hilbert (North-Holland mathematics Studies 1973).

Burgers, G., van Leeuwen, P. J., Evensen, G. (1998), *Analysis scheme in the ensemble Kalman filter*, Mon. Wea. Rev. *126*, 1719–1724.

Chapman, B., Jost, G., and Van Der Pas, R., Using OpenMP: Portable Shared Memory Parallel Programming (MIT Press. Cambridge, 2007).

Cox G., Combustion fundamentals of fire (Academic Press, London 1995).

Evensen, G., Data assimilation, The Ensemble Kalman Filter (Springer 2009).

Ferragut L., Asensio M.I., and Simon J. (2011) *High definition local adjustment model for 3D wind fields performing only 2D computations*, Int. J. Num. Methods Biomedical Eng. *27*, 510–523.

Ferragut, L., Asensio, M.I., and Monedero, S. (2007a), *A numerical method for solving convection-reaction-difussion equation in fire spread modeling*, Advances in Engineering Software, *38*, 366–371.

FERRAGUT, L., ASENSIO, M.I., and MONEDERO, S. (2007b), *Modelling radiation and moisture content in fire spread*, Communications in Numerical Methods in Engineering, *23*, 819–833.

FILIPPI, J.B., MALLET, V. and NADER, B. (2014) *Representation and evaluation of wildfire propagation simulations*, Int. J. Wildland Fire, *23*, 46–57.

JOHNS, C.J., and MANDEL, J. (2008) *A two-stage ensemble Kalman filter for smooth data assimilation*, Environmental and Ecological Statistics, *15*, 101–110.

KALMAN, R. E. (1960), *A New Approach to Linear Filtering and Prediction Problems*, Journal of Basic Engineering, *82*(Serie D), 35–45.

MANDEL, J., BENNETHUM, L.S., BEEZLEY, J.D., COEN, J.L., DOUGLAS, C.C., KIM M., and VODACEK, A. (2008) *A wildfire model with data assimilation*, Mathematics and Computers in Simulation *79*, 584–606.

MARGERIT, J., and SÉRO-GUILLAUME, O. (2002), *Modelling forest fires. Part II: reduction to two-dimensional models and simulation of propagation*, Int. J. Heat and Mass Transfer *45*, 1723–1737.

MENDES-LOPES, J.M., VENTURA J.M.P. and AMARAL, J.M.P. (2003) *Flame characteristics, temperature-time curves, and rate of spread in fires propagating in a bed of Pinus pinaster needles.* Int. J. Wildland Fire *12*, pp. 64–84.

PASTOR, E., ZARATE, L., PLANAS, E., and ARNALDOS, J. (2003), *Mathematical models and calculation systems for the study of wildland fire behaviour*, Progress in Energy and Combustion Sciene *29*(2), 139–153.

ROTHERMEL R.C., (1972), A mathematical model for predicting fire spread in wildland fires, USDA Forest Service Research Paper INT-115.

SAKOV P. and OKE P.R. (2008), A deterministic formulation of the ensemble Kalman filter: an alternative to ensemble square root filters, Tellus, 60A, 321–371.

SALTELLI, A., RATTO, M., ANDRES, T., CAMPOLONGO, F., CARIBONI, J., GATELLI, D., SAISANA, M., and TARANTOLA, S., Global Sensitivity Analysis. The Primer (John Wiley & Sons Ltd, England, 2008).

SALTELLI, A., TARANTOLA, S., CAMPOLONGO, F., and RATTO, M., Sentitivity analysis in practice: a guide to assessing scientifics models (John Wiley & Sons Ltd, England, 2004).

SIEGEL R.D J.R. HOWELL J.H., Thermal Radiation Heat Transfer, (McGraw-Hill Inc, New York 1972).

SIMLAB (2009) Version 2.2 Simulation Environment for Uncertainty and Sensitivity Analysis, developed by the Joint Research Centre of the European Commission.

SULLIVAN, A.L. (2009a), *Wildland surface fire spread modelling, 1990–2007. 1: Physical and quasi-physical models*, Int. J. Wildland Fire *18*, 349–368.

SULLIVAN, A.L. (2009b), *Wildland surface fire spread modelling, 1990–2007. 2: Empirical and quasi-empirical models*, Int. J. Wildland Fire *18*, 369–386.

SULLIVAN, A.L. (2009c), *Wildland surface fire spread modelling, 1990–2007. 3: Simulation and mathematical analogue models*, Int. J. Wildland Fire *18*, 387–403.

WEBER, R.O. (1989), *Analytical models of fire spread due to radiation*, Combustion and Flame 78, 398–408.

WEBER, R.O. (1991), *Modelling fire spread through fuel beds*, Progress in Energy and Combustion Science *17*(1), 67–82.

(Received February 28, 2014, revised June 24, 2014, accepted June 27, 2014, Published online July 22, 2014)

Pure Appl. Geophys. 172 (2015), 141–148
© 2014 Springer Basel
DOI 10.1007/s00024-014-0918-4

Computer Simulation of Packing of Particles with Size Distributions Produced by Fragmentation Processes

MIGUEL ANGEL MARTÍN,[1] FRANCISCO J. MUÑOZ,[1] MIGUEL REYES,[2] and F. JAVIER TAGUAS[1]

Abstract—Fragmentation schemes inspired by theoretical results and conjectures of Kolmogorov are applied to produce particle size distributions of different natures, depending on fragmentation parameters. A two-dimensional computer simulation method of packing is applied to the resulting distributions and the void fraction is evaluated. The relationship between the void fraction and characteristic parameters of the fragmentation process is studied.

Key words: Fragmentation, granular media, packing, void fraction, particle size distribution.

1. Introduction: Particle Size Distributions, Fragmentation and Packing

The study of particulate systems is of great interest in many fields of science and technology. Soil, sediments, powders, granular materials, colloidal and particulate suspensions are examples of systems involving many size particles. For those systems, the statistical description of the particle size distribution (PSD), that is, the mathematical distribution that defines the relative amounts of particles present, sorted according to size, is a central issue. The PSD can be important in understanding soil hydraulic properties, the geological origin of sediments, or the physical and chemical properties of granular materials and ceramics, among others.

Several probability distributions have been used as PSD models, depending on the nature of the particle system. The log-normal and Weibul distributions have been traditionally used for modelling particle size distributions generated by grinding, milling and crushing operations. The log-hyperbolic distribution was proposed in BAGNOLD and BARN-DORFF-NIELSEN (1980) to model PSD of naturally occurring sediments. Dynamical explanations for the occurrence of hyperbolic distribution are also given in that study and factors in the formation of sediments, such as transport by wind or water, are considered. The log-Laplace model was proposed in FIELLER *et al.* (1984) as an alternative for the same scenario.

An important amount of the above mentioned granular media is the result of grinding or fragmentation processes: either natural ones, as in the case of soil and rocks, or artificial ones, as occurs in ceramics or building materials. Some of the models have been postulated on the basis of geometric features observed in such materials. Thus, in the pioneering work of ANDREASEN and ANDERSEN (1930), the differential equation

$$\frac{\mathrm{d}Q}{\mathrm{d}(\log r)} = aQ \qquad (1)$$

with $a > 0$ as a constant, is proposed as a semi-empirical model for the cumulative mass-size distribution function $Q(r)$ of certain granular products with grain size below a given limit. The differential equation is formulated for those products with a grain distribution conformed in such a way that when adding a portion of larger grains, the resulting product (grain distribution) is geometrically similar to the previous one, so that an image of both products seems equal (they have the same "granulography", a term coined therein). The model and ability to predict both the PSD and the void fraction are tested using data

---

[1] Department of Applied Mathematics, E.T.S.I. Agrónomos, Universidad Politécnica de Madrid, Avda. Complutense s/n, 28040 Madrid, Spain. E-mail: miguelangel.martin@upm.es
[2] Department of Applied Mathematics, E.T.S.I. Informáticos, Universidad Politécnica de Madrid, Campus de Montegancedo s/n, Boadilla del Monte, 28660 Madrid, Spain.

obtained by materials and operations commonly used in the ceramics industry.

The log-normal distribution, the most traditional distribution used in particulate media literature, is strongly supported by important theoretical results due to the brilliant mathematician A. N. Kolmogorov. In KOLMOGOROV (1941), it is mathematically shown that the log-normal distribution for the asymptotic PSD holds when random rules for the fragmentation process are independent of the ratio between the size r of the particle and the size of the particles obtained from it.

In MANDELBROT (1982) and TURCOTTE (1986), a connection between fractals and fragmentation was made, and this was the start point of a huge number of studies using fractals to model fragment size distributions that have been shown to obey fractal or power law behaviour in many different natural scenarios. It is just to point out that, surprisingly, the concept of self-similarity or fractality is hidden in the rational basis of the model proposed in ANDREASEN and ANDERSEN (1930), a half century before MANDELBROT (1982)! In fact, the solution of the differential equation proposed therein is a power law. Several models have been proposed to rationalize such empiric evidence. Thus, multiplicative models, together with large deviation theory, have been used to understand these observations, suggesting that the observed power law distributions of fragment sizes should correspond to the superposition of probability density functions that are log-normal in the "centres" and take the power-law form in the "tail". It would be the result of a natural mixing of simple multiplicative processes that take place along the fragmentation of different particles (FRISH and SORNETTE1997; SORNETTE 2006). However, as was indicated in the above references, there is no accepted theoretical description, although fragmentation processes indeed have a crucial role in the explanation of models of PSD.

Abstract fragmentation schemes certainly may help in simulating the PSD. However, the result of fragmentation is not easy to predict. An important question was raised in KOLMOGOROV (1941), about what PSD could be expected when the ratio between the size r of the initial particle and the size of the particles obtained from it have a power law

dependence of the size r of the initial particle. This open question still remains without a theoretical answer. Recently, fragmentation algorithms inspired by Kolmogorov's question have been used in MARTÍN et al. (2009), giving results far removed from lognormality, showing, on the contrary, great complexity for simulated PSD that are in fact of fractal/multifractal type.

The packing of particles affects the physical properties of the granular system. Properties of particulate materials highly depend on the packing structure. Also, the hydraulic properties (water retention) of granular porous media, as in the case of soil, depend on the bulk density and the geometric scale arrangement of the intergranular space. Even when the particles are modelled by hard spheres, and disregarding the filling material occupying the intergranular space, models of granular media can be useful in predicting different properties of a wide number of natural and engineering systems, such as soil, ceramics, porous materials, concentrated suspensions, amorphous materials, alloys or microstructures of simple liquids. Because of that, the study of the random packing of particles has attracted researchers in many areas of science and technology (see GRAY 1968). In particular, the crucial influence of particle size distribution on the random packing structure increases the interest in relating both, either theoretically or by computational methods. SOHN and MORELAND (1968) studied the effect of Gaussian and log-normal distributions on packing density using dense random packing of sands, and ROUALT and ASSOULINE (1998) used a probabilistic approach to determine the distribution of the volume of the voids in packed spheres once their size distribution was given. ANISHCHIK and MEDVEDEV (1995) used three-dimensional Apollonian packing as a model for dense granular systems investigating the particle size distribution and the fractal nature of packings. The packing of spheres with log-normal distributions by means of computer simulation has been studied in NOLAN and KAVANAG (1993), and HE et al. (1999) used a Monte Carlo simulation for a random model of spherical particles of sizes obeying any given distribution.

The goal of this paper is to explore the connection of the characteristics of fragmentation processes with

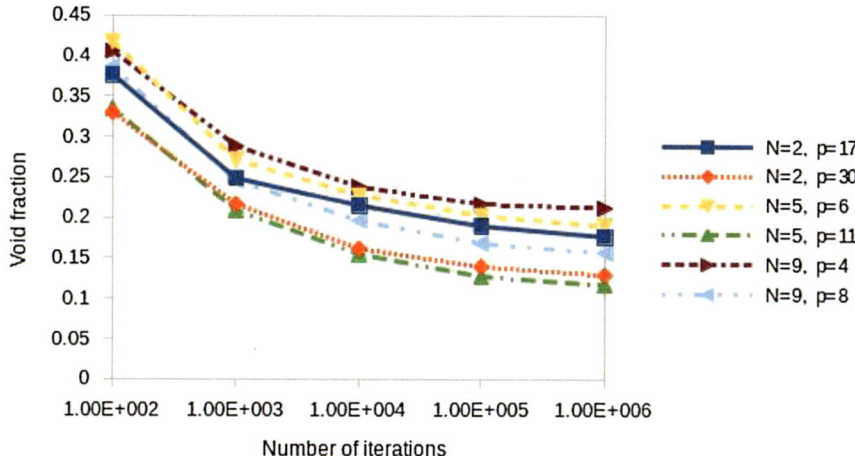

Figure 1

Influence of the number of iterations on the void fraction for some simulations. This behaviour is similar for all values of N and p used

the packing density of resulting particulate media, and examine this with respect to its important repercussions in many fields of science and technology. First, a family of fragmentation algorithms that replicate both the smooth log-normal and the fractal models is implemented. Then, a two-dimensional computer simulation method of packing of spheres is applied to the resulting distributions and the void fraction is evaluated. Finally, the relationship between the void fraction and the characteristic parameters of the fragmentation process is studied.

2. Materials and Methods

2.1. Fragmentation Algorithms

Fragmentation algorithms proposed in MARTÍN et al. (2009) provide a wide range of particle size distributions, depending of the parameters driving the fragmentation process. They vary from the log-normal model, when the fragmentation takes place with rules that are independent of the size of the fragmented particle, to the otherwise highly complex fractal/multifractal distributions.

Let N be a non-negative natural number bigger than one and $\alpha \geq 0$. Each particle of size (volume) r is divided in k smaller particles of size $\frac{r^{1+\alpha}}{k}$, k being a number randomly chosen between 1 and N, with equal probability for all the possible choices. The

Table 1

Minimum, maximum, average and standard deviation of the radii of the particles distributions of some simulations

N	p	Min	Max	Average	Standard deviation
2	17	3.12E−03	1.41E−01	1.71E−02	1.47E−02
	30	4.87E−05	2.49E−02	9.39E−04	1.19E−03
5	6	5.05E−03	1.03E−01	1.19E−02	7.71E−03
	11	1.01E−04	1.12E−02	8.14E−04	7.76E−04
9	4	8.96E−03	1.51E−01	1.72E−02	1.00E−02
	8	2.37E−04	9.73E−03	1.04E−03	7.47E−04

ratio between the size r of a particle and the size of the particles obtained from it is proportional to r^{α}, a certain power of r as suggested by KOLMOGOROV (1941).

The number of particles obtained from a particle of size r is

$$\left[\frac{r}{\frac{r^{1+\alpha}}{k}}\right] = \left[\frac{k}{r^{\alpha}}\right] \qquad (2)$$

and a remaining particle of size

$$r - \left[\frac{k}{r^{\alpha}}\right] \cdot \frac{r^{1+\alpha}}{k} \qquad (3)$$

where $[x]$ denotes the function floor of x, that is, the function that returns the greatest integer less than or equal to x.

The algorithm starts by fixing both values of N (integer bigger than one) and α (positive real

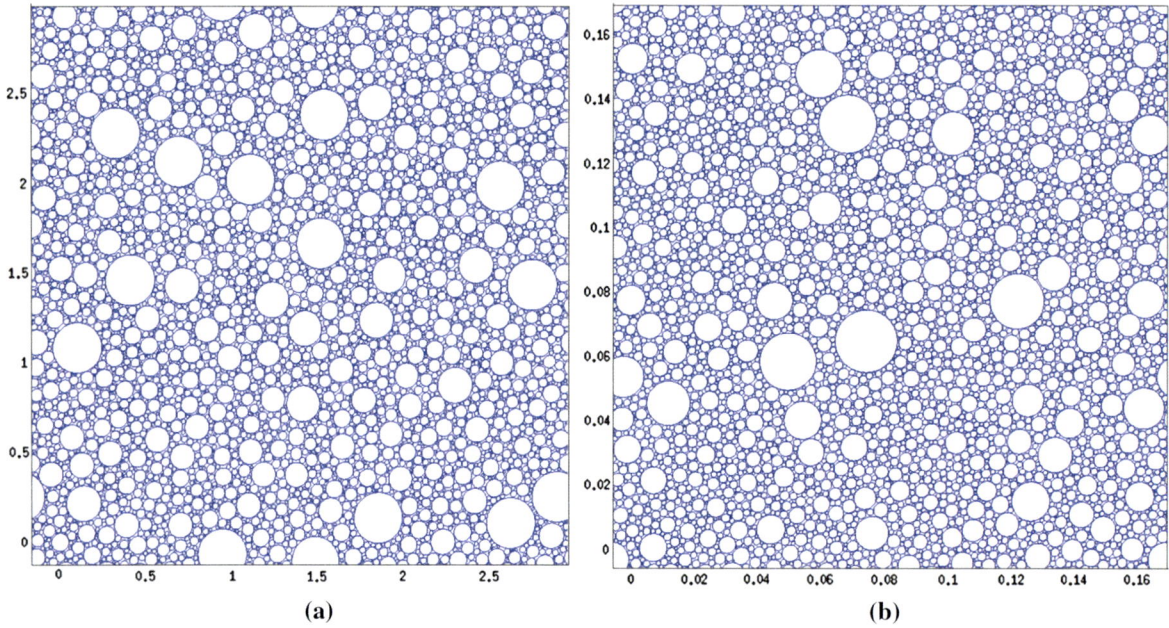

Figure 2
PSD simulated with packing algorithm with **a** $N = 2$ and **b** $N = 9$

number such that $0 \le \alpha \le 1$), for all the process, and the number of steps p of the fragmentation process. We consider an initial particle of size equal to 1, and in each step we apply, to each particle resulting from the previous step, the following procedure:

1. An integer k is randomly chosen between 1 and N, with equal probability for each one possibility.
2. The considered particle of size r is substituted by $\left[\frac{k}{r^\alpha}\right]$ particles of size $\frac{r^{1+\alpha}}{k}$ and an additional residual particle of size $r - \left[\frac{k}{r^\alpha}\right] \cdot \frac{r^{1+\alpha}}{k}$.

2.2. Computer Simulation of Packing

A Monte Carlo-based method following the model proposed in VIDAL et al. (2009) has been considered for the generation of packing processes. This method randomly positions a sequence of particles, ordered by decreasing radius, in a square domain with periodic border and edge length L_0. The center of each particle, with randomly generated coordinates (x, y), will always remain in the domain as long as it does not overlap any previously positioned particle. There exists a maximum number N_{TM} of attempts (of randomly generated (x, y)) of the positioning of each particle within the domain.

There also exists a number N_R of the repetition of the positioning of the same sequence of particles within the domain. Once all the particles have been positioned, a new domain will be defined with the length of the smaller edge L_1, where L_1 is defined as of $L_1 = \delta \cdot L_0$, with $0 < \delta < 1$ being a reduction factor of the domain. This positioning process will continue repeating in the new domain L_i, so that $L_i = \delta \cdot L_{i-1}$, until stage n, at which a new particle will exist whose number of positioning attempts reaches N_{TM}. At this point, the result of the packing simulation of the repetition of the completed N_R will be that obtained at stage $n - 1$, where all the particles have been positioned within the domain of size L_{n-1}. The void fraction, V, resulting from that packing will be

$$V = 1 - \frac{S_T}{L_{n-1}^2} \tag{4}$$

where S_T is the total surface area of all the particles of the sequence.

For the execution of the packing simulations, the initial size of the domain, L_0, has been obtained as a function of the total surface area of the particles by means of the equation $L_0 = (S_T \cdot 2.5)^{1/2}$, which produces an initial void fraction $V = 60\%$. The

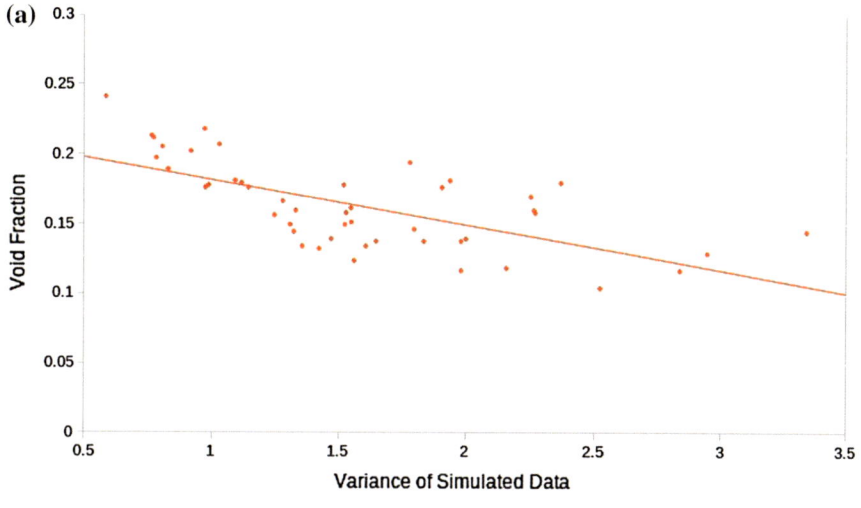

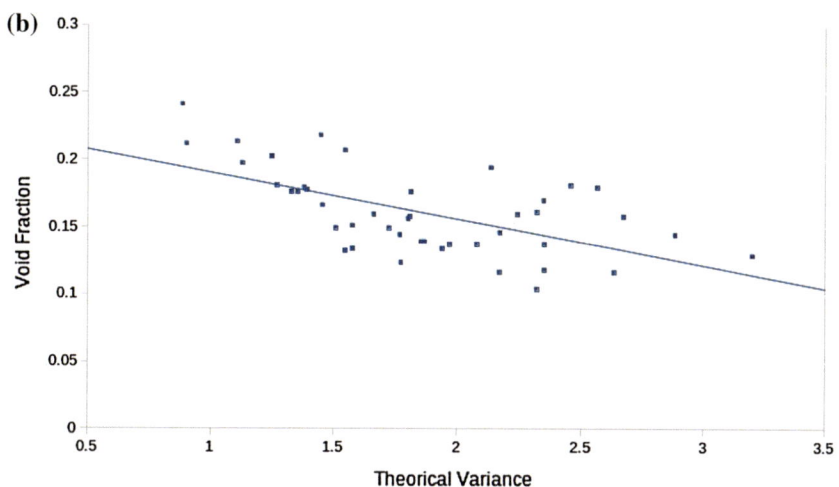

Figure 3
Scatter plots and linear fit of void fraction versus **a** theoretical variance and **b** real variance

Table 2

The average and standard deviation of the void fraction for different values of α and N

N	Fragmentation exponent	Average void fraction	Standard deviation
2	0.1	0.067	0.0038
	0.2	0.058	0.0067
	0.3	0.117	0.0149
	0.5	0.189	0.0030
3	0.1	0.080	0.0029
	0.2	0.081	0.0059
	0.3	0.137	0.0068
	0.5	0.155	0.0031

reduction factor δ has been fixed at 0.999 and the maximum number of iterations at 10^6. Each particle sequence repeats $N_R = 10$ times and the value of the ending void fraction V of the simulation for that distribution will be the lowest of these repetitions.

3. Results and Discussion

Data coming from MARTÍN *et al.* (2009) have been used to simulate the packing algorithm. A total number of 47 simulated distributions corresponding to the value $\alpha = 0$ and $N = 2, 3, 4, \ldots, 9$ have been selected. Figure 1 shows the influence of the number

145

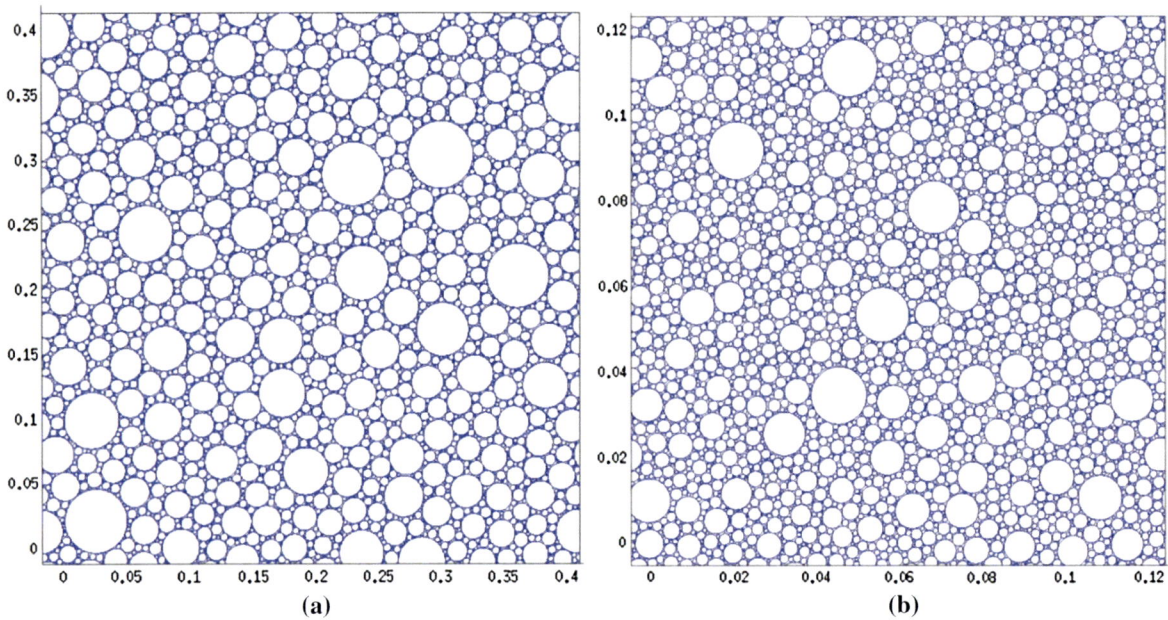

Figure 4
PSD simulated with packing algorithm with **a** $N = 3$, $\alpha = 0.1$ and **b** $N = 9$, $\alpha = 0.5$

of the iterations (in the packing algorithm) on the void fractions. It can be observed that the void fraction has been stabilized for 10^6 iterations; therefore, the number of iterations of each packing simulation in this work is 10^6.

The log-normality was tested with a Kolmogorov–Smirnov goodness-of-fit test at a confidence level of 95 %. In the case of $\alpha = 0$, the theoretical mean value and variance may be determined by algorithm parameters and they are given by the formulas

$$E = \frac{-2 \sum_{i=1}^{N} i \cdot \ln i}{N(N+1)} \cdot p$$

$$\sigma^2 = \left(\frac{2 \sum_{i=1}^{N} i \cdot (\ln i)^2}{N(N+1)} - \frac{E^2}{p^2} \right) \cdot p \qquad (5)$$

where p is the number of steps of the algorithm and $\ln x$ is the natural logarithm function. Table 1 shows a brief summary of the selected distributions.

The theoretical variance given by the algorithm and the real variance of the simulations were computed showing a great agreement (mean error between theoretical and real computed data is 0.285 and the standard deviation is 0.155).

Figure 2 shows some results of the packing algorithm, with Fig. 2a corresponding to a PSD obtained with $N = 2$ and Fig. 2b to a PSD obtained with $N = 9$.

The results with respect to the relation between the theoretical and real variance and the void fraction of the simulated packings are shown in Fig. 3. The determination coefficient R^2 of the linear fit is 0.35 and 0.42, respectively.

Results clearly indicate that there is an inverse correlation between both quantities: the greater the variance is, the lesser the void fraction. This computational result agrees with those of Sohn and Moreland (1968) for dense packings of sands.

Also, 40 simulated distributions corresponding to different values of α (between 0.1 to 0.5) have been selected to explore how the fragmentation exponent influences the result of packing. Table 2 shows averages and standard deviations of void fractions for different values of the fragmentation exponent α and parameter N.

The packing algorithm was applied to them. Figure 4 shows some examples of the packing results. Figure 4a corresponds to a PSD obtained with $N = 3$

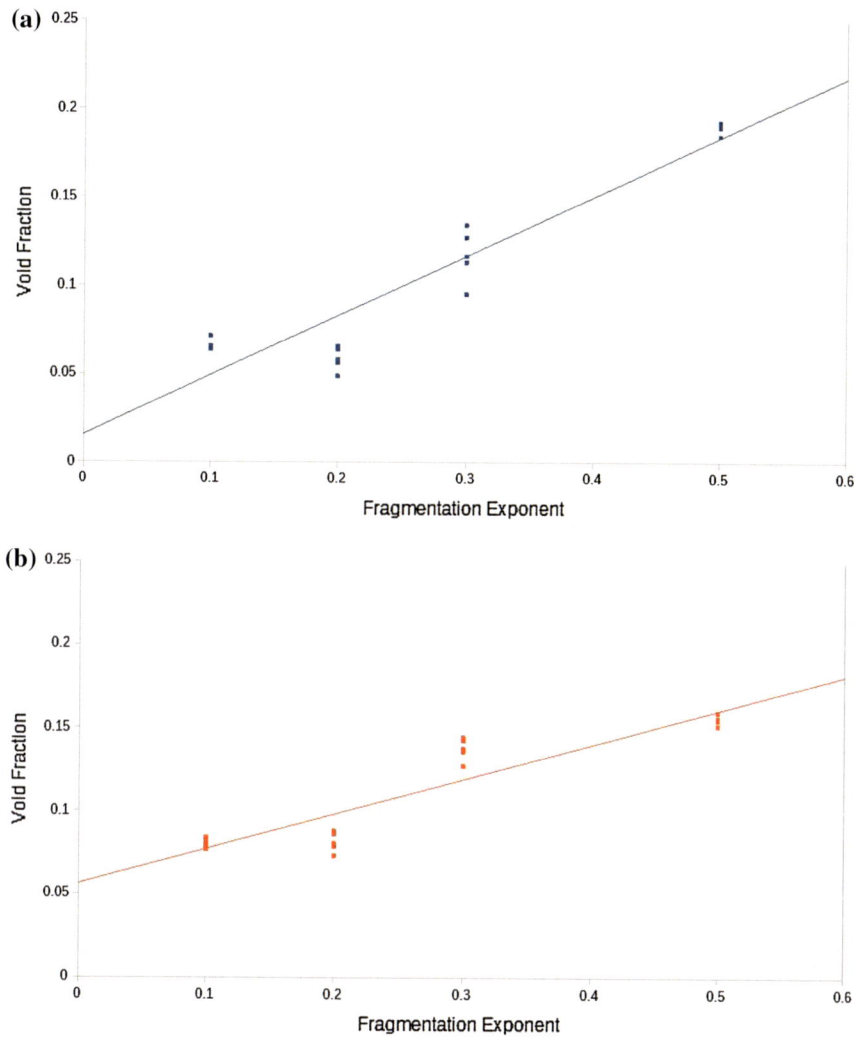

Figure 5

Plots of void fraction versus fragmentation exponent and trend lines for **a** $N = 2$ and **b** $N = 3$

and $\alpha = 0.1$, and Fig. 4b corresponds to a PSD obtained with $N = 9$ and $\alpha = 0.5$.

Figure 5 shows the relation between the value of α and the value of the void fraction for $N = 2$ and $N = 3$. The trend lines are also shown. The coefficient of determination R^2 of the corresponding fit is 0.893 and 0.835, respectively. Results indicate that the void fraction increase for increasing values of α.

A deeper insight around interesting questions that are related to the result of the fragmentation process for $\alpha > 0$ (open question posed by Kolmogorov) and the derived packing arrangement is suggested by this

result. On one hand, a dependence on the scale may be now expected, which might explain the fractal nature of the resulting distribution. Secondly, for $\alpha = 0$, the fragmentation of a given particle may give particles of any size with the same probability, and small particles in particular, which would facilitate a greater packing density. On the contrary, for $\alpha > 0$, the result of fragmentation does not have such free rules and depends on the size of the fragmented particle (scale): the size of any particle influences the emergence of new sizes; in particular, neighbouring sizes that might increase the intergranular space.

4. Conclusions

Fragmentation algorithms inspired by results and open problems raised in KOLMOGOROV (1941) provide a wide range of particle size distributions, depending on the parameters driving the fragmentation process. They vary from the log-normal model to the highly complex fractal/multifractal distributions. Because of this, they offer an interesting possibility to explore the connection of the characteristics of fragmentation processes with the packing density of resulting particulate media, and to examine this with respect to its repercussions in many scientific fields.

For the log-normal model ($\alpha = 0$), both the theoretical and the real variance of simulations inversely correlate with the void fraction: the greater the variance, the lesser the void fraction.

When $\alpha > 0$, the fragmentation algorithm produces non-smooth complex distributions for which the void fraction increases for increasing values of α. A deeper insight might indicate that, on one hand, scale dependence on the result of fragmentation may be expected, and on the other, that such dependence might facilitate a greater intergranular space.

Acknowledgment

This research work was been funded by Spain's Plan Nacional de Investigación Científica, Desarrollo e Innovación Tecnológica (I + D + I) under ref. AGL2011–25175.

REFERENCES

ANDREASEN, A. and ANDERSEN, J. (1930), *Ueber die Beziehung zwischen Kornabstufung und Zwischenraum in Produkten aus losen Körnern*, Kolloid-Zeitschrift *50*, 217–228.

ANISHCHIK, S.V. and MEDVEDEV, N.N. (1995), *Three-Dimensional Apollonian Packing as a Model for Dense Granular Systems*, Phys. Rev. Lett. *75*, 4314–4317.

BAGNOLD, R.A. and BARNDORFF-NIELSEN, O. (1980), *The pattern of natural size distributions*, Sedimentology *27* (2), 199–207.

FIELLER, N.R.J, GILBERTSON, D.D. and OLBRICHT, W. (1984), *A new method for environmental analysis of particle size distribution data from shoreline sediments*, Nature *311*, 648–651.

FRISH, U. and SORNETTE, D. (1997), *Extreme deviations and applications*, J. Phys. I France *7*, 1155–1171.

GRAY, W.A., *The packing of solid particles* (Chapman and Hall, London, 1968).

HE, D., EKERE, N.N. and CAI, L. (1999), *Computer simulation of random packing of unequal particles*, Phys. Rev. E *60*, 7098–7104.

KOLMOGOROV, A.N. (1941), *On the logarithmic normal distribution of particle sizes under grinding*. Dokl. Akad. Nauk SSSR *31*, 99–101.

MANDELBROT, B.B., *The fractal geometry of nature* (W.H. Freeman, New York 1982).

MARTÍN, M.A., GARCÍA-GUTIÉRREZ, C. and REYES, M. (2009), *Modeling multifractal features of soil particle size distributions with Kolmogorov fragmentation algorithms*, Vadose Zone J. *8*, 202–208.

NOLAN, G.T. and KAVANAG, P.E. (1993), *Computer simulation of random packings of spheres with log-normal distributions*, Powder Technol. *76*, 309–316.

ROUALT, Y. and ASSOULINE, S. (1998), *A probabilistic approach towards modeling the relationships between particle and pore size distributions: the multicomponent packed sphere case*, Powder Technol. *96*, 33–41.

SOHN, H.Y. and MORELAND, C. (1968), *The effect of particle size distribution on packing density*, Can. J. Chem. Eng. *46*, 162–167.

SORNETTE D., *Critical phenomena in Natural Sciences* (Springer-Verlag, Berlín 2006).

TURCOTTE, D.L. (1986), *Fractals and fragmentation*, J. Geophys. Res. *91*, 1921–1926.

VIDAL, D., RIDGWAY, C., PIANET. G., SCHOELKOPF, J., ROY, R. and BERTRAND, F. (2009), *Effect of particle size distribution and packing compression on fluid permeability as predicted by lattice-Boltzmann simulations*, Computers and Chemical Engineering *33*, 256–266.

(Received March 28, 2014, revised July 25, 2014, accepted July 26, 2014, Published online August 17, 2014)

Pure Appl. Geophys. 172 (2015), 149–165
© 2014 Springer Basel
DOI 10.1007/s00024-014-0900-1

❙ **Pure and Applied Geophysics**

Evaluation of the Optimal Utility of Some Investment Projects with Irreversible Environmental Effects

IÑIGO ARREGUI[1] and CARLOS VÁZQUEZ[1]

Abstract—In this work, the authors propose efficient numerical methods to solve mathematical models for different optimal investment problems with irreversible environmental effects. A relevant point is that both the benefits of the environment and the alternative project are uncertain. The cases with instantaneous and progressive transformation of the environment are addressed. In the first case, an augmented Lagrangian active set (ALAS) algorithm combined with finite element methods are proposed as a more efficient technique for the numerical solution to the obstacle problem associated with a degenerated elliptic PDE. In the second case, the mathematical model can be split into two subsequent steps: first we solve numerically a set of parameter dependent boundary value problems (the parameter being the level of progressive transformation), and secondly an evolutive nonstandard obstacle problem is discretized, thus leading to an obstacle problem at each time step. Also, an ALAS algorithm is proposed at each time step. Numerical solutions are validated through qualitative properties theoretically proven in the literature for different examples.

Key words: Investment under uncertainty, obstacle problems, finite elements, augmented Lagrangian algorithms.

1. Introduction

Physiocrats understand the earth as the main source of health, so that economics should focus on getting the most benefit from unlimited renewable Earth resources. Later, most economic schools separated the environment from economic development, thus environmental resources were understood as an exogenous factor in the economy. More recently, the discipline of environmental economics incorporates classical tools from economics to the analysis of environmental problems in order to focus economic

activity to minimize negative effects on the earth. Clearly, the tools of applied mathematics can be used for the modeling and solution of problems arising both in economics and environmental sciences, and therefore, in environmental economics. The present article can be framed in this general setting. More precisely, we focus on a problem arising in the economics of environmental management involving uncertain and irreversible effects.

From earlier civilizations, human action has changed planet Earth. Jointly with water, wind, and vegetation, human action has become one of the main external agents acting on the environment. Traditionally, the effects of these agents are incorporated in the modeling of different geophysical phenomena (FOWLER 2011). In more recent times, the extremely quick advance of technology makes human effects especially relevant. The building of new cities and the expansion of human habitat into previously unoccupied areas has also contributed to it, resulting in the construction of buildings, communication routes (roads, airports, railways, etc.), or infrastructures for the exploitation of natural resources (dams, ports, mines, terraces, etc.). Thus, in order to assess environmental policies, the impact of human actions in different geophysical processes needs to be analyzed, mathematically modeled, and numerically simulated.

In order to decide when and how best to invest in infrastructure or industrial projects, it is important to consider not only the financial aspects, but also the environmental impact of the investment. Also, an important aspect to consider is that their benefits will mainly occur in the future, and future uncertain factors may be involved. The irreversibility of some actions on the environment represents also an important factor to be considered when starting investment projects.

---
[1] Faculty of Informatics, Department of Mathematics, University of A Coruña, Campus Elviña s/n, 15071-A Coruña, Spain. E-mail: arregui@udc.es; carlosv@udc.es

In this work, we deal with the numerical solution of some models related to the opportunity of starting an industrial project that provides some uncertain benefits, but also involves some irreversible effects on the environment. In the case of instantaneous irreversible effects, the implicit uncertain profit associated with the environment stops once the project is initiated. When both the environment and the industrial project benefits are uncertain and governed by stochastic processes, an important problem is to determine if the project should be started and which will be the joint utility if started at an optimal instant. In the case of a progressive transformation of the environment, the utility also depends on the evolution of the fraction of the environment that is being transformed. In DIXIT and PINDYCK (1994) this kind of problem is studied and some analogies are established between them and the ones related to financial options pricing.

In SCHEINKMAN and ZARIPHOPOULOU (2001), a first approach to the rigorous mathematical modeling of these kinds of optimization problems is performed, its departure point being the one period example treated in ARROW and FISHER (1974) to discuss the so called quasi-option value. More precisely, in SCHEINKMAN and ZARIPHOPOULOU (2001), it is proved that the optimal utility function is the unique viscosity solution to the associated Hamilton–Jacobi–Bellman equation. Moreover, this optimal utility value function depends on the fraction of the environment that has been converted for the alternative project and from the initial profits associated with the environment and the alternative industrial project. More recently, DÍAZ et al. (2007) show that a suitable change of variable allows the formulation of the same problem in terms of two sequential PDE problems. The first one is associated with an elliptic degenerate PDE depending on a time parameter. The second can be formulated as an evolutive problem involving a multivalued operator. The model is mathematically analyzed by means of L^∞-accretive operator techniques. Although in SCHEINKMAN and ZARIPHOPOULOU (2001) some analytical solutions are discussed for particular expressions of the utility function (additive HARA utility functions), numerical methods are clearly required for the general case. In the present paper, taking advantage of the equivalent formulation

in DÍAZ et al. (2007), we propose some suitable numerical techniques to solve the general problem.

Moreover, in DÍAZ and FAGHLOUMI (2002), the case of instantaneous irreversible effects on the environment is addressed. This case corresponds to a particular choice of the utility function, so that an obstacle problem associated with a second-order elliptic operator which is not in divergence form is posed. Under certain not too restrictive hypotheses, by using the specific structure of the differential operator and the unbounded spatial domain, an appropriate change of variable is chosen in DÍAZ and FAGHLOUMI (2002) to prove the existence and uniqueness of solution. Some regularity results and qualitative properties of the solution are also stated for specific choices of the data. As these properties cannot be stated for the general case, some numerical methods to compute approximated solutions have been proposed in ACCIÓN et al. (2010). More precisely, projected Gauss–Seidel- (GLOWINSKI et al. 1976) and LIONS and MERCIER (1980)-type algorithms for the discretized finite elements problem are applied. In the second case, the method is combined with a multigrid and adaptive refinement (HOPPE 1987). The numerical methods are validated by means of examples whose solutions exhibit proved qualitative properties in DÍAZ and FAGHLOUMI (2002). However, some of the techniques can be replaced for the more efficient ones we propose in the present paper.

In the present work, we propose different numerical techniques to solve the mathematical models analyzed in DÍAZ et al. (2007) and DÍAZ and FAGHLOUMI (2002), proposed for the case of progressive and instantaneous effects on the environment, respectively. For this purpose, we consider a finite element discretization of the equivalent PDE problem posed on a suitable bounded domain, thus avoiding the required domain truncation of the original unbounded domain (KANGRO and NICOLAIDES 2000). Moreover, in the case of instantaneous effects, we propose an augmented Lagrangian active set method as a more efficient alternative to those ones proposed in ACCIÓN et al. (2010). We also validate the qualitative properties of the solution of this problem in different particular cases. The same numerical techniques are adapted for the obstacle problems appearing in the PDE model for progressive transformation of the environment.

The outline of the paper is the following. In Sect. 2 we describe the mathematical models of the optimal investment problem in the case of progressive and instantaneous effects on the environment. In Sect. 3 we describe the proposed numerical methods, mainly based on finite elements discretization and the augmented Lagrangian active set method for the discretized obstacle problems. Section 4 is devoted to the numerical tests illustrating the good performance of the numerical methods and the qualitative properties of the numerical solutions. Finally, some conclusions are presented in Sect. 5.

2. Mathematical Model

Let us consider a certain aspect related to the environment (a natural site, like a lake, a beach, a river or a mountain, for example) which can be developed into an alternative use (industrial plant, hotel, building, bridge, port, mine, tunnel, etc), the effects of which are irreversible on the environment. We assume that the benefits (per unit) of the original environmental aspect and the alternative project are random. More precisely, we assume that the benefits (per unit) of the environmental aspect (by means of tourism, for example), at time $t \geq 0$ are given by the stochastic process X_t. Also, let us consider that the random benefits (per unit) associated with the alternative project (by means of transport facilities, mine resources, additional business companies, for example) are given by the stochastic process Y_t. The dynamics of both stochastic processes are governed by the following differential equations:

$$
\begin{cases}
dX_t = \mu_1(X_t)\,dt + \sqrt{2}\sigma_1(X_t)\,dB_{1_t}, & X_0 = x \in \mathbb{R} \\
dY_t = \mu_2(Y_t)\,dt + \sqrt{2}\sigma_2(Y_t)\,dB_{2_t}, & Y_0 = y \in \mathbb{R},
\end{cases}
\tag{1}
$$

where $\{B_{i_t}\}_{t \geq 0}$ are Brownian motions defined in a certain probability space, μ_1 and μ_2 represent the drift functions of processes X_t and Y_t, and σ_1 and σ_2 are the standard deviation functions. Although we can assume that for $i = 1, 2$ the functions μ_i and σ_i are Lipschitz continuous and vanishing at the origin as in

SCHEINKMAN and ZARIPHOPOULOU (2001), DÍAZ et al. (2007), for simplicity we will consider them as constants, so that

$$
\mu_i(z) = \mu_i, \quad \sigma_i(z) = \sigma_i, \qquad i = 1, 2.
$$

We also consider a constant correlation coefficient ρ so that $dX_t dY_t = \rho dt$. In more practical applications, these and other model parameters need to be calibrated from real data. For example, the ones concerning environmental benefits could incorporate the output of some geophysical models related to the underlying particular environmental aspects that are involved.

As we consider a general case where only a part of the environment has been transformed into the industrial project, we denote by $\theta_t \in [0, 1]$ the transformed fraction of the environment at time t. We assume that

$$
\theta_t = \theta + M_t \in [0, 1],
\tag{2}
$$

where M_t is a process describing the nonnegative and nondecreasing cumulative development starting with $M_0 = 0$. Notice that the initial data $X_0 = x$, $Y_0 = y$ and $\theta_0 = \theta \in [0, 1]$ are given.

In order to pose the optimization problem we need to consider a utility function U depending on two arguments. In the general case, we impose that U is concave and nondecreasing in their arguments and that $U(0, 0) = 0$. As indicated in SCHEINKMAN AND ZARIPHOPOULOU (2001), cases where $U(0, 0)$ is not finite require further developments in the theory (and also in numerics). As only a fraction of the environment is transformed into the alternative industrial project, the joint utility at time t is given by $U((1 - \theta_t)X_t, \theta_t Y_t)$. In terms of this utility, the expected cumulative utility functional is defined by the expression

$$
J(x, y, \theta) = \mathbb{E}\left[\int_0^{+\infty} e^{-\alpha s} U((1 - \theta_s)X_s, \theta_s Y_s)\,ds \right],
\tag{3}
$$

where $\mathbb{E}$ represents the expected value, α denotes the constant continuous discount rate, X_t and Y_t follow Eq. (1) and θ_t satisfies condition (2). Moreover, let $\mathcal{A}_\theta$ denote the domain for M_t such that condition (2) holds.

Next, we pose the problem of determining the value function, v, associated with the utility functional (3) as

$$v(x, y, \theta) = \max_{\mathcal{A}_\theta} J(x, y, \theta). \qquad (4)$$

Notice that the utility function depends on the initial benefits to the environment (x) and the alternative project (y), which are assumed to be nonnegative, and also on the initial fraction of the environment (θ) that is transformed.

Additionally, in SCHEINKMAN and ZARIPHOPOULOU (2001) for the case of constant μ_i and σ_i the authors focus on additive HARA utility functions in the form:

$$U(x, y) = \frac{(x + y)^p}{p}, \qquad (5)$$

where $1 - p$ represents the constant relative risk aversion parameter (so that $p = 1$ corresponds to risk-neutrality). In this case, U has homogeneity of degree p and the linearity of the state equations for X_t and Y_t implies the homogeneity of the value function and a possibility to reduce the dimension of the equation. Notice that the limit case $p \to 0$ corresponds to logarithmic utility that fails out of the hypothesis $U(0,0) = 0$ (or even $U(0,0)$ finite), thus requiring further research. In the numerical results section, we mainly address examples with the utility function defined by (5).

In the next section, we recall the PDE models corresponding to two situations: first, when a fraction of the environment is progressively transformed (DÍAZ et al. 2007) and, secondly, when the environment is instantaneously transformed (DÍAZ and FAGHLOUMI 2002) into the alternative industrial problem.

2.1. PDE Model with Progressive Transformation of the Environment

By using the theory of viscosity solutions in CRANDALL et al. (1992) or the arguments in DÍAZ et al. (2007), the following Hamilton–Jacobi–Bellman (HJB) equation can be obtained:

$$\min[-\mathcal{L}v + \alpha v - U((1 - \theta)x, \theta y), -v_\theta] = 0, \quad \text{in } \Omega, \qquad (6)$$

where $\Omega = (0, +\infty) \times (0, +\infty)$ and $\mathcal{L}$ is the second order differential operator defined by

$$\mathcal{L}v = \sigma_1^2 x^2 v_{xx} + \sigma_2^2 y^2 v_{yy} + 2\sigma_1 \sigma_2 \rho xy v_{xy} + \mu_1 x v_x + \mu_2 y v_y. \qquad (7)$$

Actually, by introducing the matrix A and the vector $\mathbf{b}$ given by

$$A(x, y) = \begin{pmatrix} \sigma_1^2 x^2 & \sigma_1 \sigma_2 \rho xy \\ \sigma_1 \sigma_2 \rho xy & \sigma_2^2 y^2 \end{pmatrix},$$

$$\mathbf{b}(\mathbf{x}, \mathbf{y}) = \begin{pmatrix} (\mu_1 - 2\sigma_1^2 - \sigma_1 \sigma_2 \rho)\mathbf{x} \\ (\mu_2 - 2\sigma_2^2 - \sigma_1 \sigma_2 \rho)\mathbf{y} \end{pmatrix},$$

we can write (7) in the following divergence form:

$$\mathcal{L}v = \text{div}(A \nabla v) + \mathbf{b} \cdot \nabla \mathbf{v}.$$

Notice that at the boundaries $x = 0$ and $y = 0$, the operator (7) degenerates and homogeneous Neumann-like boundary conditions are implicit. Moreover, we introduce the function

$$\hat{U}(x, y, \theta) = U((1 - \theta)x, \theta y)$$

and notice that when we reach $\theta = 1$ all the environment has been transformed into the alternative problem so that we have

$$v(x, y, 1) = \mathbb{E}\left[\int_0^{+\infty} e^{-\alpha s} U((0, Y(s)) \, ds\right].$$

So, the HJB Eq. (6) jointly with homogeneous Neumann boundary condition and the value with $\theta = 1$ can be written in the form

$$\begin{cases} \min[-\mathcal{L}v + \alpha v - \hat{U}, -v_\theta] = 0, & \text{in } \Omega \times (0, 1), \\ A\nabla v \cdot \mathbf{v} = 0, & \text{in } \partial\Omega \times (0, 1), \\ v(x, y, 1) = \mathbb{E}\left[\int_0^{+\infty} e^{-\alpha s} U(0, Y(s)) \, ds\right], & \text{in } \Omega, \end{cases}$$
$$(8)$$

where $\mathbf{v}$ denotes the unit normal vector to $\partial\Omega$.

Moreover, in DÍAZ et al. (2007), the change of variable $t = 1 - \theta$ and the function

$$\overline{U}(x, y, t) = \hat{U}(x, y, 1 - t) = U(tx, (1 - t)y)$$

are introduced, so that the following auxiliary problem is considered:

$$\begin{cases} -\mathcal{L}f + \alpha f = \overline{U}(\cdot,\cdot,t), & \text{in } \Omega \times (0,1), \\ A\nabla f \cdot \boldsymbol{v} = 0, & \text{in } \partial\Omega \times (0,1). \end{cases} \quad (9)$$

By using the solution of (9) to introduce the new unknown

$$u(x,y,t) = v(x,y,1-t) - f(x,y,t), \quad (10)$$

problem (8) can be written as the following evolutive problem

$$\begin{cases} \min[u_t + f_t, -\mathcal{L}v + \alpha v - \hat{U}] = 0, & \text{in } \Omega \times (0,1), \\ A\nabla u \cdot \boldsymbol{v} = 0, & \text{in } \partial\Omega \times (0,1), \\ u(x,y,0) = \mathbb{E}\left[\int_0^{+\infty} e^{-\alpha s} U(0,Y(s))\,\mathrm{d}s\right] - f(x,y,0), & \text{in } \Omega. \end{cases} \quad (11)$$

In Díaz et al. (2007), a suitable multivalued operator is introduced to pose an equivalent formulation to (11) with a homogeneous initial condition so that existence of solutions is obtained by using L^∞ accretive operators theory.

In the section of numerical methods we propose a strategy to solve problem (6) by sequentially solving problems (9) and (11).

2.2. PDE Model with Instantaneous Transformation of the Environment

In this section, we assume instantaneous irreversible effects on the environment. If the project is developed at time T, then the expected cumulative utility functional is given by

$$J(x,y;T) = \mathbb{E}\left[\int_0^T e^{-\alpha s} X(s)\,\mathrm{d}s + \int_T^{+\infty} e^{-\alpha s} Y(s)\,\mathrm{d}s\right]$$
$$= \mathbb{E}\left[\int_0^{+\infty} \left(\mathbb{I}_{[0,T]} e^{-\alpha s} X(s) + \mathbb{I}_{[T,+\infty)} e^{-\alpha s} Y(s)\right)\,\mathrm{d}s\right],$$

where $\mathbb{I}_C$ represents the characteristic function of set C. In order to model more general problems, we can consider the following utility function:

$$J(x,y;T) = \mathbb{E}\left[\int_0^{+\infty} e^{-\alpha s} \left(\mathbb{I}_{[0,T]} f(X(s)) + \mathbb{I}_{[T,+\infty)} Y(s)\right)\,\mathrm{d}s\right] \quad (12)$$

for a given function f.

Thus, analogously to the general case, we pose the problem of determining the value, v, which is obtained when the alternative project is started at the optimal instant, T, that is:

$$v(x,y) = \max_T J(x,y;T).$$

Notice that the value function depends on the initial benefits of the environment (x) and the alternative project (y), which are assumed to be nonnegative.

Next, using techniques in Bensoussan and Lions (1978), it is easy to prove that function v is the solution of the following linear complementarity problem, posed on the domain $\Omega = (0,+\infty) \times (0,+\infty)$:

$$-\mathcal{L}v + \alpha v \geq f, \quad v \geq h, \quad (-\mathcal{L}v + \alpha v - f)(v - h) = 0, \quad (13)$$

where $\mathcal{L}$ is given by (7) and the lower obstacle, h, is the obtained utility when the project is started at time $T = 0$ (i.e., $h(x,y) = J(x,y,0)$). Thus, the analytical expression of function h is given by:

$$h(x,y) = \mathbb{E}\left[\int_0^{+\infty} e^{-\alpha s} Y(s)\,\mathrm{d}s\right] = \frac{y}{(\lambda-1)(\sigma_2^2\lambda + \mu_2)}, \quad (14)$$

with

$$\lambda = \frac{1}{2}\left[\left(1 - \frac{\mu_2}{\sigma_2}\right) - \sqrt{\left(1 - \frac{\mu_2}{\sigma_2}\right)^2 + \frac{4\alpha}{\sigma_2^2}}\right]. \quad (15)$$

On the other hand, if we introduce a new unknown $u = v - h$, the equivalent problem to obtain u can be written as

$$-\mathcal{L}u + \alpha u \geq G, \qquad u \geq 0, \qquad (-\mathcal{L}u + \alpha u - G)u = 0, \tag{16}$$

which is still posed on the unbounded domain Ω. Notice that $G = f + \mathcal{L}h - \alpha h$ and the associated obstacle function are identically zero. Moreover, due to the expression of matrix A, the homogeneous Neumann boundary condition $(A\nabla u) \cdot \mathbf{v} = 0$ is again naturally obtained on the boundaries $x = 0$ and $y = 0$.

In order to overcome the difficulty due to the unboundedness of Ω, we use the following change of variable proposed in Díaz and Faghloumi (2002):

$$\theta = \arctan x \quad \beta = \arctan y, \tag{17}$$

that maps the domain Ω into $\mathcal{F}(\Omega) = (0, \pi/2) \times (0, \pi/2)$, where $\mathcal{F}(x, y) = (\theta(x, y), \beta(x, y))$. Thus, the

$$S(\theta, \beta) = \begin{pmatrix} \frac{\sigma_1^2}{4}\sin^2 2\theta & \frac{\sigma_1\sigma_2\rho}{4}\sin 2\theta \sin 2\beta \\ \frac{\sigma_1\sigma_2\rho}{4}\sin 2\theta \sin 2\beta & \frac{\sigma_2^2}{4}\sin^2 2\beta \end{pmatrix},$$

$$\mathbf{p}(\theta, \beta) = \begin{pmatrix} \frac{\sin 2\theta}{2}(\mu_1 - 2\sigma_1^2 \cos^2 \theta - \sigma_1\sigma_2\rho \cos 2\beta) \\ \frac{\sin 2\beta}{2}(\mu_2 - 2\sigma_2^2 \cos^2 \beta - \sigma_1\sigma_2\rho \cos 2\theta) \end{pmatrix}.$$

Following Díaz and Faghloumi (2002), we assume that there exist some parameters $m_1 > 1$ and $m_2 > 1$ such that

$$\omega^{1/2}G = (1 + x^2)^{-m_1/2}(1 + y^2)^{-m_2/2}G \in L^2(\Omega), \tag{19}$$

which is equivalent to assume that $\widehat{\omega}^{1/2}\widehat{G} \in L^2(\mathcal{F}(\Omega))$, with $\widehat{\omega} = \omega \circ \mathcal{F}^{-1}$.

Next, we consider the weighted Sobolev spaces:

$$L_\omega^2(\mathcal{F}(\Omega)) = \left\{ \varphi : \widehat{\omega}^{1/2}\varphi \in L^2(\mathcal{F}(\Omega)) \right\}$$

$$H_\omega^1(\mathcal{F}(\Omega), S) = \left\{ \varphi \in L_\omega^2(\mathcal{F}(\Omega)) : (\sin 2\theta)\varphi_\theta \in L_\omega^2(\mathcal{F}(\Omega)), (\sin 2\beta)\varphi_\beta \in L_\omega^2(\mathcal{F}(\Omega)) \right\}$$

equivalent complementarity problem consists in finding $\widehat{u}$, defined by $\widehat{u} = u \circ \mathcal{F}^{-1}$, such that

$$\begin{aligned} &-\mathcal{L}_{(\theta,\beta)}\widehat{u} + \alpha\widehat{u} \geq \widehat{G}, \qquad \widehat{u} \geq 0, \\ &(-\mathcal{L}_{(\theta,\beta)}\widehat{u} + \alpha\widehat{u} - \widehat{G})\widehat{u} = 0, \end{aligned} \tag{18}$$

where

$$\begin{aligned} &\mathcal{L}_{(\theta,\beta)}\widehat{u} = \mathrm{div}_{(\theta,\beta)}(S\nabla_{(\theta,\beta)}\widehat{u}) + \mathbf{p} \cdot \nabla_{(\theta,\beta)}\widehat{\mathbf{u}}, \\ &\widehat{G} = G \circ \mathcal{F}^{-1}, \end{aligned}$$

and the subindex in the differential operators refers to the involved spatial variables. Moreover, matrix S and vector $\mathbf{p}$ are given by

and the convex set:

$$\widehat{\mathcal{K}} = \left\{ \varphi \in H_\omega^1(\mathcal{F}(\Omega), S) : \varphi \geq 0 \right\}.$$

Thus, problem (18) admits the following variational inequality formulation:

Find $u \in \widehat{\mathcal{K}}$, such that

$$a(\widehat{u}, \varphi - \widehat{u}) \geq L(\varphi - \widehat{u}), \quad \forall \varphi \in \widehat{\mathcal{K}}, \tag{20}$$

where the bilinear form a and linear operator L are given by

$$a(\widehat{u}, \varphi) = \int_{\mathcal{F}(\Omega)} \widehat{\omega} \left\{ \nabla \varphi \cdot S \nabla \widehat{u} \right.$$

$$+ \left[\left(\frac{1}{\widehat{\omega}} \nabla \widehat{\omega} \cdot S - \mathbf{p} \cdot \right) \nabla \widehat{u} \right] \varphi + \alpha \widehat{u} \varphi \right\} d\theta \, d\beta$$

$$L(\varphi) = \int_{\mathcal{F}(\Omega)} (\widehat{\omega} \, \widehat{G} \, \varphi) \, d\theta \, d\beta \,.$$

The existence and uniqueness of solution for (20) are proven in DÍAZ and FAGHLOUMI (2002) for the case $\mu_1 = \mu_2 = 0$ and in FAGHLOUMI (2003) without this assumption.

By using formulation (20), in ACCIÓN et al. (2010) several numerical techniques to solve this problem have been proposed. In the next section, we propose new ones which improve the efficiency in the numerical solution and can also be used for the model with progressive transformation of the environment.

3. Numerical Methods

3.1. Numerical Solution of the Model for Instantaneous Transformation

First, in order to discretize the continuous problem (20), let us consider a family of finite element triangular meshes (τ_h) of $\mathcal{F}(\Omega)$ for $h > 0$. Each mesh is formed by elements of diameter less than or equal to h and we denote its set of nodes by

$$\Sigma_h = \bigcup_{T \in \tau_h} \Sigma_T = \bigcup_{T \in \tau_h} \{a_i \mid 1 \leq i \leq N\} \,,$$

where Σ_T denotes the set of nodes in the mesh triangle T. Next, we define the space of piecewise linear Lagrange finite elements:

$$V_h = \left\{ \varphi_h \in \mathcal{C}^0(\overline{\Omega}) : \varphi_h|_T \in P_1, \quad \forall T \in \tau_h \right\},$$

where P_1 denotes the space of polynomials of degree less or equal than one, and the convex set:

$$K_h = \left\{ \varphi_h \in V_h : v_h(a_i) \geq 0, \quad \forall a_i \in \Sigma_h \right\}. \quad (21)$$

Next, we pose the following discretized variational inequality problem:

Find $\widehat{u}_h \in K_h$ such that

$$a(\widehat{u}_h, \varphi_h - \widehat{u}_h) \geq L(\varphi_h - \widehat{u}_h), \qquad \forall \varphi_h \in K_h \,. \tag{22}$$

In order to introduce a matrix formulation of problem (22), we can choose the finite element space basis:

$$B = \left\{ \psi_i \in V_h, \quad \psi_i(a_j) = \delta_{ij}, \quad 1 \leq i, j \leq N \right\}.$$

So, every function $\varphi_h \in V_h$ can be uniquely expressed in terms of the basis elements as

$$\varphi_h = \sum_{i=1}^{N} \varphi_h(a_i) \psi_i \,.$$

Moreover, we introduce the vector of node values, $\bar{\varphi}_h$, defined by $(\bar{\varphi}_h)_i = \varphi_h(a_i)$, $1 \leq i \leq N$ and the finite element discretization matrix and vector

$$A_{ij} = a(\psi_j, \psi_i), \qquad b_i = L(\psi_i), \qquad 1 \leq i, j \leq N.$$

Thus, problem (22) can be written in the form

$$A_h \bar{u}_h \geq b_h, \qquad \bar{u}_h \geq 0, \qquad (A_h \bar{u}_h - b_h) \bar{u}_h = 0 \,, \tag{23}$$

with the vector and matrix notations $A_h = (A_{ij})$ and $b_h = (b_i)$, respectively.

Different classical techniques exist for the numerical solution of discretized elliptic variational inequalities with unilateral constraints (see GLOWINSKI et al. 1976, for example). Among them, projected relaxation methods applied to the discretized problem can be chosen, having in mind that the convergence of these methods highly depends on the value of the relaxation parameter. In ACCIÓN et al. (2010), we analyzed a comparison between projected Gauss–Seidel and Lions–Mercier algorithms(LIONS and MERCIER 1980), this last one combined with a multigrid technique and adaptive refinement. In the present paper, to solve problem (23), we propose the use of the more recent augmented Lagrangian active set (ALAS) algorithm (KÄRKKÄINEN et al. 2003). The method has been previously used for pricing pension plans (see CALVO-GARRIDO et al. 2013).

More precisely, after the finite element discretization the discrete problem can be written in the form

$$A_h \bar{u}_h + P_h = b_h, \tag{24}$$

with the discrete complementarity conditions

$$\bar{u}_h \geq 0, \quad P_h \leq 0, \quad \bar{u}_h P_h = 0, \tag{25}$$

where P_h denotes the vector of the multiplier values associated to the inequality constraint.

The basic iteration of the ALAS algorithm consists of two steps. In the first one, the domain is decomposed into active and inactive parts (depending on whether the constraints are active or not), and in the second step, a reduced linear system associated with the inactive part is solved. Thus, we use the algorithm for unilateral problems, which are based on the augmented Lagrangian formulation.

First, for any decomposition $\mathcal{N} = \mathcal{I} \cup \mathcal{J}$, where $\mathcal{N} := \{1, 2, \ldots N_{\mathrm{dof}}\}$, let us denote by $[A_h]_{\mathcal{I}\mathcal{I}}$ the principal minor of matrix A_h and by $[A_h]_{\mathcal{I}\mathcal{J}}$ the co-diagonal block indexed by $\mathcal{I}$ and $\mathcal{J}$. Thus, the ALAS algorithm computes not only $\bar{u}_h$ and P_h, but also a decomposition $\mathcal{N} = \mathcal{J} \cup \mathcal{I}$ such that

$$\begin{aligned}
A_h \bar{u}_h + P_h &= b_h, \\
[P_h]_j + \beta[\bar{u}_h]_j &\leq 0, & \forall j \in \mathcal{J}, \quad (26) \\
[P_h]_i &= 0, & \forall i \in \mathcal{I},
\end{aligned}$$

for a given positive parameter β. In the above equations, $\mathcal{I}$ and $\mathcal{J}$ are the *inactive* and the *active* sets, respectively. More precisely, the iterative algorithm builds sequences $\{\bar{u}_h^m\}_m$, $\{P_h^m\}_m$, $\{\mathcal{I}^m\}_m$ and $\{\mathcal{J}^m\}_m$, converging to $\bar{u}_h$, P_h, $\mathcal{I}$ and $\mathcal{J}$, by means of the following steps:

1. Initialize $\bar{u}_h^0 = 0$ and $P_h^0 = \min\{b_h - A_h \bar{u}_h^0, 0\} \leq 0$. Choose $\beta > 0$. Set $m = 0$.
2. Compute

$$\begin{aligned}
Q_h^m &= \min\{0, P_h^m + \beta(\bar{u}_h^m - 0)\}, \\
\mathcal{J}^m &= \{j \in \mathcal{N}, [Q_h^m]_j < 0\}, \\
\mathcal{I}^m &= \{i \in \mathcal{N}, [Q_h^m]_i = 0\}.
\end{aligned}$$

3. If $m \geq 1$ and $\mathcal{J}^m = \mathcal{J}^{m-1}$ then convergence is achieved.
4. Let $\bar{u}$ and P be the solution of the linear system:

$$\begin{aligned}
A_h \bar{u} + P &= b_h, \\
P = 0 \quad \text{on} \quad \mathcal{I}^m \quad &\text{and} \quad \bar{u} = 0 \quad \text{on} \quad \mathcal{J}^m.
\end{aligned} \quad (27)$$

Set $\bar{u}_h^{m+1} = V$, $P_h^{m+1} = \min\{0, P\}$, $m = m + 1$ and go to Step 2.

It is important to notice that instead of solving the full linear system in (27), for $\mathcal{I} = \mathcal{I}^m$ and $\mathcal{J} = \mathcal{J}^m$ the following reduced one on the inactive set is solved:

$$\begin{aligned}
[A_h]_{\mathcal{I}\mathcal{I}} [\bar{u}]_{\mathcal{I}} &= [b_h]_{\mathcal{I}}, \\
[\bar{u}]_{\mathcal{J}} &= 0_{\mathcal{J}}, \quad (28) \\
P &= b - A_h \bar{u}.
\end{aligned}$$

In KÄRKKÄINEN *et al.* (2003), the convergence of the algorithm in a finite number of steps is proved for a Stieltjes matrix (i.e. a real symmetric positive definite matrix with negative off-diagonal entries) and a suitable initialization (the same we consider in this paper). They also prove that $\mathcal{I}^m \subset \mathcal{I}^{m+1}$. A Stieltjes matrix is obtained for linear elements.

Also, we have implemented a refinement criterium that selects the elements where

$$u_h(a_i) > 0 \quad \text{and} \quad u_h(a_j) = 0, \quad \text{for } i \neq j \text{ in } \{1, 2, 3\},$$

a_i being the vertices of the triangle. So, elements close to the free boundary are refined. Once one element has been selected for refinement, we implement the refinement technique proposed in RIVARA (1984).

In Sect. 4.1. some examples illustrate the performance of the described numerical strategy for the case of instantaneous transformation of the environment. Also, the method is compared with projected Gauss-Seidel, especially in terms of computational time.

3.2. Numerical Solution of the Model for Progressive Transformation

In order to solve problem (11) numerically, we start writing the first equation in (11) in the following equivalent form:

$$\begin{aligned}
u_t + f_t \geq 0, \quad -\mathcal{L}u + \alpha u \geq 0, \\
(u_t + f_t)(-\mathcal{L}u + \alpha u) = 0, \qquad \text{in } \Omega \times (0, 1).
\end{aligned} \quad (29)$$

In order to discretize in variable t the previous equation, we consider a natural number $N > 0$, the step size $\Delta t = 1/(N+1)$, and the discrete values $t^n = n\Delta t$, for $t = 0, 1, \ldots, N+1$. Moreover, we introduce the functional notation $h^n = h(\cdot, t = t^n))$ for all functions depending on $(x, y) \in \Omega$ and $t \in [0, 1]$ and propose the following approximation:

$$(u_t + f_t)(\cdot, t^n) = \frac{u^n - u^{n-1}}{\Delta t} + \frac{f^n - f^{n-1}}{\Delta t}. \quad (30)$$

Thus, by introducing approximation (30) in Eq. (29), after initializing $u^0 = u(\cdot, 0)$ we sequentially compute u^n as the solution of the complementarity problem

$$u^n \geq \psi^n, \quad -\mathcal{L}u^n + \alpha u^n \geq 0, \quad (u^n - \psi^n) \\ (-\mathcal{L}u^n + \alpha u^n) = 0, \qquad \text{in}\,\Omega, \tag{31}$$

jointly with homogeneous Neumann boundary conditions, where

$$\psi^n = u^{n-1} - f^n + f^{n-1} \tag{32}$$

represents the obstacle function at each step of variable t.

Notice that for each step n, problem (31) is analogous to problem (16) appearing in the instantaneous transformation case, although with a nonzero obstacle function ψ^n and a null second member of the equation instead of G.

The terms f^n are previously computed by solving problem (9) for the different values $t = t^n$. For the numerical solution of problem (9) we use the same finite element discretization as in the instantaneous transformation case, so that the discretized problem takes the form

$$A_h \bar{f}_h^n = b_h^n,$$

where b_h^n denotes the second member associated to the function $\overline{U}(\cdot, t^n)$ appearing in (9). The linear systems are solved by means of a classical LU factorization direct method.

For the numerical solution of (31), after the change of variable (17) to work in a bounded computational domain, we also propose the same finite elements discretization as in the other problems. Thus, the mixed formulation of the discretized problem can be written in the form:

$$A_h \bar{u}_h^n + P_h^n = 0, \tag{33}$$

with the discrete complementarity conditions

$$\bar{u}_h^n \geq \overline{\psi}_h^n, \quad P_h^n \leq 0, \quad \left(\bar{u}_h^n - \overline{\psi}_h^n\right) P_h^n = 0, \tag{34}$$

where P_h^n denotes the vector of the multiplier values and $\overline{\psi}_h^n$ denotes the vector of the nodal values defined by function ψ^n.

We apply the ALAS algorithm for solving formulation (34). More precisely, for each value t^n, the ALAS algorithm computes not only $\bar{u}_h^n$ and P_h^n, but also a decomposition $\mathcal{N} = \mathcal{J}^n \cup \mathcal{I}^n$ such that

Table 1

Financial data set for Example 1

σ_1	σ_2	μ_1	μ_2	ρ	α	c
0.18	0.0	0.0	0.0	0.0	1.0	1.0

$$\begin{aligned}
A_h \bar{u}_h^n + P_h^n &= 0, \\
\left[P_h^n\right]_j + \beta\left[\bar{u}_h^n - \overline{\psi}\right]_j &\leq 0, &\forall j \in \mathcal{J}^n, \\
\left[P_h^n\right]_i &= 0, &\forall i \in \mathcal{I}^n,
\end{aligned} \tag{35}$$

for a given positive parameter β. In the above equations, $\mathcal{I}^n$ and $\mathcal{J}^n$ are the *inactive* and the *active* sets at t_n, respectively. More precisely, the iterative algorithm builds sequences $\left\{\bar{u}_{h,m}^n\right\}_m$, $\left\{P_{h,m}^n\right\}_m$, $\left\{\mathcal{I}_m^n\right\}_m$ and $\left\{\mathcal{J}_m^n\right\}_m$, converging to $\bar{u}_h^n$, P_h^n, $\mathcal{I}^n$ and $\mathcal{J}^n$, by means of the following steps:

1. Initialize $\bar{u}_{h,0}^n = \overline{\psi}_h^n$ and $P_{h,0}^n = \min\{b_h^n - A_h V_{h,0}^n, 0\} \leq 0$. Choose $\beta > 0$. Set $m = 0$.
2. Compute

$$\begin{aligned}
Q_{h,m}^n &= \min\left\{0, P_{h,m}^n + \beta\left(\bar{u}_{h,m}^n - \overline{\psi}_{h,m}^n\right)\right\}, \\
\mathcal{J}_m^n &= \left\{j \in \mathcal{N}, \left[Q_{h,m}^n\right]_j < 0\right\}, \\
\mathcal{I}_m^n &= \left\{i \in \mathcal{N}, \left[Q_{h,m}^n\right]_i = 0\right\}.
\end{aligned}$$

3. If $m \geq 1$ and $\mathcal{J}_m^n = \mathcal{J}_{m-1}^n$ then convergence is achieved.
4. Let $\bar{u}$ and P be the solution of the linear system

$$A_h \bar{u} + P = 0, \\
P = 0 \quad \text{on} \quad \mathcal{I}_m^n \quad \text{and} \quad \bar{u} = \overline{\psi}_{h,m}^n \quad \text{on} \quad \mathcal{J}_m^n. \tag{36}$$

Set $\bar{u}_{h,m+1}^n = \bar{u}$, $P_{h,m+1}^n = \min\{0, P\}$, $m = m + 1$ and go to Step 2.

As in the instantaneous transformation case, instead of solving the full linear system in (36), for $\mathcal{I} = \mathcal{I}_m^n$ and $\mathcal{J} = \mathcal{J}_m^n$ the following reduced one on the inactive set is solved:

$$\begin{aligned}
\left[A_h\right]_{\mathcal{I}\mathcal{I}}[\bar{u}]_{\mathcal{I}} &= -\left[A_h\right]_{\mathcal{I}\mathcal{J}}\left[\overline{\psi}\right]_{\mathcal{J}}, \\
[\bar{u}]_{\mathcal{J}} &= \left[\overline{\psi}\right]_{\mathcal{J}}, \\
P &= -A_h \bar{u}.
\end{aligned} \tag{37}$$

157

Table 2

Mesh refinement, computational time, iterations, and errors obtained for Example 1

L	ALAS					Projected Gauss–Seidel				
	NE	NN	t_A	I_A	e_A	NE	NN	t_{GS}	I_{GS}	e_{GS}
1	512	289	4.39	2	6.92×10^{-3}	512	289	45.20	56	6.41×10^{-3}
2	704	386	7.10	4	6.03×10^{-3}	714	391	90.55	62	5.68×10^{-3}
3	1,136	605	14.94	6	6.07×10^{-3}	1,130	601	234.66	69	4.91×10^{-3}
4	2,096	1,087	40.51	7	4.80×10^{-3}	2,153	1,115	1,819.61	160	4.58×10^{-3}
5	3,840	1,982	100.86	10	3.61×10^{-3}	4,070	2,076	8,064.63	218	3.79×10^{-3}

L, NE, NN, *t*, *I* and *e* represent the level of refinement, number of elements, number of nodes, computational time in seconds, number of iterations and relative error, respectively. When appearing, subindex *A* refers to ALAS algorithm and GS to projected Gauss-Seidel

 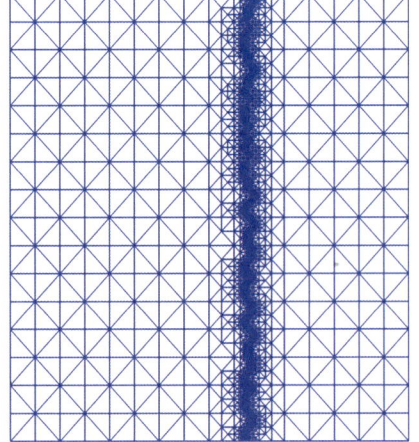

Figure1

Adaptive meshes of the domain $\mathcal{F}(\Omega)$ for ALAS (*left*) and projected Gauss-Seidel (*right*), after five refinement steps for Example 1

4. Numerical Results

4.1. Examples with Instantaneous Transformation

As in ACCIÓN *et al.* (2010), in the present paper, we first validate the proposed numerical methods by solving one example with a known analytical solution. Next, we consider an example for which theoretically stated qualitative properties have been proved in DÍAZ and FAGHLOUMI (2002) and FAGHLOUMI (2003). The third example corresponds to a more realistic function *h* given by (14) in terms of the financial data.

Example 1　In this example we choose $f(x) = x$ and $h(x, y) = c/\alpha$, *c* being a constant, so that $G(x, y) = x - c$. So, we assume a linear behaviour of

Table 3

Financial data set for Example 2

σ_1	σ_2	μ_1	μ_2	ρ	α	c
0.18	0.40	0.0	0.0	0.20	1.0	1.0

the utility associated with the environment with respect to the benefit (per unit) to the environment and that the utility of the alternative project is constant. Table 1 shows the set of model parameters. With these parameters, the exact solution (which only depends on *x*) is given by

$$v(x, y) = \begin{cases} \dfrac{c^{1-\gamma}}{\alpha(1-\gamma)^{1-\gamma}|\gamma|^{\gamma}}x^{\gamma} + \dfrac{x}{\alpha} & \text{if } x \geq x_f \\ \dfrac{c}{\alpha} & \text{if } x \leq x_f, \end{cases}$$

Table 4

Mesh refinement, computational time, iterations, and errors obtained for Example 2

	ALAS				Projected Gauss–Seidel			
L	NE	NN	t_A	I_A	NE	NN	t_{GS}	I_{GS}
1	512	289	5.21	3	512	289	68.38	86
2	732	401	8.63	4	732	401	130.13	86
3	1,275	675	19.41	6	1,288	682	463.04	108
4	2,367	1,224	52.41	9	2,445	1,263	2,743.57	190
5	4,466	2,277	133.67	11	4,766	2,428	1,4419.40	271

L, NE, NN, t and I represent the level of refinement, number of elements, number of nodes, computational time in seconds, and number of iterations, respectively. When appearing, subindex A refers to the ALAS algorithm and GS refers to the projected Gauss–Seidel

with

$$\gamma = \frac{1}{2}\left(1 - \sqrt{1 + \frac{4\alpha}{\sigma_1^2}}\right), \quad x_f = \frac{\gamma}{\gamma - 1}c.$$

Notice that the free boundary that separates the region where $v > h$ from the one with $v = h$ is the straight line $x = x_f$.

very close in both algorithms, the ALAS algorithm reports much shorter computational time (t) and less iterations (I).

Figure 1 shows the meshes of the domain $\mathcal{F}(\Omega)$ after five refinement levels ($L = 5$). Notice that the one corresponding to ALAS seems to better follow the free-boundary of the problem. However, the error in both approximations is of the same order.

Example 2 We choose $f(x) = x$ and $h(x,y) = y/\alpha$ so that $G(x,y) = x - y$ and utilities associated with environment and alternative projects linearly depend on their respective benefits (per unit). The chosen parameters are shown in Table 3. Notice that in this example both volatilities and the correlation coefficient are different from zero.

Although we cannot obtain the expression of the exact solution, for the case $\mu_1 = \mu_2 = 0$ and $c > 0$ the following upper bound for the solution is stated in DÍAZ and FAGHLOUMI (2002):

$$\bar{u}(x,y) = \begin{cases} \dfrac{y}{\alpha}, & \text{if}(x,y) \in \Omega_1 \\[3mm] \dfrac{(\delta - 1)^{\delta-1}}{\alpha c^{\delta-1}\delta^{\delta}}y^{\delta} + \dfrac{c}{\alpha}, & \text{if}(x,y) \in \Omega_2 \\[3mm] \dfrac{c^{1-\gamma}}{\alpha(1-\gamma)^{1-\gamma}\mid\gamma\mid^{\gamma}}x^{\gamma} + \dfrac{x}{\alpha} + \dfrac{(\delta - 1)^{\delta-1}}{\alpha c^{\delta-1}\delta^{\delta}}y^{\delta}, & \text{if}(x,y) \in \Omega_3 \\[3mm] \dfrac{c^{1-\gamma}}{\alpha(1-\gamma)^{1-\gamma}\mid\gamma\mid^{\gamma}}x^{\gamma} + \dfrac{x}{\alpha} + \dfrac{y - c}{\alpha}, & \text{if}(x,y) \in \Omega_4 \end{cases}$$

Table 2 summarizes the numerical results for up to five refinement levels starting from a uniform initial mesh with 289 nodes and 512 triangular elements. For both projected Gauss–Seidel and ALAS algorithms, the stopping test in the relative quadratic error between two iterations is set to 10^{-5}, while for ALAS the parameter $\beta = 10,000$ is chosen. In Table 2 we can observe that the relative error (e) with respect to the analytical solution is

where

$$\Omega_1 = \left\{(x,y) \in \Omega : x \leq \frac{\gamma}{\gamma-1}c \ \text{and} \ y \geq \frac{\delta}{\delta-1}c\right\}$$

$$\Omega_2 = \left\{(x,y) \in \Omega : x \leq \frac{\gamma}{\gamma-1}c \ \text{and} \ y \leq \frac{\delta}{\delta-1}c\right\}$$

$$\Omega_3 = \left\{(x,y) \in \Omega : x \geq \frac{\gamma}{\gamma-1}c \ \text{and} \ y \leq \frac{\delta}{\delta-1}c\right\}$$

$$\Omega_4 = \left\{(x,y) \in \Omega : x \geq \frac{\gamma}{\gamma-1}c \ \text{and} \ y \geq \frac{\delta}{\delta-1}c\right\}.$$

159

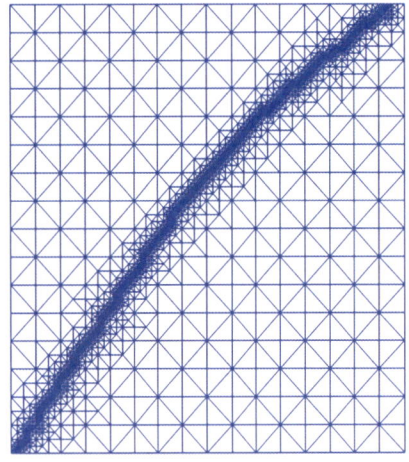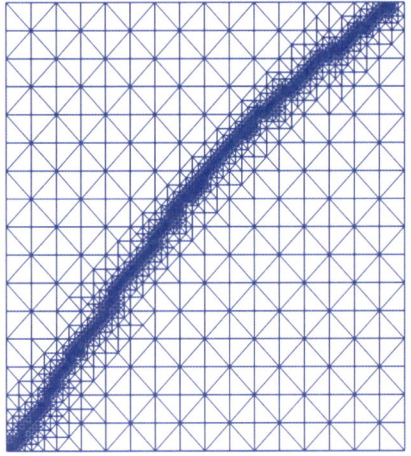

Figure2
Adaptive meshes of the domain $\mathcal{F}(\Omega)$ for ALAS (*left*) and projected Gauss–Seidel (*right*), after five refinement steps for Example 2

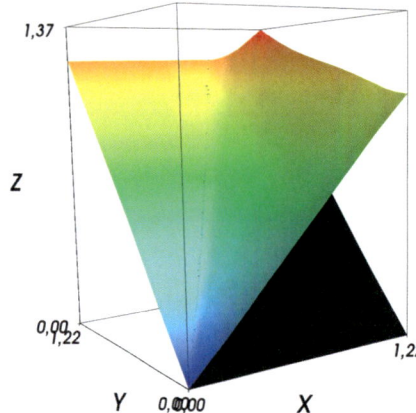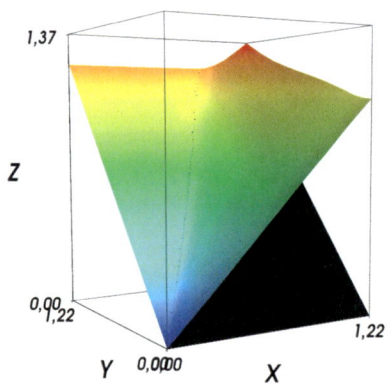

Figure3
Numerical solution for ALAS (*left*) and projected Gauss–Seidel (*right*), after five refinement steps for Example 2

Table 5

Financial data set for Example 3

σ_1	σ_2	μ_1	μ_2	ρ	α
0.18	0.40	0.08	0.15	0.20	0.25

Moreover, as stated in FAGHLOUMI (2003), we can identify the following subset, S_C, of the coincidence set characterized by the condition:

$$S_C = \{(x, y) \in \Omega, y \geq \gamma x\}$$
$$\subset \{(x, y) \in \Omega, u(x, y) = h(x, y)\}$$

with

$$\gamma = \frac{\left(1 + \sqrt{1 + \frac{4\alpha}{\sigma_1^2}}\right)\left(1 + \sqrt{1 + \frac{4\alpha}{\sigma_2^2}}\right)}{\left(\sqrt{1 + \frac{4\alpha}{\sigma_1^2}} - 1\right)\left(\sqrt{1 + \frac{4\alpha}{\sigma_2^2}} - 1\right)} \cdot$$

Both theoretically stated qualitative properties, the supersolution, and coincidence subset, have been verified by the numerical tests.

Table 4 summarizes the numerical results for five refinement levels from a uniform initial mesh with 289 nodes and 512 triangular elements. Moreover, the same stopping test in the relative quadratic error between two iterations and the same parameter β as in Example 1 have been chosen. As the exact solution cannot be computed, errors cannot be reported. We

Table 6

Mesh refinement, computational time, iterations, and errors obtained for Example 3

	ALAS				Projected Gauss–Seidel			
L	NE	NN	t_A	I_A	NE	NN	t_{GS}	I_{GS}
1	512	289	5.19	3	512	289	114.93	150
2	745	409	8.75	6	745	409	327.24	214
3	1,296	688	19.66	9	1,299	689	1,506.50	350
4	23,79	1,231	52.97	12	2,488	1,287	7,582.44	511
5	4,583	2,321	139.81	15	4,686	2,390	3,6982.91	725

L, NE, NN, t and I represent the level of refinement, number of elements, number of nodes, computational time in seconds, and number of iterations, respectively. When appearing, subindex A refers to the ALAS algorithm and GS refers to projected Gauss–Seidel

can observe that the ALAS algorithm exhibits much shorter computational time (t) and less iterations (I).

Figure 2 shows the meshes after five refinement levels. As in Example 1 the one corresponding to ALAS seems to follow a bit better the free boundary (optimal investment boundary) of the problem. Notice that the region above and at the left of the free boundary corresponds to the part of the domain where $v = h$, while the region located below and at the right of the free boundary corresponds to $v > h$. Thus, for each value of the benefit (per unit) of the environment, there exists a different critical value of the benefit (per unit) of the industrial project above which the optimal utility is equal to the one obtained if we start the investment at the initial time.

The associated numerical solutions at this refinement level with both numerical methods are shown in Fig. 3 and are very close each other. The plane in black behind the solution represents the obstacle function h.

Example 3 In this last example of the instantaneous irreversible case, the function h is given by (14), so that for the choice $f(x) = x$ the function G is

$$G(x, y) = x - \frac{(\alpha - \mu_2)y}{(\lambda - 1)(\sigma_2^2 \lambda + \mu_2)},$$

with λ given by (15). The financial parameters are shown in Table 5 so that an optimal investment boundary appears.

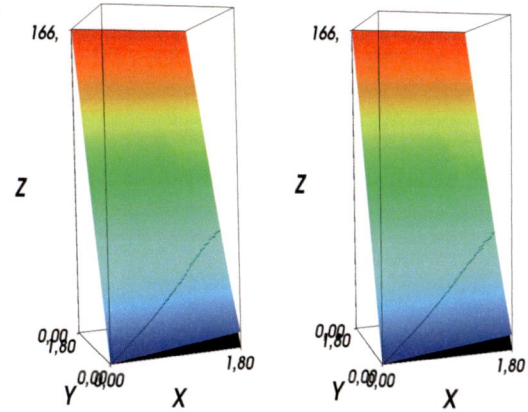

Figure5
Numerical solution for ALAS (*left*) and projected Gauss–Seidel (*right*), after five refinement steps for Example 3

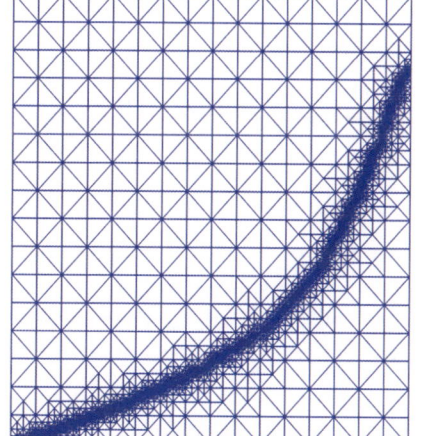
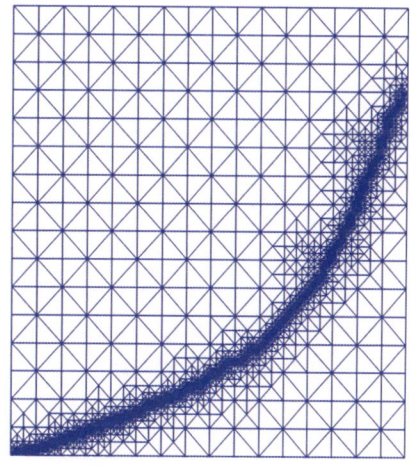

Figure4
Adaptive meshes of the domain $\mathcal{F}(\Omega)$ for ALAS (*left*) and projected Gauss–Seidel (*right*), after five refinement steps for Example 3

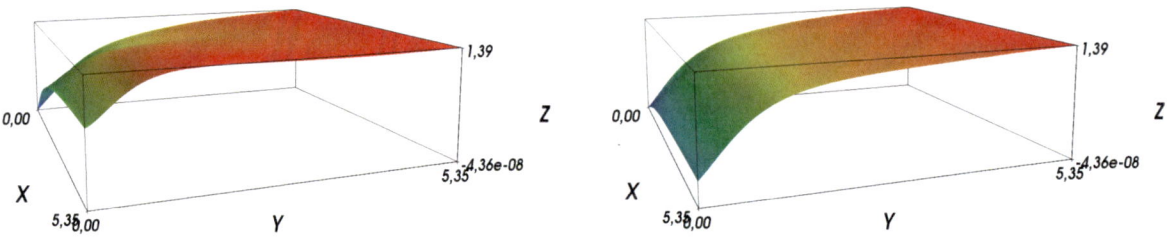

Figure6
Numerical solution for $\theta = 0.25$ (*left*) and $\theta = 0.75$ (*right*) with $p = 1$ in the progressive transformation case

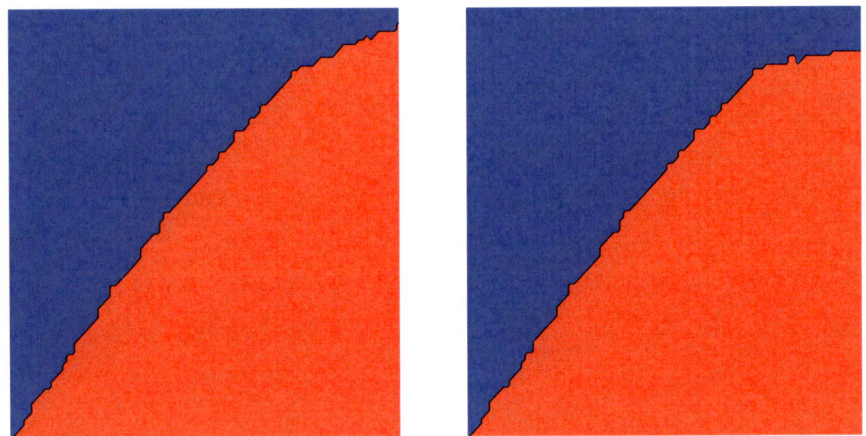

Figure7
Free boundary for $\theta = 0.25$ (*left*) and $\theta = 0.75$ (*right*) with $p = 1$ in the progressive transformation case

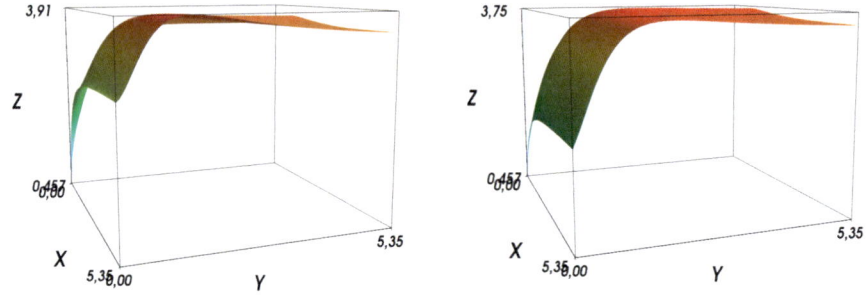

Figure8
Numerical solution for $\theta = 0.25$ (*left*) and $\theta = 0.75$ (*right*) with $p = 0.3$ in the progressive transformation case

Table 6 summarizes the numerical results for five refinement levels from a uniform initial mesh with 289 nodes and 512 triangular elements. Moreover, the same stopping test in the relative quadratic error between two iterations and the same parameter β as in Example 1 are chosen. Again, the ALAS algorithm exhibits a much shorter computational time (t) and less iterations (I).

Figure 4 shows the meshes after five refinement levels. As in previous examples, the one corresponding to ALAS seems to better follow the free boundary of the problem. The associated numerical solutions at this refinement level are shown in Fig. 5. As in the previous examples, the mesh corresponds to the bounded domain ($\mathcal{F}(\Omega)$), while the solution is presented, jointly with the obstacle, over a bounded

Figure 9
Free boundary for $\theta = 0.25$ (*left*) and $\theta = 0.75$ (*right*) with $p = 0.3$ in the progressive transformation case

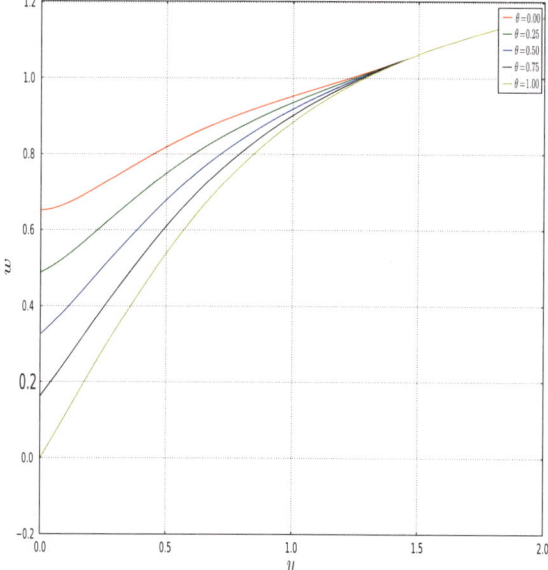

Figure 10
Solution $w(\cdot, \theta) = v(1, \cdot, \theta)$ for different θ with $p = 1$ in the progressive transformation case

subdomain of the unbounded original domain Ω. In this case, the only theoretically stated qualitative properties are the bounds indicated in FAGHLOUMI (2003), ACCIÓN *et al.* (2010), which have been verified by the computed numerical solutions.

4.2. Examples with Progressive Transformation

In this section, we consider several examples in which the effect on the environment is not instantaneous. Following SCHEINKMAN and ZARIPHO-POULOU (2001), we consider the additive HARA utility functions U given by expression (5) for different values of p, where $1 - p$ represents the constant relative risk aversion parameter.

We consider the same parameters as in Example 2 in the instantaneous case (see Table 3). In this setting, we first consider the case $p = 1$, which corresponds to risk-neutrality and a linear utility function. For this case, Fig. 6 shows the computed numerical solution v for the values $\theta = 0.25$ and 0.75, associated with different initial transformed fractions of the environment. For each value of θ the corresponding regions $v_\theta = 0$ (blue) and $v_\theta < 0$ (red) are shown in Fig. 7, with the free boundary separating both regions. Actually, we obtain these regions by considering $t = 1 - \theta$ and representing the regions where $u(\cdot, t) = \psi(\cdot, t)$ (blue) and $u(\cdot, t) > \psi(\cdot, t)$ (red).

The results have been obtained for a discretization time step $\Delta t = 0.025$ and a uniform mesh with 4,225 nodes and 8,192 triangular elements. The parameters for the ALAS algorithm are the same as in the case of instantaneous effects.

Next, Figs. 8 and 9 show the analogous computed results for the case $p = 0.3$, thus corresponding to a nonlinear HARA utility function and the presence of risk aversion.

On the other hand, as indicated in SCHEINKMAN AND ZARIPHOPOULOU (2001), the homogeneity of degree p of U implies the same homogeneity for the value function v. Thus, for utilities given by (5) this homogeneity allows us to reduce the dimension of the problem and analyze some qualitative properties of the function

$$w(y, \theta) = v(1, y, \theta), \qquad (38)$$

that satisfies $v(x, y, \theta) = x^p w(y/x, \theta)$ for $x > 0$. We have used some of these properties to validate the numerical techniques proposed here. In particular, the free boundary that separates the region where $w_\theta < 0$ from the region where $w_\theta = 0$ can be parameterized by a curve $y = g(\theta)$. The first region is characterized by $y \leq g(\theta)$ and no action should be taken in the associated singular control problem, while the second region corresponds to $y > g(\theta)$, and the optimal policy is to jump to θ_0 such that $y = g(\theta_0)$. Moreover, the authors prove that

$$w(y, \theta) = w(y, 1), \qquad y \in [g(1), +\infty), \qquad (39)$$

for all $\theta \in (0, 1)$. This means that for all values of θ, the functions $w(., \theta)$ meet at the point $y = g(1)$ and take the same value after this point, as it is clearly illustrated in Fig. 10 for $p = 1$. Moreover, the observed results are in agreement with the estimated value of $g(1) = 1.4945$ in SCHEINKMAN and ZARIPHO-POULOU (2001). In the region where $y > g(1)$ the optimal policy leads to full conversion of the environment. On the other hand, when $y < g(0)$ the alternative project should not start until their benefits do not exceed those ones of the environment by a certain amount. For the risk-neutrality case ($p = 1$) the identity $g(0) = g(1)$ holds.

5. Conclusions

In this paper, we address the numerical solution of investment problems with instantaneous and progressive irreversible effects on the environment. In the instantaneous case, an obstacle problem associated with a degenerated elliptic equation is numerically solved. As illustrated by the numerical tests, more efficient methods than in the previous work of ACCIÓN et al. (2010) are proposed. More precisely, the computational time of the ALAS algorithm is much shorter than the projected Gauss–Seidel. In the progressive case, following the formulation proposed in DÍAZ et al. (2007), the problem is formulated in terms

of the subsequent solution of parameter dependent elliptic boundary value problems and an evolutive obstacle problem. After the time discretization, at each time step, an obstacle problem associated with a degenerated elliptic equation is posed and solved with the same numerical techniques as in the instantaneous case. Numerical examples are partially validated with the qualitative properties obtained in SCHEINKMAN and ZARIPHOPOULOU (2001) for the case of additive HARA utility functions. These properties are mainly based on a certain homogeneity property. Notice that the proposed numerical techniques can be applied to more general cases in which this reduction cannot be applied. However, the limit case $p \to 0$ corresponding to unbounded utility at the origin will require further treatment from the theoretical and numerical point of view.

Acknowledgment

This work is partially supported by MICINN (MTM2010-21135-C02-01) and by Xunta de Galicia (Ayuda CN2011/004 cofinanced with FEDER funds).

REFERENCES

A. ACCIÓN, I. ARREGUI, C. VÁZQUEZ,, *A Numerical solution of a free boundary problem associated to investments with instantaneous irreversible effects.* Appl. Math. Comput., *215* (2010), 3461–3472.

K.J. ARROW, A.C. FISHER, *Environmental preservation, uncertainty and irreversibility,* Quaterly J. Econ., *88* (1974), 312–319.

A. BENSOUSSAN, J. L. LIONS, Application des Inégalités Variationnelles en Control Stochastique, Dunod, Paris, 1978.

A. BRANDT, C. W. CRYER, *Multigrid algorithms for the solution of linear complementarity problems arising from free boundary problems,* SIAM J. Sci. Stat. Comput., *4* (1983), 655–684.

M. C. CALVO-GARRIDO, A. PASCUCCI, C. VÁZQUEZ, *Mathematical analysis and numerical methods for pricing pension plans allowing early retirement,* SIAM J. Appl. Math., *73* (2013), 5, 1747–1767.

M.G. CRANDALL, H. ISHII, P.L. LIONS, *User's guide to viscosity solutions of second order partial differential equation,* Bull. Amer. Math. Soc., *27* (1992), 1–42.

G. DÍAZ, J. I. DÍAZ, C. FAGHLOUMI, *On an evolution problem associated to modelling of incertitude into the environment,* Nonlinear Analysis: Real World Applications, *8* (2007), 399–404.

J. I. DÍAZ, C. FAGHLOUMI, *Analysis of a degenerate obstacle problem on an unbounded set arising in the environment,* Appl. Math. Optim., *45* (2002), 251–267.

K. Dixit, R. S. Pindyck,, *A Investment under Uncertainty*, Princeton University Press, Princeton. NJ, 1994.

A. C. Faghloumi, *Modelización y tratamiento matemático de algunos problemas en medio ambiente*, Ph. D. Dissertation, Universidad Complutense de Madrid, 2003.

A.C. Fowler, *Mathematical Geoscience*, Springer, 2011.

R. Glowinski, J. L. Lions, R. Tremolières, Analyse Numérique des Inéquations Variationnelles, Dunod, Paris, 1976.

H. W. Hoppe,, *R Multigrid methods for variational inequalities*. SIAM J. Numer. Anal., *24* (1987), 1046–1065.

R. Kangro, R. Nicolaides, *Far field boundary conditions for Black-Scholes equations*, SIAM J. Numer. Anal., *38* (2000), 1357–1368.

T. Kárkkáinen, K. Kunisch, P. Tarvainen, *Augmented Lagrangian active set methods for obstacle problems*, JOTA, *119* (2003), 499–533.

D. Kinderlehrer, G. Stampacchia, *An Introduction to Variational Inequalities and their Applications*, Academic Press, New York, 1980.

R. Kornhüber, *Monotone multigrid methods for elliptic variational inequalities*. Numer. Math., *69* (1994), 167–184.

P. L. Lions, B. Mercier, *Approximation numérique des équations de Hamilton-Jacobi-Bellman*, RAIRO Anal. Numér., *14* (1980), 369–393.

M. C. Rivara, *Algorithms for refining triangular grids suitable for adaptive and multigrid techniques*, Int. J. Numer. Meth. Engrg., *20* (1984), 745–756.

J. A. Scheinkman, Th. Zariphopoulou, *Optimal environment management in the presence of irreversibilities*, J. Econom. Theory *96* (2001), 180–207.

(Received March 29, 2014, revised July 8, 2014, accepted July 10, 2014, Published online August 9, 2014)

Reprinted from the journal

Pure Appl. Geophys. 172 (2015), 167–179
© 2014 Springer Basel
DOI 10.1007/s00024-014-0897-5

Volume, Surface, Connectivity and Size Distribution of Soil Pore Space in CT Images: Comparison of Samples at Different Depths from Nearby Natural and Tillage Areas

F. J. Muñoz-Ortega,[1] F. San José Martínez,[1] and F. J. Caniego Monreal[1]

Abstract—The study of soil structure, i.e., the pores, is of vital importance in different fields of science and technology. Total pore volume (porosity), pore surface, pore connectivity and pore size distribution are some (probably the most important) of the geometric measurements of pore space. The technology of X-ray computed tomography allows us to obtain 3D images of the inside of a soil sample enabling study of the pores without disturbing the samples. In this work we performed a set of geometrical measures, some of them from mathematical morphology, to assess and quantify any possible difference that tillage may have caused on the soil. We compared samples from tilled soil with samples from a soil with natural vegetation taken in a very close area. Our results show that the main differences between these two groups of samples are total surface area and pore connectivity per unit pore volume.

1. Introduction

A soil matrix is commonly viewed as a hierarchical system made of primary and secondary particles (e.g., aggregates) of different sizes that shape soil pore space. Its structure is believed to be among the main factors controlling soil processes and functioning (Dexter 1988; Revil and Cathles 1999). In particular, important physical and biological processes within the soil–plant–microbial system, such as microbial population dynamics, nutrient cycling, diffusion, mass flow and nutrient uptake by roots, are affected by soil pore structure (Young and Crawford 2004).

Land use and management have long been known to affect soil structure (Gale and Cambardella 2000; Six *et al.* 2000; Lal 2002). Recent studies showed

that contrasting land use and management practices can also lead to marked differences in soil pore structure (Peth *et al.* 2008; Kravchenko *et al.* 2011; Wang *et al.*, 2012). Such differences can potentially affect processes within the soil matrix. It is well known that differences in land use and management generate notable changes in soil physical and hydraulic properties, including changes in soil organic matter content, soil porosity, hydraulic conductivity and water retention (Brye and Pirani, 2005).

Recent advances in X-ray CT of undisturbed soil columns provide 3D images of the soil interior (Saucier and Muller 1999; Gantzer and Anderson 2002; Posadas *et al.* 2003; Dathe *et al.* 2006; Gibson *et al.* 2006; Nunan *et al.* 2006; Chun *et al.* 2008; San José *et al.* 2010, 2013). This is a significant step in characterizing heterogeneity of soil environments at micro-scales, as it allows obtaining exact information of the physical structure of the soil matrix.

Image analysis may be seen as a body of tools that facilitates extraction of quantitative information from images such as X-ray tomograms. Image analysis may be subdivided into five steps: display, filtering, segmentation, transformation and measurement of image features (Glasbey and Horgan, 1995). The display renders an array of pixels values as a picture on a computer screen. Filters enhance images by applying transformations based on groups of pixels to reduce noise and emphasize edges. Segmentation divides an image into regions that correspond to different objects or parts of objects. Morphological transformations are used to study the shape of objects. And measurements extract quantitative information from the enhanced images. Mathematical morphology (Serra 1982) is a theory that is based in the assumption that images consist of structures that

---

[1]   Dpt. of Applied Mathematics in Agronomic Engineering, Universidad Politécnica de Madrid (UPM), Madrid, Spain. E-mail: f.j.munoz.ortega@upm.es; fernando.sanjose@upm.es; j.caniego@upm.es

may be handled by set theory. It provides useful morphological transformations and measurement procedures for quantifying the geometry of images (SAN JOSÉ et al. 2013 and references therein).

The objectives of this study are to characterize the geometry of the soil pores by using several geometric 3D image measurements. The second objective is to evaluate whether different soil management (natural vegetation vs. tillage) as well as the depth at which sample was taken, exhibit significantly different geometric measurements. Finally, we study how these measures are related and how these measures relate to soil type and depth at which samples were taken.

2. Theory: Morphological Image Analysis

The theory of mathematical morphology provides a number of tools that allow us to characterize geometric objects, such as the background components of a binary image (the set of pixels with value 1), called objects. The fundamental concepts of mathematical morphology are morphological transformations; they change the size or the shape of the object under study. A quantitative description of the object is obtained from the measurement of a functional (e.g., volume) over the object properly transformed. Morphological transformations are formulated in the language of set theory in which, for a binary image, these sets correspond to the objects (set of pixels with value 1) in the image. To morphologically transform a set A another set B is needed, usually a sphere, which is called a structuring element. The basic morphological transformations are two: dilation and erosion. If we call B^d as the structuring element with size d then the dilation of A is defined as:

$$\mathrm{Dil}(A)_d = \{x - y : x \text{ in } A, \ y \text{ in } B^d\}$$

The main effect of dilation of A is an "expansion" of the boundary of A. The erosion of A is defined as:

$$\mathrm{Ero}(A)_d = \{x \text{ in } R^n : (B^d + x) \text{ subset } A\}$$

where R^n is the n-dimensional Euclidean space where the objects are defined. The main effect of erosion on A is the "contraction" of the border of A. Another

transformation is morphological opening which is a combination of the two basic operations. The opening of A by B^d is defined as:

$$\mathrm{Op}(A)_d = \mathrm{Dil}(\mathrm{Ero}(A)_d)_d$$

The effects of the opening of A is the smoothing of contours, the breaking of narrow necks and the removal of small objects (objects with size smaller than the size of the structuring element). If the set on which the opening operation is performed is formed by components of different sizes (like the pores on a porous medium image) then a sequential use of the morphological opening with increasing sizes of the spherical structuring element simulates a process of "sieving," since for each structuring element size all the smaller components disappear after the operation (see Fig. 1). This procedure is usually called morphological granulometry and, following VOGEL (2002), in the study of porous media images the size criteria used almost perfectly reflects the idea of the hydraulic diameter of a pore. A functional measurement such as the volume, expressed in relation to the size of the structuring element d, provides a measure of the distribution of the functional in the set. Therefore, for an image of a porous medium a function of pore size distribution can be defined (VOGEL, 2002) as:

$$F(d) = \mathrm{Vol}(\mathrm{Op}(A)_d)$$

for differents values of d. Where Vol means volume.

3. Materials and Methods

3.1. Soil Study and Sampling

Undisturbed soil samples were taken in December 2008 in two nearby areas within the experimental farm "El Encín" this farm belongs to the Instituto Madrileño de Investigación y Desarrollo Rural Agrario y Alimentario (IMIDRA) located in Alcala de Henares, Madrid, Spain. The first sampled area (labeled T), with coordinates (WGS84) 40°31'29"N, 3°17'30"W, is an agricultural area which at the time of sampling had a cotton crop (Gossypium sp.). Soil in this area was identified (INIA, 1977) as Calcic Haploxeralf, adapted to the USDA classification (SOIL

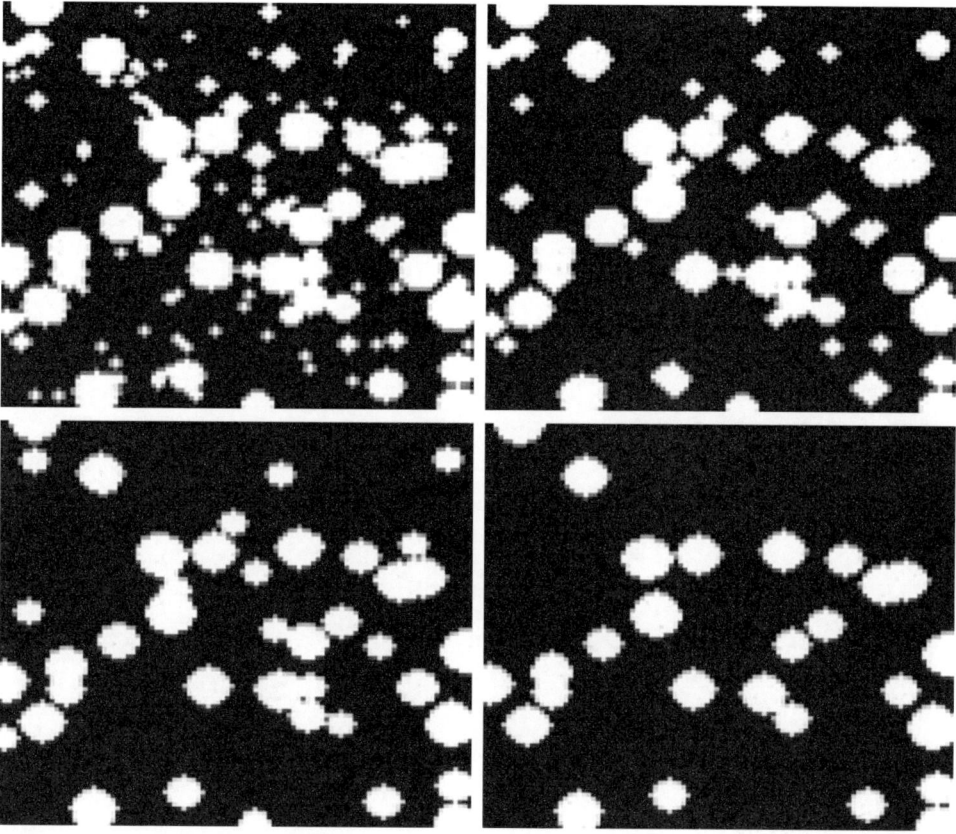

Figure 1

Example of the effect of morphological granulometry on an image of a medium consisting of objects of different sizes. In the different stages
of the process it can be seen that the objects disappear according to its increasing size

SURVEY STAFF 2010); the textural family of the soil is clay loam with the following proportions: 3.4 % sand, 35.8 % silt, 60.8 % clay and 1.62 % organic matter. The second sampled area (labeled *N*), with coordinates (WGS84) 40°31′6″N, 3°17′21″W, is a natural site located on a terrace of the Henares river with riparian vegetation (Tamarix sp., Ulmus sp.). Soil in this second area was identified (INIA, 1977) as typic xerofluvent, adapted to the USDA classification (SOIL SURVEY STAFF 2010); the textural family of the soil is silty clay loam with the following proportions: 12.3 % sand, 60.8 % silt, 26.9 % clay and 0.85 % organic matter.

In order to extract and preserve the undisturbed samples methacrylate cylindrical containers were used and they were introduced directly into the soil by manual drilling. The dimensions of the containers are: 3 cm high, 2.6 cm inner diameter and 2 mm thickness. Soil samples were taken at three depths

about 10 cm apart, the first group of samples being flush with the surface. In total 24 samples were extracted, 12 of each soil type with a total of 4 replicates in each of the three depths sampled in each area.

3.2. Image Acquisition and Processing

Soil samples were scanned using an X-ray CT scanner at the Gregorio Marañon Hospital's Laboratory of Medical Imaging and Experimental Medicine in Madrid (Spain). The tomograph used an energy of 50 keV for scanning the samples. The output data obtained after the reconstruction phase is a set of 3D unsigned 16-bit RAW images. The size of the images are 1,392 slices, each slice with 1,600 × 1,600 pixels. The spatial resolution of the images is 0.03 mm/pixel side. The spacing between sections is 0.03 mm so the 3D image has the same resolution in all three

Table 1

Labeling of images obtained according to the sampling area

	Area *T* (tillage)	Area *N* (natural)
0 cm (surface)	T1_1, T1_2, T1_3	N1_1, N1_2, N1_3, N1_4
−10 cm	T2_1, T2_2, T2_3, T2_4	N2_1, N2_2, N2_3, N2_4
−20 cm	T3_1, T3_2, T3_3, T3_4	N3_1, N3_2, N3_3

spatial directions. Due to an error in the acquisition/reconstruction of data only 22 of the 24 samples have a final reconstructed image. Table 1 summarizes labeling of the images obtained according to the sampling area. This labeling is of the form $Ln1\_n2$ where L is the treatment: T (tillage) or N (natural), $n1$ is the depth 1 (shallow), 2 (intermediate), 3 (the deepest) and $n2$ is the repetition.

Image processing was performed with the public domain program ImageJ version 1.47 v developed at the National Institute of Health (RASBAND 1997). The image format was changed from RAW to TIFF.

The first image processing was thresholding (segmentation) in order to classify the entire set of pixels of the image into two regions, the region of interest (foreground) and the region of background. The foreground in this work corresponds to the porous phase of the sample and the background region corresponds to the solid phase.

A local (adaptive) thresholding algorithm was used to implement the process of thresholding. It has been shown that these types of algorithms are stable

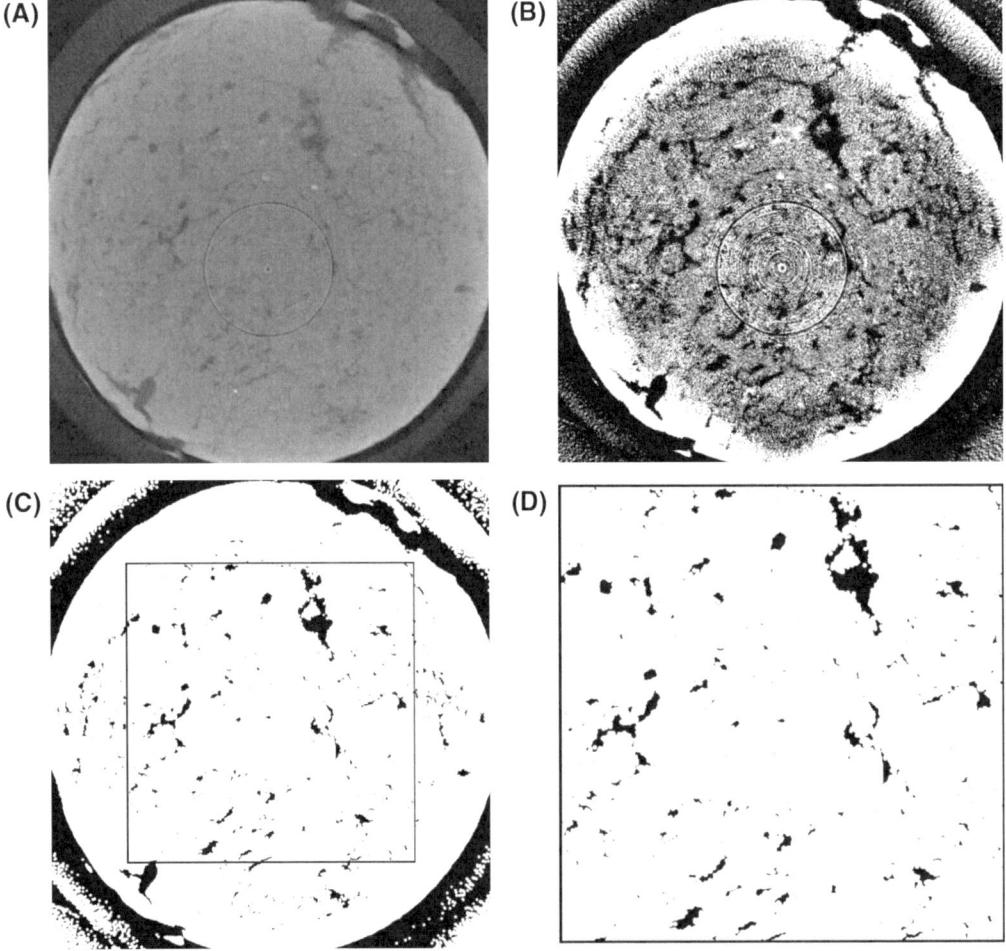

Figure 2
Procedure performed on *T2_2* sample. **a** Image acquisition; **b** local threshold; **c** maximum filtering; **d** resampling

and are precise (IASSONOV 2009). The local segmentation algorithm used is based on the mean operation. Thresholding works as follows: for each pixel of the image a threshold is calculated; if the pixel value is below the threshold then the pixel is classified as background, otherwise it is classified as an object (foreground). Calculation of the threshold for each pixel is made by examining pixel values of the surroundings of the pixel to be replaced. For this work the surroundings have been chosen for a given radius of 100 pixels which corresponds to an actual sample size of 3 mm. Figure 2b shows the result of applying this thresholding on the original image of Fig. 2a.

Images were filtered in order to reduce Gaussian noise and artifacts. These imperfections arise during the tomographic scan due to defects in the X-ray beam and the detector used (observe the concentric ring-shape structures on the inside of the sample in Fig. 2a). The filter chosen was the spatial filtering that uses the maximum function with a radius of action r equal to 2 pixels. The filtering operation operates on each image pixel as follows: each pixel x in the original image is replaced by the maximum of the values of the pixels covered by the ball of radius r when it is centered on the pixel x. Figure 2c shows the result of this filter where it can see that the noise and ring artifacts have disappeared.

An inner region with cubic shape of size $512 \times 512 \times 512$ pixels was selected from each 3D picture for the study of the geometric characteristics. This region corresponds to a cube whose edge has an actual length of 15.36 mm. This size was chosen to avoid taking pixels of the sample container or of soil at the perimeter. Figure 2c shows a section of the region where it will resample the marked square. Figure 2d shows the results of resampling.

3.3. Computation of Geometrical Features

Total pore surface, specific pore surface, total pore connectivity, specific pore connectivity, proportion of macropores (POM) and pore size distribution was calculated for each image.

For the computation of porosity, pore surface (total and specific) and connectivity (total and specific) we used computer code developed in

MICHIELSEN and DE RAEDT 2001; the authors describe a method for calculating the volume, surface and connectivity of an object in a 3D digital image. The authors compute the number of pixels (n_c), the number of vertices (n_v), the number of edges (n_e) and the number of faces (n_f) of the object in the image, so that:

$$\text{Volume} = n_c$$
$$\text{Surface} = -6 \cdot n_c + 2 \cdot n_f$$
$$\text{Connectivity} = -n_c + n_f - n_e + n_v$$

In this way porosity is obtained by dividing the volume of the object (in our case the object is the set of pores) by the total volume of the image (the product of its dimensions $512 \times 512 \times 512$). The total pore surface (mm^2) was obtained by multiplying the surface calculated in this method by the square of the resolution surface, and the specific pore surface (mm^{-1}) by dividing total pore surface by the volume of the object expressed in mm^3 (i.e., multiplied by the cube of the resolution). Total connectivity corresponds to the connectivity obtained in the program and specific connectivity (mm^{-3}) is obtained by dividing the total connectivity by the volume of the object expressed in mm^3.

For the distribution of pore sizes in this work we followed the process of morphological granulometry (VOGEL 2002), where the size of a spherical structuring element that has been used follows the sequence: 1, 3, 5, 7, ... (pixels) that correspond to equivalent pore diameters given in μm of 30, 90, 150, 210, ...

POM is defined as the percentage of pores with a diameter >75 μm (BREWER 1964); in this paper POM corresponds to the sum of the relative frequencies of pore volumes with a diameter greater than or equal to 3 pixels.

3.4. Statistics

Statistical analyzes were performed on the variables porosity, POM, total pore surface (mm^2), specific pore surface (mm^{-1}), total pore connectivity (dimensionless) and specific pore connectivity (mm^{-3}) taking treatment (soil use) and sample depth into account. Statistical analysis utilized Statgraphics Centurion XVI (StatPoint Technologies, Inc.) has

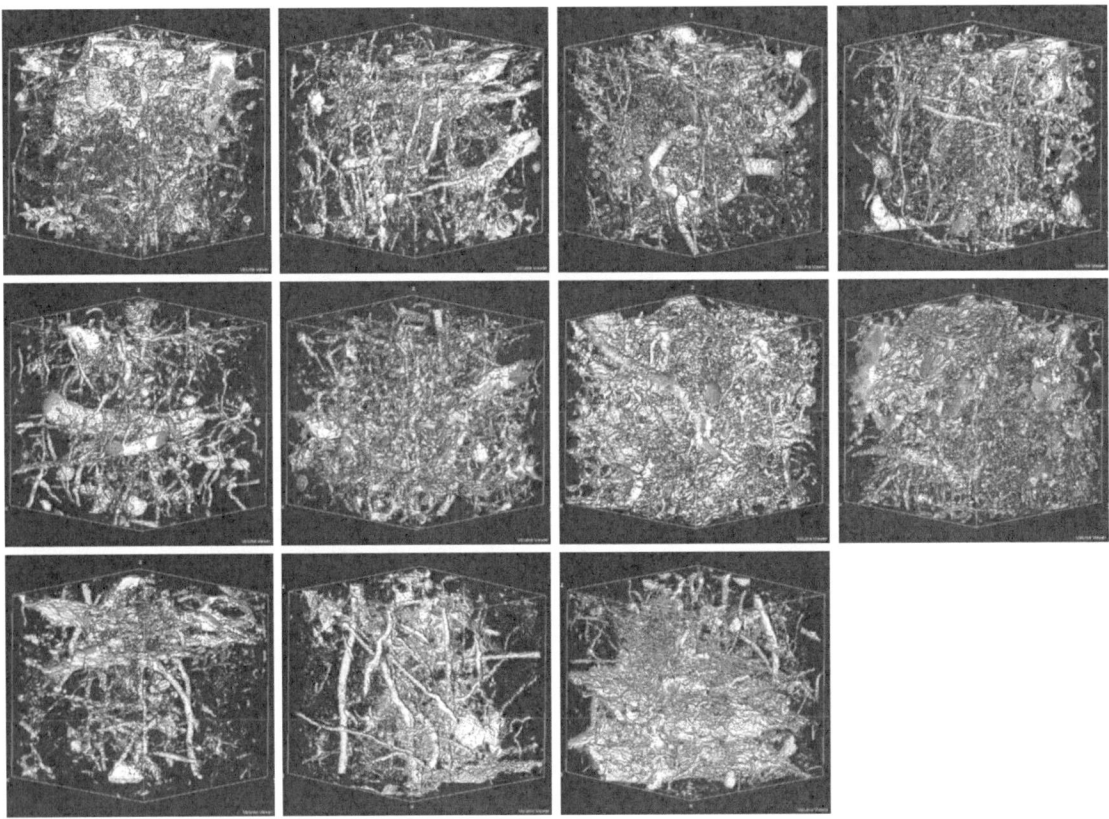

Figure 3
3D reconstructions of natural soil samples (*N*)

been used. Significance tests for the evaluation of the effects of the treatment factors and depth have been performed using the ANOVA multifactorial analysis for all variables. The ANOVA test evaluates factors that have a statistically significant effect on each dependent variable. ANOVA analysis for the contribution of each factor is measured by the sum of squares Type III which is measured by removing the effects of other factors. Pearson correlations between each pair of variables have been measured for the entire set of data (not rated by factors) in order to evaluate whether there are correlations significantly different from zero.

Comparisons between pairs of pore size distributions was performed using the Kolmogorov–Smirnov test for paired data sets, using the program MATLAB R2010a (MathWorks, Inc.). This test assumes as a null hypothesis that the two distributions of data sets come from the same continuous distribution and measures the maximum distance between the cumulative distributions for each value of the distribution.

4. Results and Discussion

4.1. Visualization of Pore Space

Figures 3 and 4 show three-dimensional reconstructions of pore space in 22 soil samples. In general, it seems to us that *N* samples have more pores than *T*. In addition, in the natural soil (Fig. 3) the samples *N1_1*, *N2_3*, *N2_4* and *N3_3* show a high porosity; the *N* samples indicate a large number of tube-shaped pores crossing the sample mainly from top to bottom; the pores clearly appear to have a biological origin. Tilled soil samples *T1_1*, *T2_2*, and *T3_1* (Fig. 3) have a high porosity; however, *sample T2_3* appears to have very little as many tubular pores are observed, similar to the natural soil. The

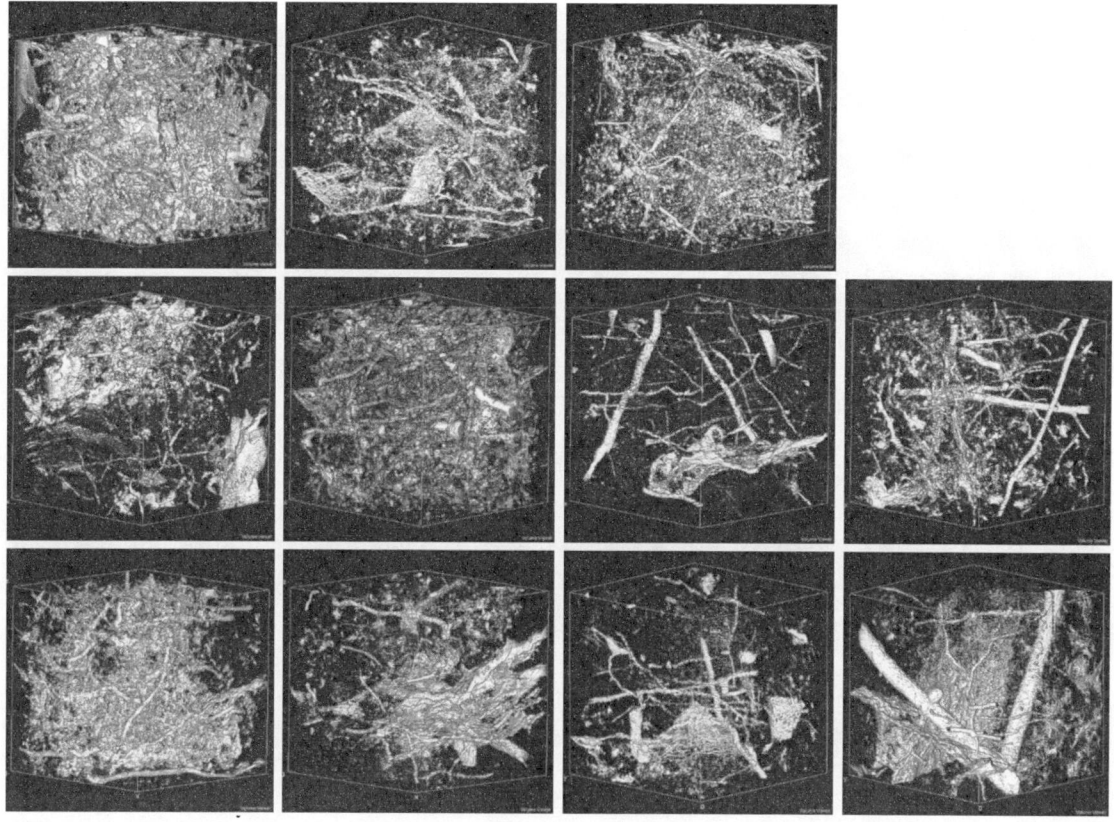

Figure 4
3D reconstructions of tilled soil samples (*T*)

largest pores were observed in *sample T3_4*. The direction of the pores in tilled soil does not seem as clear as in the natural soils.

4.2. Porosity, Surface, Connectivity and Proportion of Macropores

Table 2 lists the characteristics [porosity, POM, total pore surface (mm^2), specific pore surface (mm^{-1}), total pore connectivity (dimensionless) and specific pore connectivity (mm^{-3})] for all soil samples with the mean values and standard deviations of each group of data.

In the second column of Table 2 corresponding to porosity it can be seen that greater porosity occurs in the sample *T1_1* (11.071 %). However, on average, pores are greater for *N* samples (5.333 %) compared to *T* samples (3.071 %). As observed by visual analysis of the reconstructions (Sect. 4.1), high porosity natural treatment samples are *N1_1*, *N2_3*,

N2_4 and *N3_3* and *N1_3*; *T1_1*, *T2_2* and *T3_1* have the greatest porosity among the tillage treatment samples. Small differences in porosity show a decrease with depth; for *T* the porosity is higher in the surface samples (4.766 %) than in the other depths (2.424 and 2.444 %). In *N* samples the average porosity is 3/10 higher in the intermediate depth (5.910 %) than in the surface (5.686 %) and these two are greater than the mean of the deepest group (4.093 %).

If we relate data porosity with POM it can be seen that the samples with greater porosity (e.g., *T1_1*) have a high POM (0.897). Moreover, this sample has the greatest POM; therefore, it seems that higher porosity is due to pores having a large size and not to a high amount of smaller pores (this can be checked with the value of total pore connectivity which increases with the number of pores and decreases with the number of tunnels). Smaller values of POM are at about 0.5 (samples *T1_2*, *T2_3*, *T2_4* and *T3_3*).

Table 2

Characteristics of all soil samples analyzed for the two treatments and the three different levels of depth

Sample	Porosity (%)	POM	Sup total (mm^2)	Sup/volPoro (mm^{-1})	Total connectivity (adim)	Specific connectivity (mm^{-3})
T1_1	11.071	0.897	8,011.390	19.969	13,283.000	33.110
T1_2	0.810	0.520	1,811.018	61.677	27,851.000	948.509
T13	2.416	0.640	4,676.020	53.405	39,536.000	451.541
Mean T1	4.766	0.686	4,832.809	45.017	26,890.000	477.720
SD T1	5.519	0.192	3,103.158	22.083	13,152.857	458.261
T2_1	2.958	0.762	3,721.219	34.709	16,770.000	156.419
T2_2	4.450	0.777	6,066.432	37.614	24,029.000	148.990
T2_3	0.915	0.592	1,732.554	52.224	9,740.000	293.593
T2_4	1.373	0.563	2,821.415	56.689	19,019.000	382.135
Mean T2	2.424	0.673	3,585.405	45.309	17,389.500	245.284
SD T2	1.610	0.112	1,843.066	10.784	5,933.994	112.889
T31	3.292	0.692	5,691.022	47.700	22,772.000	190.868
T3_2	2.982	0.730	4,095.443	37.902	13,415.000	124.151
T3_3	0.924	0.585	1,761.242	52.600	11,823.000	353.099
T3_4	2.578	0.771	3,280.464	35.110	15,267.000	163.399
Mean T3	2.444	0.694	3,707.043	43.328	15,819.250	207.879
SD T3	1.055	0.080	1,638.611	82.07	4,844.108	100.610
Mean T	3.071	0.684	3,969.838	44.509	19,409.545	295.074
SD T	2.896	1.260	2,015.107	12.390	8,705.074	250.801
N1_1	9.137	0.821	9,453.758	28.552	7,626.000	23.032
N1_2	3.729	0.700	5,814.484	43.026	28,244.000	208.998
N1_3	5.487	0.782	6,637.091	33.379	23,543.000	118.402
N1_4	4.389	0.750	5,953.207	37.429	27,939.000	175.656
Mean N1	5.686	0.763	6,964.635	35.596	21,838.000	131.522
SD N1	2.412	0.051	1,697.926	6.140	9,715.057	81.430
N2_1	3.432	0.750	4,493.565	36.130	17,403.000	139.926
N2_2	4.406	0.710	6,810.233	42.655	20,338.000	127.385
N2_3	8.601	0.800	10,056.866	32.264	1,226.000	3.933
N2_4	7.199	0.776	9,049.522	34.687	13,765.000	52.762
Mean N2	5.910	0.759	7,602.547	36.434	13,183.000	81.002
SD N2	2.402	0.039	2,477.293	4.443	8,412.510	64.189
N3_1	4.546	0.806	5,081.254	30.846	27,936.000	169.584
N3_2	2.778	0.723	4,124.254	40.961	20,925.000	207.824
N3_3	4.956	0.768	6,236.258	34.723	22,459.000	125.049
Mean N3	4.093	0.766	5,147.255	35.510	23,773.333	167.486
SD N3	1.157	0.041	1,057.548	5.104	3,685.666	41.427
Mean N	5.333	0.762	6,700.954	35.877	19,218.545	122.959
SD N	2.095	2.067	2,001.703	4.758	8,706.446	69.978

It should be noted that these are *T* samples and indicate a ratio of approximately 50 % of macropores. For *N* samples, the values are in the range 0.700–0.821, so the minimum POM in these samples is 70 %; therefore, as can be observed in the group mean values, POM in *N* samples is greater than in *T* samples.

Surface samples indicate that, although in terms of absolute total pore surface (mm^2) the highest values on average correspond to *N* samples (6,700.9 in *N* vs. 3,969.8 in *T*), in relative terms the size of pores and specific pore surface (mm^{-1}) in *T* samples

are higher (44.509 in *T* vs. 35.877 in *N*); all mean values in *T* samples are greater than 43 mm^{-1} while all averages values for *N* are below 37 mm^{-1}. This suggests that, although there are more pore surfaces in *N*, the specific surface area per pore volume is greater in *T*; that is, for the same volume of pores, pores in *T* samples provide the largest surface area. This can be interpreted as the pores in *T* samples are irregular or there are a great number of small pores (except for *T*1_1). The latter interpretation is in agreement with the POM data; connectivity data can be used to test it.

Table 3

Pearson correlations between each pair of variables to the entire set of samples

	Porosity	POM	Total pore surface	Specific pore surface	Total pore connectivity	Specific pore connectivity
Porosity	1	0.826***	0.9143***	−0.8182***	−0.3723*	−0.6632***
POM	0.826***	1	0.743***	−0.9841***	−0.2462	−0.8294***
Total pore surface	0.9143***	0.743***	1	−0.6931***	−0.2569	−0.6759***
Specific pore surface	−0.8182***	−0.9841***	−0.6931***	1	0.3405	0.8264***
Total pore connectivity	−0.3723	−0.2462	−0.2569	0.3405	1	0.4575**
Specific pore connectivity	−0.6632***	−0.8294	−0.6759***	0.8264***	0.4575**	1

* <0.1, ** <0.05, *** <0.001

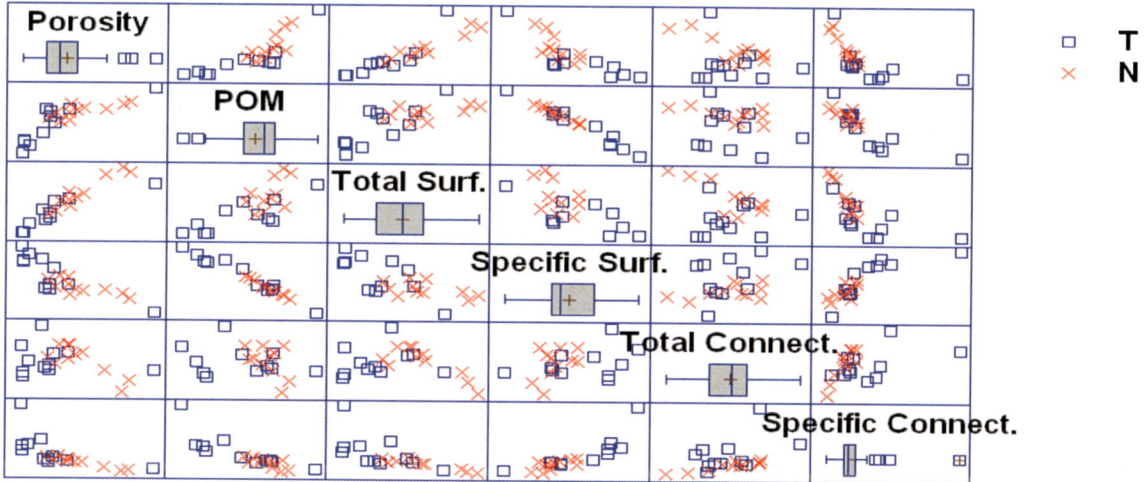

Figure 5
Dispersion matrix that shows possible correlations between groups of samples

In absolute terms, the two treatments on average have approximately the same total pore connectivity (dimensionless), around 19,000. To know which is more connected it is recalled that a sample will be more connected, have more tunnels, branches and closed loops when its connectivity value is lower. The $T1\_3$ sample exhibits an extreme value of this total connectivity, resulting in a high positive value (39,536.00). This indicates a high number of connected components (isolated pores) and/or a small value of tunnel branches and loops. The $N2\_3$ sample has the minimum value of total pore connectivity (dimensionless) (1,226.00) In the same way that high porosity is visualized in the sample we think that all this porosity forms a low value of connected components and most of the branches and loops are connected. For $T1\_1$, which has a high POM and a high value of porosity, since its connectivity is low it may indicate that porosity is due to a few number of large pores (which can have many ramifications and tunnels) and is not due to a higher amount of smaller pores (which would increase the value of connectivity).

Regarding surface values, pores in T samples are either more irregular or are smaller and more numerous (this second option is more consistent with the POM data) and specific connectivity in T samples is greater than in N samples (i.e., for an equal volume of pores, pores in N either have fewer components or have more branches versus pores in T samples in which many disconnected (without branching structures form) pores appear (without branching structures).

Table 3 shows the relationship between each pair of variables measured for the entire set of samples quantified by Pearson correlations. Figure 5 shows

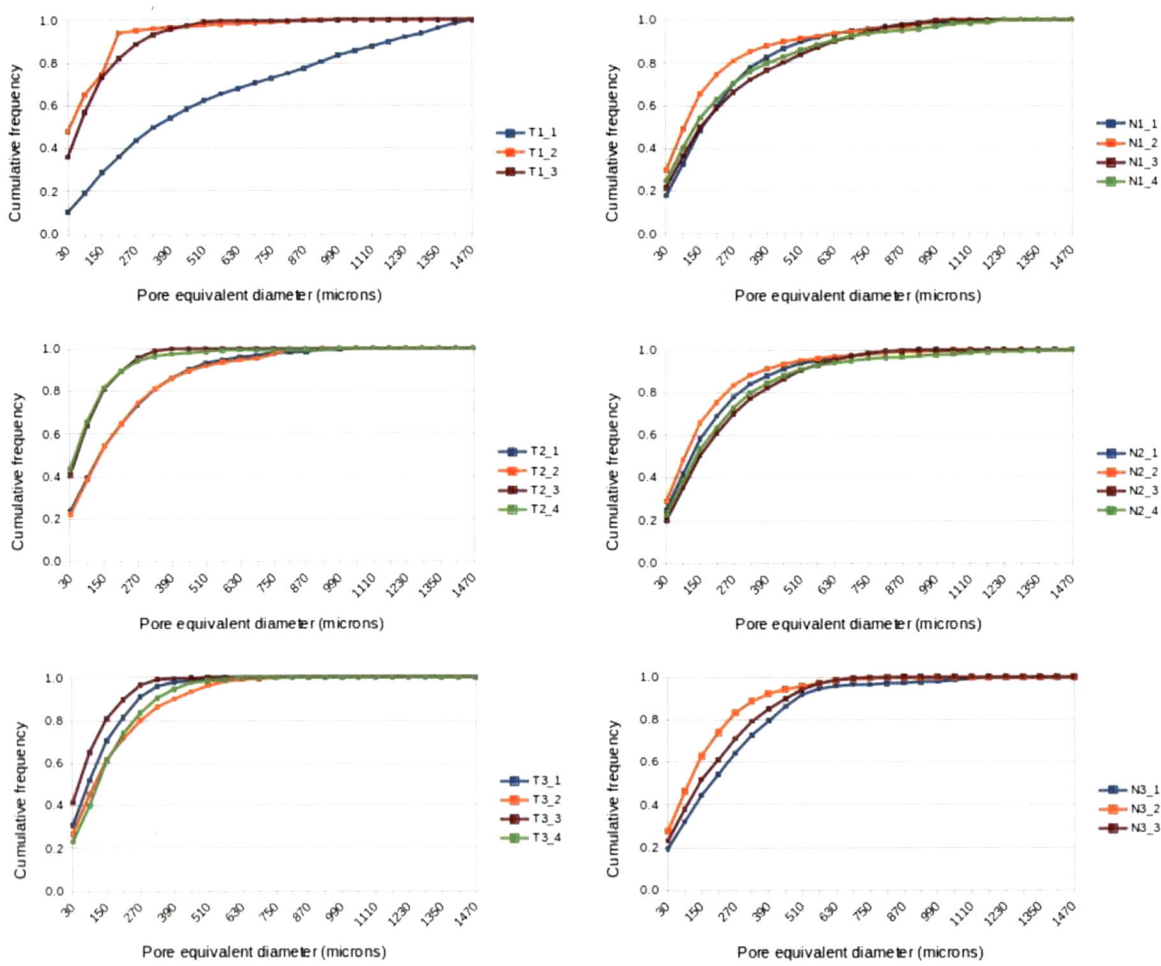

Figure 6
Distribution of cumulative relative frequencies of pore size for each group of samples, expressed in terms of equivalent diameter (μm)

the scatter plots of these relationships, identifying samples of each treatment. Significant relationships found at a confidence level of 95 % are listed following Table 3.

- Porosity and total pore surface: as expected, the higher porosity, the higher the pore surface.
- Porosity and specific pore surface: this relation is negative; the greater porosity, the lower the pore surface. This relationship is also expected because when the pores are larger the surface area per unit volume of pore is lower. Moreover when the number of pores is increased it begins to coalesce and share walls so that the surface pores decreases.
- Specific pore connectivity and specific pore surface: for a given pore, the more irregular the pore

is, the more branches and loops it has and, hence, greater surface.
- Porosity and POM: a higher porosity in absolute terms correlates to a higher proportion of large pores.
- Total pore surface and POM: if there is greater surface area of pores there is a higher proportion of large pores.
- Total pore surface and specific pore surface: this relation is negative; a higher total pore surface area correlates with a smaller pore surface area per unit area. That is, if there are more pores they are more regular and/or larger and, hence, its specific surface area is smaller.
- Total pore connectivity and specific pore connectivity: the less connected a pore is the less connected the total is.

- Specific pore surface and POM: this relation is negative; when there are pores with a high specific surface value then small pores are more abundant. That is, the greater specific surface areas are small.

4.3. Pore Size Distribution

Figure 6 shows the cumulative relative frequency distribution of pore sizes in each group of samples. These distributions are expressed as a function of equivalent pore diameters obtained from the morphological opening radius used at each stage of morphological granulometry. It can be seen that the N treatment samples exhibit similar behavior: All start between 20 and 30 %, and all reach 75 % of the volume of pores with a diameter about 250 μm and 100 % at 1000 μm. In contrast, T samples have more diversity; T1_1 starts about 10 %, T1_2 at 50 % and the remaining T samples are between 20 and 40 %. The group within T that seems to show less variability is T3, where 75 % is reached in about 200–250 μm. The T1 group shows the singular behavior of the sample T1_1; this sample does not reach 100 % of the pores until it reaches a diameter of 1,470 μm. The sample T1_2 has a singularity; it appears to lack pores between 30 and 150 μm, there is an abrupt change making the curve not grow in the same smooth way as the rest. T2 samples are grouped into two behaviors, one starting at 20 % and another at 40 %.

Table 4 shows statistical results of the Kolmogorov–Smirnov test that compares two sample distributions. Comparisons of all possible pairs of data are shown. Bold type and shading highlight pairs of samples with a statistically significant difference for a confidence level of 95 % or higher. It can be seen that sample T1_1 is statistically different from all others; the rest of the statistical differences are mostly between T and N samples. Not taking sample T1_1 into account, there are 37 of 110 pairs of samples statistically different between groups, and within the same group (either N or T) there are four pairs of samples statistically different from the 100 possible. Furthermore, differences between N and T samples seem to correlate with depth, and so there are 23 over 40 in N1, 16 over 40 in N2 and 5 over 30 in N3.

Table 4

Statistical results of the Kolmogorov–Smirnov test in comparison of pairs of distributions with * if P < 0.1, ** if P < 0.05 and *** if P < 0.01

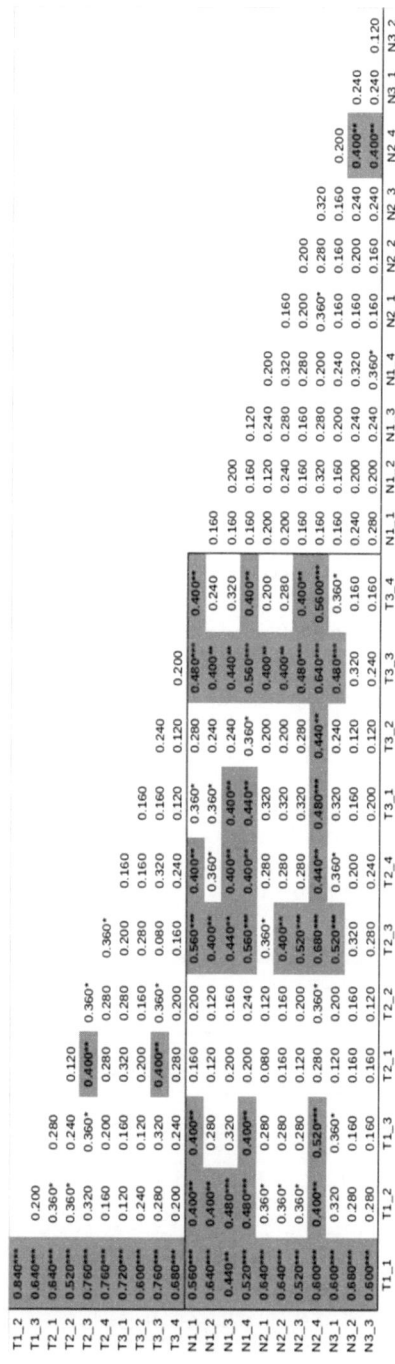

Table 5

*Mean squares (MS) of the multifactorial ANOVA analysis with * if P < 0.1, ** if P < 0.05 and *** if P < 0.01*

	Porosity	POM	Total pore surface	Specific pore surface	Total pore connectivity	Specific pore connectivity
Soil use	23.836*	0.033*	3.69E7***	412.162*	2.18E + 6	181,017.000**
Depth	0.421	0.000	3.59E + 7	4.295	1.43E + 6	36,234.600

4.4. Effects of Soil Use and Depth

Table 5 shows the mean squares values for the multifactorial ANOVA analysis. The value of total pore surface has a P value <0.05 suggesting that factor treatment produces a statistically significant difference for this variable. Furthermore, the difference in specific pore connectivity between T and N samples is also statistically significant.

5. Summary and Conclusions

This work shows the utility of performing different geometric measures of soil pore space using 3D images in order to characterize the soil structure. We characterized in quantitative terms several geometric factors of great interest in the study of biological and physical processes, such as porosity, pore surface area, connectivity of the pores and pore size distribution. For this latter purpose, we needed to use tools of mathematical morphology, which reflect almost perfectly pore size criteria (VOGEL, 2002). Measurements were made directly on the samples without altering them, this being an advantageous technique employed by further research. Our comparisons indicate total pore surface and specific pore connectivity show the greatest differences between the two soil types (natural vegetation and tillage). Iin addition, we quantified the relationships between different geometric measurements for all soil samples demonstrating that most of them are related via high correlation coefficients.

Acknowledgments

This research work was partially funded by Spain's Plan Nacional de Investigación Científica, Desarrollo e Innovación Tecnológica (I + D + I) under ref. AGL2011-25175.

REFERENCES

BREWER, R. 1964 Fabric and mineral analysis of soils. John Wiley and Sons.

BRYE, K.R., and A.L. PIRANI. 2005 Native soil quality and the effects of tillage in the grand prairie region of eastern Arkansas. Am. Midl. Nat. 154, 28 41. doi:10.1674/0003-0031(2005)154 [0028:NSQATE]2.0.CO;2.

CHUN, H.C., D. GIMÉNEZ, and S.W. YOON. 2008 Morphology, lacunarity and entropy of intra-aggregate pores: Aggregate size and soil management effects. Geoderma 146, 83–93. doi:10.1016/j.geoderma.2008.05.018.

DATHE, A., TARQUIS, A.M., and PERRIER, E. 2006 Multifractal analysis of the pore-and solid-phases in binary two-dimensional images of natural porous structures. Geoderma 134(3), 318–326.

DEXTER, A.R. 1988 Advances in characterization of soil structure. Soil Till. Res. 11, 199–238.

GALE, W.J., CAMBARDELLA, C.A. 2000 Carbon dynamics of surface residue and root-derived organic matter under simulated no-till. Soil Sci. Soc. Am. J. 64, 190–195.

GANTZER, C.J., ANDERSON, S.H. 2002 Computed tomographic measurement of macroporosity in chisel-disk and no-tillage seedbeds. Soil Till. Res. 64, 101–111.

GIBSON, J.R., LIN, H., BRUNS, M.A. 2006 A comparison of fractal analytical methods on 2- and 3-dimensional computed tomographic scans of soil aggregates. Geoderma 134, 335–348.

GLASBEY, C. A. and HORGAN, G. W. 1995 Image Analysis for Biological Sciences. John Wiley and Sons. Chichester, England.

IASSONOV, P., GEBRENEGUS, T., AND TULLER, M. 2009 Segmentation of X-ray computed tomography images of porous materials: A crucial step for characterization and quantitative analysis of pore structures. Water Resources Research 45(9).

INIA. 1977 El Encín, suelo y clima. INIA-MAPA, Madrid, 213 pp.

KRAVCHENKO, A.N., WANG, W., SMUCKER, A.J.M., RIVERS, M.L. 2011 Long-term differences in tillage and land use affect intra-aggregate pore heterogeneity. Soil Sci. Soc. Am. J. 75, 1658–1666.

LAL, R. 2002 Soil carbon dynamics in cropland and rangeland. Environ. Pollut. 116, 353–362.

MICHIELSEN, K., DE RAEDT, H. 2001 Integral-geometry morphological image analysis. Physics Reports 347(6), 461–538.

NUNAN, N., RITZ, K., RIVERS, M., FEENEY, D.S., YOUNG, I.M. 2006 Investigating microbial micro-habitat structure using X-ray computed tomography. Geoderma 133, 398–407.

PETH, S., HORN, R., BECKMAN, F., DONATH, T., FISHER, J., SMUCKER, A.J.M. 2008 Three-dimensional quantification of intra-aggregate *pore space features using synchrotron radiation-based microtomography*. Soil Sci. Soc. Am. J. *72*, 897–907.

POSADAS, A.N.D., GIMÉNEZ, D., QUIROZ, R., PROTZ, R. 2003 *Multifractal characterization of soil pore systems*. Soil Science Society of America Journal *67*, 1361–1369.

REVIL, A., CATHLES, L.M. 1999 *Permeability of shaly sands*. Water Resour. Res. *35*, 651–662.

RASBAND, W. S. 1997 ImageJ, US National Institutes of Health, Bethesda, Maryland, USA.

SAUCIER, A., MULLER, J. 1999 *Textural analysis of disordered materials with multifractals*. Physica A *267*, 221–238.

SAN JOSÉ MARTÍNEZ. F., M.A. MARTÍN, F.J. CANIEGO, M. TULLER, A. GUBER, Y. PACHEPSKY, C. GARCÍA-GUTIÉRREZ. 2010 *Multifractal analysis of discretized X-ray CT images for the characterization of soil macropore structures*. Geoderma *156*, 32–42.

SAN JOSÉ MARTÍNEZ, F., F.J. MUÑOZ, F.J. CANIEGO, F. PEREGRINA. 2013 *Morphological Functions to Quantify Three-Dimensional Tomograms of Macropore Structure in a Vineyard Soil with Two Different Management Regimes*. Vadose Zone J. *12*(3).

SERRA, J. 1982 Image analysis and mathematical morphology. Academic Press Inc. Orlando, Florida, USA.

SIX, J., PAUSTIAN, K., ELLIOTT, E.T., COMBRINK, C. 2000 *Soil structure and organic matter. I. Distribution of aggregate-size classes and aggregate associated carbon*. Soil Science Society of America Journal *64*(2), 681–689.

SOIL SURVEY STAFF. 2010 Keys to Soil Taxonomy, 11th ed. USDA-Natural Resources Conservation Service, Washington, DC.

VOGEL, H. J. 2002 Topological characterization of porous media. In Morphology of condensed matter (pp. 75–92). Springer Berlin Heidelberg.

WANG, W., KRAVCHENKO, A.N., SMUCKER, A.J.M, RIVERS, M.L. 2012 *Intra-aggregate pore characteristics: X-ray computed microtomography analysis*. Soil Sci. Soc Am J *76*, 1159–1171.

YOUNG, I.M. and CRAWFORD, J.W. 2004 *Interactions and self-organization in the soil–microbe complex*. Science *304*(5677).

(Received March 29, 2014, accepted July 8, 2014, Published online August 8, 2014)

Pure Appl. Geophys. 172 (2015), 181–196
© 2014 Springer Basel
DOI 10.1007/s00024-014-0889-5

Disperse Two-Phase Flows, with Applications to Geophysical Problems

Luigi C. Berselli,[1] Matteo Cerminara,[2] and Traian Iliescu[3]

Abstract—In this paper, we study the motion of a fluid with several dispersed particles whose concentration is very small (smaller than 10^{-3}), with possible applications to problems coming from geophysics, meteorology, and oceanography. We consider a very dilute suspension of heavy particles in a quasi-incompressible fluid (low Mach number). In our case, the Stokes number is small and—as pointed out in the theory of multiphase turbulence—we can use an Eulerian model instead of a Lagrangian one. The assumption of low concentration allows us to disregard particle–particle interactions, but we take into account the effect of particles on the fluid (two-way coupling). In this way, we can study the physical effect of particles' inertia (and not only passive tracers), with a model similar to the Boussinesq equations. The resulting model is used in both direct numerical simulations and large eddy simulations of a dam-break (lock-exchange) problem, which is a well-known academic test case.

Key words: Dilute suspensions, Eulerian models, direct and large eddy simulations, slightly compressible flows, dam-break (lock-exchange) problem.

1. Introduction

One of the characteristic features of geophysical flows (see for instance Cushman-Roisin and Beckers 2011) is stratification (the other one is rotation). In this manuscript, we study some problems related to suspensions of heavy particles in incompressible—or slightly compressible—fluids. Our aim is a better understanding of mixing phenomena between the two phases, the fluid and solid one. We especially study this problem because (turbulent) mixing with stratification plays a fundamental role in the dynamics of both oceanic and atmospheric flows. In this study, we perform the analysis of some models related to the transport of heavy dilute particles, with special emphasis on their mixing. Observe that mixing is very relevant near the surface and the bottom of the ocean, near topographic features, near polar and marginal seas, as well as near the equatorial zones (Kantha and Clayson 2000). Especially in coastal waters, precise analysis of transport and dispersion is needed to study biological species, coastal discharges, and also transport of contaminants. The other main motivation of our study is a better understanding of transport of particles (e.g., dust and pollution) in the air. This happens—for instance—in volcanic eruptions or more generally by natural and/or human generation of jets/plumes of particles in the atmosphere.

Following Balachandar and Eaton (2010), in the physical regimes we will consider, it is appropriate to use the Eulerian approach, that is, the solid-phase (the particles) will be modeled as a continuum. This choice is motivated by the presence of a huge number of particles and because we are analyzing the so-called "fine particle" regime (that is, the Stokes number is much smaller than one). In this regime, a Lagrangian approach could be computationally expensive, and the Eulerian approach may offer more computationally efficient alternatives. We will explain the precise assumptions that make this *ansatz* physically representative, and we will also study numerically the resulting models, with and without large-scale further approximation. In particular, we will model the particles as dust, investigating a model related to *dusty gases*, and which belongs to the hierarchy of reduced multiphase models, as reviewed by Balachandar and Eaton (2010). These models

---
[1] Dipartimento di Matematica, Università di Pisa, Pisa, Italy. E-mail: berselli@dma.unipi.it
[2] Scuola Normale Superiore di Pisa, Istituto Nazionale di Vulcanologia e Geofisica, Pisa, Italy. E-mail: matteo.cerminara@sns.it
[3] Department of Mathematics, Virginia Tech, Blacksburg, VA, USA. E-mail: iliescu@vt.edu

represent a good approximation when the number of fine particles to be traced is very large and a direct numerical simulation (DNS) of the fluid with a Lagrangian tracer for each particle would be too expensive. As well explained in BALACHANDAR and EATON (2010), the point-like Eulerian approach for multiphase fluid-particle dynamics becomes even more efficient in the case of large eddy simulations (LES), because the physical diameter of the particles has to be compared with the large eddy length-scale and not with the smaller Kolmogorov one. We will use the dusty gas model in a physical configuration that is very close to that modeled by the Boussinesq system, and this explains why we compare our numerical results with those reported in ÖZGÖKMEN et al. (2007), BERSELLI et al. (2011). Observe that the dusty gas model reduces to the Boussinesq system with a large Prandtl number if: (1) the fluid velocity is divergence-free; and (2) the relative ratio of solid and fluid bulk densities is very small [see Sect. 2 and Eq. (6)].

The approach we will use for multiphase fluids is well-described in MARBLE (1970). More precisely, when the *Stokes time*—which is the characteristic time of relaxation of the particle velocity with respect to the surrounding fluid—is small enough and the number of particles is very large, it could be reasonable to use the Eulerian approach (instead of the Lagrangian). In Eulerian models, both the carrier and the dispersed phase are treated as interpenetrating fluid media, and consequently both the particulate solid-phase and fluid-phase properties are expressed by a continuous field representation. Originally, we started studying these models in order to simulate ash plumes coming from volcanic eruptions, see ESPOSTI ONGARO et al. (2007), CERMINARA et al. (2013), CERMINARA (2014), VALADE et al. (2014), but here we will show that the same approach could be also used to study some problems coming from other geophysical situations, at least for certain ranges of physical parameters.

Our model is evaluated in a two-dimensional *dam-break problem*, also known as the *lock-exchange problem*. This problem, despite being concerned with (1) a simple domain; (2) nice initial and boundary conditions; and (3) smooth gravity external forcing, contains shear-driven mixing, internal waves, interactions with boundaries, and convective motions. The dam-break problem setup has long served as a paradigm configuration for studying the space–time evolution of gravity currents (cf. DÖRNBRACK 1998; FERNANDO 2000; RILEY and LELONG 2000; SIEGEL and DOMARADZKI 1994). Consequently, we set up a canonical benchmark problem, for which an extensive literature is available: The vertical barrier separating fluid and fluid with particles is abruptly removed, and counter-propagating gravity currents initiate mixing. The time evolution can be quite complex, showing shear-driven mixing, internal waves interacting with the velocity, and gravitationally unstable transients. This benchmark problem has been investigated experimentally and numerically, for instance in BRITTER and SIMPSON (1978), HACKER et al. (1996) and HALLWORTH et al. (1996, 1993). Both the impressive amount of data and the physical relevance of the problem make it an appropriate benchmark and a natural first step in the thorough assessment of any approximate model to study stratification. The results we obtain validate the proposed model as appropriate to simulate dilute suspensions of ash in the air. In addition, we found that new peculiar phenomena appear, which are generated by compressibility. Even if the behavior of the simulations is qualitatively very close to that of the incompressible case, the (even very slightly) compressible character of the fluid produces a more complex behavior, especially in the first part of the simulations. To better investigate the efficiency and limitations of the numerical solver, the numerical tests will be performed by using both DNS and LES. Complete discussion of the numerical results will be given in Sect. 3.

Plan of the Paper In Sect. 2, we present the reduced multiphase model we will consider, with particular attention to the correct evaluation of physical parameters that make the approximation effective. In Sect. 3, we present the setting of the numerical experiments we performed. Particular emphasis is posed on the initial conditions and on the interpretation and comparison of the results with those available in the literature.

2. On Multiphase Eulerian Models

In order to study multiphase flows and especially (even compressible) flows with particles, some approximate and reduced models have been proposed in the literature. In the case of dilute suspensions, a complete hierarchy of approximate models is available (see BALACHANDAR and EATON 2010) on the basis of two critical parameters determining the level of interaction between the liquid and solid phase: the fractional volume occupied by the dispersed-phase and the mass loading (that is the ratio of mass of the dispersed to carrier phase). When they are both small, the dominant effect on the dynamics of the dispersed-phase is that of the turbulent carrier flow (*one-way coupled*). When the mass of the dispersed phase is comparable with that of the carrier-phase, the back-influence of the dispersed phase on the carrier phase dynamics becomes relevant (*two-way coupled*). When the fractional volume occupied by the dispersed phase increases, interactions between particles become more important, requiring a *four-way coupling*. In the extreme limit of very large concentration, we encounter the granular flow regime.

Here, we consider rather heavy particles such that $\widehat{\rho}_s \gg \widehat{\rho}_f$ (air), or $\widehat{\rho}_s \sim \widehat{\rho}_f$ (liquid), where in the sequel the subscript "$_s$" stands for solid, while "$_f$" stands for fluid. Here a hat $\widehat{\cdot}$ denotes material densities (as opposed to bulk densities): In particular, we suppose $\widehat{\rho}_s \sim 400 - 3{,}000\,\mathrm{kg/m^3}$. A rather small particle/volume concentration must be assumed (to have dilute suspensions), that is

$$\epsilon_s := \frac{V_s}{V} < 10^{-3},$$

where V_s is the volume occupied by the particles over the total volume V. When ϵ_s is smaller than 10^{-3}, particle–particle collisions and interactions can be neglected and the particle-phase can be considered as a pressure-less and non-viscous continuum. In this situation, the particles move approximately with the same velocity of the surrounding fluid, and the theory has been developed by CARRIER (1958) (see a review in MARBLE 1970). With these assumptions the bulk densities $\rho_f := (1 - \epsilon_s)\widehat{\rho}_f$ and $\rho_s := \epsilon_s\widehat{\rho}_s$ are of the same order of magnitude, about $1\mathrm{kg/m^3}$ in the case of

dust-in-air (two-way coupling). In the case of water with particles, the ratio ρ_s/ρ_f is of the order of 10^{-3}; hence, particles behave very similarly to passive tracers (almost one-way coupling).

Another assumption required by Marble's analysis is that particles can be considered *point-like*, if their typical diameter d_s is smaller than the smallest scale of the problem under analysis, that is, the Kolmogorov length η (DNS), or the smallest resolved LES length-scale ξ (LES).

To describe the gas/fluid-particle drag, we observe that it depends in a strong nonlinear way on the local flow variables and especially on the relative Reynolds number:

$$Re_s = \frac{\widehat{\rho}_f|u_s - u_f|d_s}{\mu},$$

where μ is the gas dynamic viscosity coefficient and u_f and u_s are the fluid and solid phase velocity field, respectively. On the other hand, for a point-like single particle and in the hypothesis of small velocities difference ($Re_s < 1$), the drag force (per volume unit) acting on a single particle depends just linearly on the difference of velocities:

$$f_d = \frac{\rho_s}{\tau_s}(u_s - u_f), \qquad \text{with} \qquad \tau_s := \frac{(2\widehat{\rho}_s/\widehat{\rho}_f + 1)\,d_s^2}{36\nu},$$

where τ_s is the *particle relaxation time or Stokes time*, which is the time needed for a particle to equilibrate to a change of fluid velocity (BALACHANDAR and EATON 2010), and $\nu := \mu/\widehat{\rho}_f$ is the fluid kinematic viscosity. In particular, in the case of water with particles, we have $\widehat{\rho}_s/\widehat{\rho}_f \sim 1$, while in the case of a gas $\widehat{\rho}_s/\widehat{\rho}_f \gg 1$ and hence

$$\tau_s \sim \begin{cases} \dfrac{d_s^2}{12\nu} & \text{(water)}, \\[2ex] \dfrac{\widehat{\rho}_s\,d_s^2}{18\mu} & \text{(air)}. \end{cases} \tag{1}$$

In order to measure the lack of equilibrium between the two phases, we have to compare τ_s with the smallest time of the dynamics. In the turbulent regime, the smallest time is the Kolmogorov smallest eddy's turnover time $\tau = \tau_\eta = \eta^2/\nu$ (DNS) (cf. FRISCH 1995) or analogously $\tau = \tau_\xi = \tau_\eta (\xi/\eta)^{\frac{2}{3}}$ (LES). It is possible to characterize this situation by using as non-dimensional parameter—the Stokes number—which is defined by comparing the Stokes time with the fastest time-scale of the problem under analysis $St := \tau_s/\tau$. If $St < 10^{-3}$ (the "fine particle regime"), we say that we have *kinematic equilibrium* between the two phases and so we can use the dusty gas model in a consistent way. In order to also have *thermal equilibrium* between the two phases, one has to assume that the *thermal relaxation time* (cf. MARBLE 1970) is small, that is:

$$\tau_T := \frac{\hat{\rho}_s C_s}{k_s} \frac{d_s^2}{4} \ll 1.$$

Comparing the kinetic and thermal relaxation times, we get the Stokes thermal time

$$St_T := \frac{\tau_T}{\tau} = \frac{\tau_T}{\tau_s}\frac{\tau_s}{\tau} = \frac{3}{2}\frac{C_s\mu}{k_s} \quad St = \frac{3}{2} \; Pr_s \; St, \quad (2)$$

i.e., the particle Prandtl number, where C_s is the solid-phase specific heat-capacity at constant volume and k_s is its thermal conductivity. To ensure that the dusty gas model is physically reasonable, both kinematic and thermal equilibrium must hold, that is, both Stokes numbers should be less than 10^{-3}. This implies that we have a single velocity $u = u_f = u_s$ for both phases and also a single temperature field $T = T_f = T_s$.

To check that our assumptions are fulfilled, we first show that if the Stokes number is small, then also the thermal Stokes number remains small. Indeed, using the typical value of the dynamic viscosity $\mu = 10^{-3}$ Pa s (water) or $\mu = 10^{-5}$ Pa s (air), specific heat capacity $C_s = 10^3$ J kg^{-1} K^{-1} and thermal conductivity $k_s \sim 1$ W m^{-1} K^{-1}, we can evaluate the particle Prandtl number in both cases:

$$Pr_s = \frac{\mu C_s}{k_s} \sim \begin{cases} \dfrac{10^3 \times 10^{-3}}{1} \sim 1 & \text{(water)}, \\[2ex] \dfrac{10^3 \times 10^{-5}}{1} \sim 10^{-2} & \text{(air)}. \end{cases}$$

Hence, formula (2) shows that $St_T \lesssim St$.

Summarizing, we used the following assumptions:

(a) Continuum assumption for both the gaseous and solid phase;

(b) The solid-phase is dispersed ($\epsilon_s < 10^{-3}$); thus, it is pressure-less and non-interacting;

(c) The relative Reynolds number between the solid and gaseous phases is smaller than one, so that it is appropriate to use the Stokes law for drag;

(d) The Stokes number is smaller than one, so that the Eulerian approach is appropriate;

(e) All the phases, either solid or gaseous, have the same velocity and temperature fields $u(x, z, t)$, $T(x, z, t)$ (local thermal and kinematic equilibrium). We showed that this assumption is accurate if the Stokes number is much smaller than one.

In this regime, the equations for the balance of mass, momentum, and energy are:

$$\begin{cases} \partial_t \rho + \nabla \cdot (\rho\, u) = 0, \\ \partial_t \rho_s + \nabla \cdot (\rho_s\, u) = 0, \\ \partial_t(\rho\, u) + \nabla \cdot (\rho\, u \otimes u + p\,\mathbb{I} - \mathbb{T}) = \rho\, g, \\ \partial_t(\rho\, e) + \nabla \cdot (\rho\, u e) + p\,\nabla \cdot u = \mathbb{T} : \nabla u - \nabla \cdot q, \end{cases}$$
$$(3)$$

where $\rho := \rho_f + \rho_s$ is the mixture density, $e := \frac{C_v \rho_f + C_s \rho_s}{\rho}$ is the internal mixture energy, and g is the gravity acceleration pointing in the downward vertical direction. The stress-tensor is

$$\mathbb{T} := 2\mu(T)\left[\frac{\nabla u + \nabla u^T}{2} - \frac{1}{D}(\nabla \cdot u)\,\mathbb{I}\right],$$

with $\mu(T)$ the dynamic viscosity, possibly depending on the temperature T, and D the spatial dimension of the problem. The Fourier law for the heat transfer assumes $\vec{q} = -k\,\nabla T$, where k is the fluid thermal conductivity. We denote by C_v and C_s the fluid and solid phase specific heat-capacity at constant volume, respectively. System (3) is completed by using the constitutive law $p = p(\rho, \rho_s, T)$. In the case of air and particles (the one for which we will present the simulations), $p = \rho_f RT$, where R is the air gas constant.

Remark 1 The correct law would be $p = \frac{\rho_f RT}{1 - \epsilon_s}$, but in our dilute setting ϵ_s is very small, which justifies the approximation $p = \rho_f RT$. A different constitutive law must be used in the presence of water or other fluids.

Remark 2 Note that the constant particle pressure $\nabla p_s = 0$ is justified by the lack of particle–particle forces. Note that in the case of uniform particle distribution ($\rho_s/\rho_f = C$), the equations (3) reduce to the compressible Navier–Stokes equations, with density multiplied by a factor C. Some numerical experiments (with $\rho_s/\rho_f \neq C$) were performed in SUZUKI et al. (2005), where the dusty gas model was applied to volcanic eruptions, i.e., a flow with vanishing initial solid density ρ_s and particles injected into the atmosphere from the volcanic vent.

Denoting by $y_s = \rho_s/\rho$ the solid-phase mass-fraction, we can rewrite the system (3) with just one flow variable (ρu) as follows:

$$
\begin{cases}
\partial_t \rho + \nabla \cdot (\rho u) = 0, \\
\partial_t (\rho y_s) + \nabla \cdot (\rho u y_s) = 0, \\
\partial_t (\rho u) + \nabla \cdot (\rho u \otimes u + p\,\mathbb{I} - \mathbb{T}) = \rho g, \\
\partial_t (\rho e) + \nabla \cdot (\rho u e) + p \nabla \cdot u = \mathbb{T} : \nabla u - \nabla \cdot q.
\end{cases}
\tag{4}
$$

In the following, we will also assume to have an iso-entropic flow with a perfect gas (which is a reasonable approximation for the air, see for example MORTON et al. (1956). We can thus substitute the energy Eq. (4) by the constitutive

$$
p(x(t), z(t), t) = p_0(x(0)) \left(\frac{\rho(x(t), z(t), t)}{\rho(x(0), z(0), 0)} \right)^{\gamma(x(t), z(t), t)},
$$

where $\gamma(x(t), z(t), t) = \frac{1 - y_s(x(t), z(t), t)R}{(1 - y_s(x(t), z(t), t))C_v + y_s(x(t), z(t), t)C_s}$ and $(x(t), z(t))$ is the streamline starting at $(x(0), z(0))$ for $t = 0$ [we have not been able to find this expression in the literature; for its full derivation see CERMINARA (2014)]. In particular, a simple calculation shows that $\gamma(x(t), z(t), t) = \gamma(x(0), z(0), 0) \sim \gamma$. Moreover, since $T(x(0), z(0), 0)/\rho(x(0), z(0), 0)^{\gamma(x(0), z(0), 0)} = a(x(0), z(0), 0) \sim a$ (where $a(x(0), z(0), 0) \sim a$ and $\gamma(x(0), z(0), 0) \sim \gamma$ are motivated by the small density variations compared with a constant temperature), we can consequently study the following system (with $p = a\rho^\gamma$; and a, γ are constants determined from the initial conditions):

$$
\begin{cases}
\partial_t \rho + \nabla \cdot (\rho u) = 0, \\
\partial_t (\rho y_s) + \nabla \cdot (\rho y_s u) = 0, \\
\partial_t (\rho u) + \nabla \cdot (\rho u \otimes u + p\,\mathbb{I} - \mathbb{T}) = \rho g.
\end{cases}
\tag{5}
$$

Here the iso-entropic assumption is justified. Indeed, since the Reynolds number is typically much greater

than 1, and the Prandtl number is of the order of 10, the two dissipation terms $\mathbb{T} : \nabla u$ and $\nabla \cdot q$ (corresponding to the conduction of heat and its dissipation by mechanical energy) can be neglected. Moreover, since $C_v \sim C_s$ and the temperature fluctuations are small, we can disregard the heat transfer from solid to fluid phase.

Observe that if $\rho_f = $ constant, $T = $ constant, and if we use the Boussinesq approximation, we get from (4) the following system:

$$
\begin{cases}
\nabla \cdot u = 0, \\
\partial_t \rho_s + (u \cdot \nabla) \rho_s = 0, \\
\partial_t u + \nabla \cdot (u \otimes u + p\,\mathbb{I} - \mathbb{T}) = \rho_s g,
\end{cases}
\tag{6}
$$

which is exactly the Boussinesq equations, except that there is no diffusion for the density perturbation (i.e., infinite Prandtl number). Thus, numerical results concerning (5) are comparable with results from the classical Boussinesq equations, see ÖZGÖKMEN et al. (2007), BERSELLI et al. (2011).

3. Numerical Results

To validate the Eulerian model for multiphase flows (5), we use it to perform both DNS and LES of a dam-break (lock-exchange) problem.

3.1. Model Configuration

Since we want to compare our results with accurate results available in the literature, we use a setting that is very close to that in ÖZGÖKMEN et al. (2007), in terms of both equations and initial conditions. In particular, we consider a two dimensional rectangular domain $-L/2 \leq x \leq L/2$ and $0 \leq z \leq H$ with an aspect ratio large enough ($L/H = 5$) in order to obtain high shear across the interface, and to create Kelvin–Helmholtz (KH) instability. We use this setting because in a domain with large aspect ratio, the density interface has more space to tilt and stretch.

For this test case, the typical velocity magnitude is (for further details see e.g., ÖZGÖKMEN et al. 2007) $U_0 = \sqrt{g\rho_s h(H - h)/\rho_0 H}$, where H is the layer thickness and h the volumetric fraction of denser material times H. From now on, with a slight abuse of notation, we denote by g the modulus of the gravity acceleration. In our simulation, we set $h = H/2$, from which we get

$$U_0 = \frac{1}{2}\sqrt{\frac{g\rho_s H}{\rho_0}}.$$

We use the characteristic length-scale ℓ to non-dimensionalize all the equations in (5). In order to have $\tau = \ell/U_0$ when $H = 2\ell$, we need to set $\rho_0/g = \rho_s/2$. Moreover, we choose a dimensional system where the initial solid bulk density is $\rho_{s,0} = 1$, which yields $g = 2\rho_0$. The Froude number is $2^{-1/2}$ for all the simulations, so we are free to choose a ρ_0 such that $\rho_0 \gg \rho_s$. We set $\rho_0 = 100$. In these non-dimensional units, the Reynolds number is $Re = (\rho + \rho_s)U_0\ell/\mu = (\rho_0 + 1)/\mu$, we set the dynamic viscosity $\mu = 0.02348837$ such that the maximum Reynolds number we consider is $Re = 4{,}300$. One of the inherent time-scales in the system is the (Brunt-Väisälä) buoyancy period

$$T_b = 2\pi\sqrt{\frac{\rho_0 H}{g\rho_s}} = 2\pi,$$

which is the natural time related to gravity waves. In order to have a quasi-incompressible flow, we set $\mathrm{Ma} = U_0/c = 0.01$. Using our non-dimensional variables, the perfect gas relationship is $p_0 = \rho_0 R$ and the speed of sound is $c = \gamma R$. We want $c = 100$ and $\gamma = 1.4$, so we set $R = 7{,}142.857143$ and $p_0 = 7.142857143 \times 10^5$. Experiments are performed at different resolutions (from about 10^4, up to about 10^6 grid cells), see the next section for details.

The initial condition is a state of rest, in which the fluid with particles on the left is separated from the fluid (without particles) on the right by a sharp transition layer. Since the tilting of the density interface puts the system gradually into motion, the system can be started from a state of rest. Due to the (slight) compressibility of the fluid some peculiar phenomena occur close to the initial time. These effects are not present in the incompressible case, cf. the discussion below.

We consider the isolated problem, so that the isoentropic approximation is valid and consequently we supplement system (5) with the following boundary conditions: the boundary condition for the density perturbation y_s is no-flux, while free-slip for the velocity is:

$$\begin{cases} u \cdot n = 0, \\ n \cdot (\mathbb{T} - p\,\mathbb{I}) \cdot \tau = 0, \end{cases} \quad \text{and} \quad n \cdot \nabla y_s = 0,$$

where n is the unit outward normal vector, while τ is a tangential unit vector on $\partial\Omega$. In the two-dimensional setting we use for the numerical simulation (the two dimensional rectangular domain $\Omega =]-L/2, L/2[\times]0, H[$), the boundary conditions become:

$$\begin{cases} \dfrac{\partial u_1}{\partial z} = 0, \quad u_2 = 0, \quad \dfrac{\partial y_s}{\partial z} = 0, \quad \text{at } z = 0, H, \ -\dfrac{L}{2} < x < \dfrac{L}{2}, \\[2mm] \dfrac{\partial u_2}{\partial x} = 0, \quad u_1 = 0, \quad \dfrac{\partial y_s}{\partial x} = 0, \quad \text{at } x = \pm\dfrac{L}{2}, \ 0 < z < H. \end{cases}$$

3.2. On the Initial Conditions

We considered as initial datum the classical situation used in the dam-break problem, with all particles confined in the left half of the physical domain (with uniform distribution), while a uniform fluid fills the whole domain. Moreover, we have an initial uniform temperature $T(x, z, 0) = T_0$ and pressure distribution $p(x, z, 0) = p_0$. Suddenly the wall dividing the two phases is removed and we observe the evolution.

Even if our numerical code is compressible, we started with this setting, widely used to study incompressible cases, since we are in the physical regime of quasi-incompressibility. The compressibility is mostly measured by the Mach number. For air, we have a typical velocity $U_0 \sim 4\,m/s$; hence, the Mach number of air in this condition is around 0.01, as we choose for our simulations. On the other hand, for water we would obtain $U_0 \sim 0.04\,m/s$ and $\mathrm{Ma} \sim 2.5\ 10^{-5}$. Nevertheless, as we will see especially in Fig. 6, even this very small perturbation creates a new instability and new phenomena for times very close to $t = 0$. In particular, new effects appear for $0 < t < T_b$. These effects seem limited to the beginning of the evolution. The characteristic time of the stratification (for a DNS) is defined as (see CUSHMAN-ROISIN and BECKERS 2011, Section 11)

$$T_a = 2\pi\sqrt{\frac{\rho H}{g\Delta\rho_f}},$$

where $\Delta\rho_f$ is the density difference between the ground level and the height H of the upper boundary wall. In particular, we know that for the gaseous phase, the stable solution is the barotropic stratification, due to the gravity acceleration:

$$T(z) = T_0 - \frac{g\,z}{\gamma\,C_v}, \quad \rho(z) = \rho_0 \left(\frac{T(z)}{T_0}\right)^{\frac{1}{\gamma-1}},$$

$$p(z) = \rho_0 \left(\frac{T(z)}{T_0}\right)^{\frac{\gamma}{\gamma-1}},$$

and in the case of perfect gases, we recover the fact that the typical stratification height for the atmosphere ($R \sim 287$) is

$$z_{\text{gas}} = \frac{1}{\eta_{\text{gas}}} = \frac{\gamma\,R\,T}{g} \sim 10^4 m,$$

while for water in the iso-thermal case we would obtain

$$z_{\text{water}} = \frac{1}{\eta_{\text{water}}} = \frac{1}{\alpha\rho_0 g} \sim 10^5 m.$$

Since η is small in both cases, we can use the following approximation:

$$\rho(z) \sim \rho_0 \left(1 - \frac{g\,z}{\gamma\,R\,T_0}\right) := \rho_0(1 - \eta\,z).$$

For a domain with volume V and mass m, in the incompressible case, the stable stationary configuration is with vanishing velocity and $\rho_{\text{homog.}} = \frac{m}{V}$. On the contrary, in our slightly compressible case, the stable stationary configuration is:

$$\frac{\rho(z)}{\rho_{\text{homog.}}} = \frac{1 - \eta\,z}{1 - \eta\,\frac{H}{2}}.$$

The length η has to be compared with the height of the domain H, in order to evaluate the importance of stratification. For instance, if we use realistic values of density, pressure, and gravity acceleration for air (to come back to dimensional variables), we get that the height of the domain is $H_{air} \sim 600m$, while for water we get $H_{\text{water}} = 0.6m$. In the case of air, we obtain density variations due to gravity that are of the order of 5%, while for water they should be of the order of 0.0003%. This explains that in the case of water, the dominant variations of density, which

are of the order of 1%, are those imposed by the initial configuration of particles. On the other hand, in the case of particles in air, which we are mostly interested in, the two phenomena create fluctuations that are comparable in magnitude, and this can be seen in Fig. 7. In particular, in Fig. 7, one can see that the fluctuations created by the non-stratified initial condition affect the behavior of the background potential energy defined below. In the case of air, we have that $T_a < T_b$; and thus the effects of these instabilities (due to the initial heterogeneity) will be observed before the mixing effects, which are dominant in the rest of the evolution. On the other hand, this effect can not be seen by analyzing just the mixed fraction; see Fig. 4 and the discussion below.

We will also compare the results obtained from DNS with those obtained by different LES models, as discussed later on. The accuracy of the LES models is evaluated through a posteriori testing. The main measure used is the background/reference potential energy (RPE), which represents an appropriate measure for mixing in an enclosed system (WINTERS et al. 1995). RPE is the minimum potential energy that can be obtained through an adiabatic redistribution of the masses. To compute RPE, we use the approach in WINTERS et al. (1995) directly, since the problem is two-dimensional and the computations do not require too much time

$$\text{RPE}(t) := g \int_\Omega \rho_s(x, z, t)\, z_r(x, z, t)\, dxdz,$$

where $z_r(\rho')$ is the height of fluid of density ρ' in the minimum potential energy state. To evaluate $z_r(\rho')$, we use the following formula:

$$z_r(x, z, t) = \frac{1}{L} \int_\Omega \mathcal{H}(\rho_s(x', z', t) - \rho_s(x, z, t))\, dx'dz',$$

where $\mathcal{H}$ is the Heaviside function. It is convenient to use the non-dimensional background potential energy

$$\text{RPE}^*(t) := \frac{\text{RPE}(t) - \text{RPE}(0)}{\text{RPE}(0)}, \qquad (7)$$

which shows the relative increase of the RPE with respect to the initial state by mixing. Further

Table 1

The dimensional maximum particle diameter fulfilling the dusty gas hypothesis

	Ultra-res	Mid-res	Low-res
Water	6.3 μm	10 μm	18 μm
Gas	82 μm	140 μm	240 μm

Table 2

N is the number of nodes of the different homogeneous meshes for our simulations

Low-res	$N = 10,580$
Mid-res	$N = 264,500$
Ultra-res	$N = 1,058,000$

discussion of the energetics of the dam-break problem can be found in CUSHMAN-ROISIN and BECKERS (2011), ÖZGÖKMEN *et al.* (2009a, b, 2007).

With these considerations, we are now able to compute the maximum particle diameter fulfilling our hypothesis ($St < 10^{-3}$). First, we must evaluate the smallest time-scale of the dynamics. As described in Table 2, we used three different resolutions. The ultra-res resolution can be considered as a DNS, so the smallest time-scale of the simulation is the Kolmogorov time $\tau_\eta = Re^{-\frac{1}{2}} = 1.525 \times 10^{-2}$, while the smallest length-scale is $\eta = Re^{-\frac{3}{4}} = 1.883 \times 10^{-3}$. The other two resolutions have been used for LES: We have $\xi = 8.696 \times 10^{-3}$ and $\xi = 4.348 \times 10^{-2}$ for the mid-res and low-res resolutions, respectively. By using the relationship $\tau_\xi = \tau_\eta(\xi/\eta)^{\frac{2}{3}}$, we found $\tau_\xi = 4.229 \times 10^{-2}$ and $\tau_\xi = 1.237 \times 10^{-1}$, respectively. In Table 1, we report the dimensional maximum particle diameter for which the dusty gas hypothesis is fulfilled [cf. Eq. (1)] at various resolutions.

3.3. Numerical Methods and Results

We tested our numerical code on a well-documented test case. At the initial time, the particles occupy only one side of the computational domain. Then—abruptly—the wall dividing the fluid with particles from the fluid without particles is removed and the two fluids start mixing under the effect of gravity. The situation is complex, even in the two-dimensional case. Results of numerical simulations with the DNS and also LES models are presented in

this section. All simulations are obtained by using OpenFOAM®, which is an Open Source computational fluid dynamics code used worldwide. The numerical algorithm we used is PISO (Pressure Implicit with Splitting of Operators) (FERZIGER and PERIĆ 1999; ISSA 1986), which allows the user to choose the numerical scheme and order for both the time and space discretization. In particular, we choose a second order unbounded and conservative scheme for the Laplacian terms; a central second order scheme for interpolation from cell center to cell faces; a second order scheme for the gradient terms; and a bounded second central scheme for the divergence term (JASAK 1996). On the other hand, we choose a second order bounded and implicit time scheme (Crank-Nicolson), with an adaptive time stepping based on the maximum initial residual of the previous time step (KAY *et al.* 2010), and on a threshold that depends on the Courant number ($C < 0.2$).

The linear system is solved by using the PbiCG solver (Preconditioned bi-Conjugate Gradient solver for asymmetric matrices) and the PCG (Preconditioned Conjugate Gradient solver for symmetric matrices), respectively, preconditioned by a Diagonal Incomplete Lower Upper decomposition (DILU) and a Diagonal Incomplete Cholesky (DIC) decomposition. The tolerance has been set to 0.01 for the initial residual and to 10^{-15} for the final one.

The high-resolution DNS, denoted ultra-res in the remainder of the paper, were performed on a HPC architecture (BLUGENE/Q system installed at CINECA) with 1,024 cores. These ultra-res runs took about 5 days. The medium-resolution simulations, denoted by mid-res, were performed on 62 cores (using the HPC infrastructure of INGV, Pisa section) for about 2 days. Since many options for LES of compressible multiphase flows are available, we chose to compare the ones that OpenFOAM has built in, to detect the most promising for our test case. In Figs. 3, 5, and 6 we especially address this topic. More specifically, the LES runs were performed using either the compressible Smagorinsky model or the one equation eddy model, that is, in Eq. (5) the stress tensor $\mathbb{T}$ is replaced by

$$\mathbb{T}_{LES} := 2(\mu(T) + \mu_{SGS})\left[\frac{\nabla u + \nabla u^T}{2} - \frac{1}{D}(\nabla \cdot u)\,\mathbb{I}\right].$$

In both cases we define a subgrid-scale (SGS) stress tensor as in FUREBY (1996) by

$$\mathbb{B} = \frac{2}{D} k \mathbb{I} - 2 C_k \sqrt{k}\, \delta \mathrm{dev}(\mathbb{D}),$$

where k is the SGS kinetic energy, $C_k = 0.02$, δ is the grid-scale, $\mathbb{D} = \mathrm{sym}(\nabla \mathbf{u})$, and $\mathrm{dev}(\mathbb{D}) = \mathbb{D} - \mathrm{Tr}(\mathbb{D})\mathbb{I}/D$. In the Smagorinsky model, k is obtained by using the equilibrium assumption

$$\rho \mathbb{D} : \mathbb{B} + \frac{C_e \rho}{\delta} k^{3/2} = 0,$$

where $C_e = 1.048$. Finally, the SGS viscosity is $\mu_{\mathrm{SGS}} = C_k \rho \, \delta \sqrt{k}$.

On the other hand, in the one equation eddy viscosity model (which is the compressible counterpart of the so called TKE model (CHACÓN REBOLLO and LEWANDOWSKI 2014), k is obtained through the following balance law:

$$\partial_t (\rho k) + \nabla \cdot (\rho u k) - \nabla \cdot ((\mu + \mu_{\mathrm{SGS}}) \nabla k)$$
$$= - \left(\rho \mathbb{D} : \mathbb{B} + \frac{C_e \rho}{\delta} k^{3/2} \right),$$

keeping $\mu_{\mathrm{SGS}} = C_k \rho \, \delta \sqrt{k}$. We perform our simulations at three different resolutions; see Table 2.

Together with the DNS simulation done on the ultra-res mesh and the four LES done on low-res and mid-res meshes, we also performed two under-resolved simulations without the SGS model, denoted by low-res DNS* and mid-res DNS*.

To illustrate the complexity of the mixing process that we investigate, in Fig. 1, we present snapshots of DNS for the density ρ_s of particles' concentration at different times (it is represented in a linear color scale for $0 \leq \rho_s \leq 1$). We notice that the results are similar to those obtained in BERSELLI et al. (2011), ÖZGÖKMEN et al. (2009b). Thus, the DNS time evolution of the density perturbation will be used as benchmark for other numerical simulations, since (as in ÖZGÖKMEN et al. 2007) the number of grid points is large enough to resolve all the relevant scales and to consider simulations at ultra-res as a DNS.

We study this problem varying both the mesh resolution (cf. Table 2) and the SGS LES model (Smagorinsky and one equation eddy model). Figure 2 displays snapshots of the solid-phase bulk densities at time $t = 4$ for the three different mesh

resolutions: DNS at ultra-res, DNS* at mid-res, and DNS* at low-res. Figure 3 displays snapshots of the solid-phase bulk density at time $t = 7$. To generate the plots in Fig. 3, we use two LES models (the Smagorinsky and the one equation eddy model) at two coarse resolutions (mid-res and low-res). To assess the quality of the LES results, we used the DNS at ultra-res as benchmark. Figure 3 shows that the LES models yield similar results. From Fig. 3, we can deduce that, even if the overall qualitative behavior is reproduced in four LES simulations, the results obtained at low-res are rather poor and only the bigger vortices are reproduced. On the other hasnd, the LES results at mid-res are in good agreement with the DNS, and the one equation eddy model seems to be better performing when looking at the smaller vortices. The two LES models required a comparable computational time, and a comparison based on more quantitative arguments will be discussed later on, see Fig. 5 and 6 and discussion therein.

Figures 1, 2, and 3 show that, just as in the case of the Boussinesq equations, the system rapidly generates the Kelvin–Helmholtz billows along the interface of gravity waves, which are counter-propagating. These waves are reflected by the side walls, and gradually both billows grow by entraining the surrounding fluid. Later, the mixing increases so much that individual billows cannot be seen anymore.

In order to check whether our DNS results are an appropriate benchmark for the LES results, in Fig. 4, we compare our ultra-res DNS results with those from ÖZGÖKMEN et al. (2007). Since we chose analogous initial conditions and since our two-phase model is comparable with the Boussinesq equations [cf. Eq. (6)], we expect similar qualitative results for all the flow variables.

using the mixed mass fraction, which is a quantity measuring the mixing, the mixed mass fraction is defined as the fraction of volume where the density perturbation is partially mixed. In particular, in our simulations with homogeneous meshes, it is obtained evaluating the percentage of cells such that $1/3 < \rho_s < 2/3$ (cf. ÖZGÖKMEN et al. 2007). The plots in Fig. 4 show that the two simulations yield similar results, as expected. The main difference is in the time interval $2 < t/T_b < 4$, where our simulation seems to mix slightly more than the simulation

Figure 1
Snapshots of the solid-phase bulk density at **a** $t/T_b = 0.637$, **b** $t/T_b = 1.114$, **c** $t/T_b = 4.297$, **d** $t/T_b = 8.276$, in ultra-res DNS at $Re = 4.300$

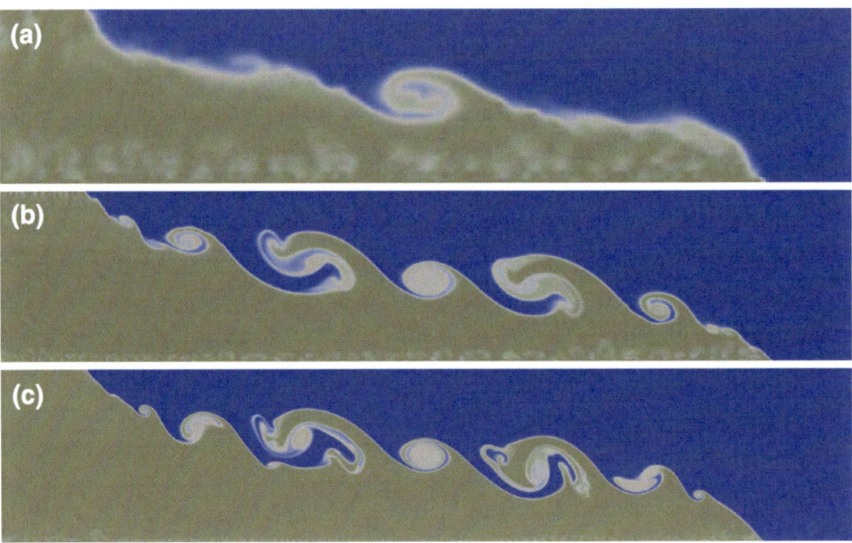

Figure 2
Snapshots of the solid-phase bulk density at $t/T_b = 0.637$ evaluated with different resolutions, **a** low-res DNS*, **b** mid-res DNS*, **c** ultra-res DNS

from ÖZGÖKMEN *et al.* (2007). As we will discuss later, this is probably due to the mixing induced by the creation of stratification.

In Fig. 5, we plot the evolution of the mixed mass fraction for all our simulations. Figure 5 yields the following conclusions: at the low-res, the one

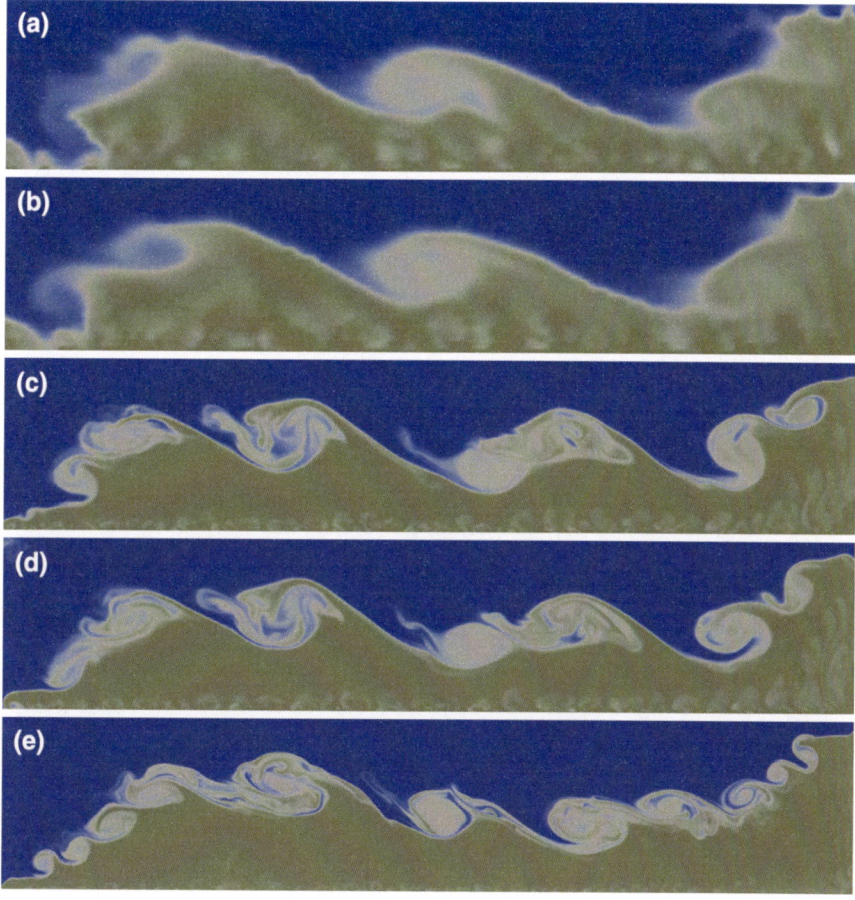

Figure 3
Snapshots of the solid-phase bulk density at $t/T_b = 1.114$ evaluated with different LES models: **a** low-res Smagorinsky, **b** low-res one eq. eddy, **c** mid-res Smagorinsky, **d** mid-res one eq. eddy, **e** ultra-res DNS

equation eddy model performs the best, followed by the DNS*, and the Smagorinsky model (in this order). At the mid-res, the Smagorinsky model performs the best, followed by the one equation eddy model, and the DNS* (in this order).

The main measure used in the assessment of the accuracy of the models employed to predict mixing in the dam-break problem is the non-dimensional background potential energy RPE* defined in (7), cf. Winters et al. (1995).

Figure 6 plots the background energy of the various LES models. The DNS results serve as a benchmark. Figure 6 yields the following conclusions: at the low-res, the one equation eddy model performs the best, followed by the DNS*, and the Smagorinsky model (in this order). At the mid-res,

the one equation eddy model again performs the best, followed by the DNS*, and the Smagorinsky model (in this order).

Apart from the above LES model assessment, we also observe that new important phenomena appear in the compressible case: While in the incompressible case the RPE is monotonically increasing, in our investigation it is initially decreasing, it then reaches a minimum, and it finally starts to increase monotonically, as expected. In order to better understand this phenomenon, we have to compare the background energy of the homogeneous initial condition with that of the stratified initial condition. Evaluating the initial potential energy (PE$_0$), the available energy (APE$_0$), and the background energy (RPE$_0$) for the homogeneous initial density of the solid-phase, we get:

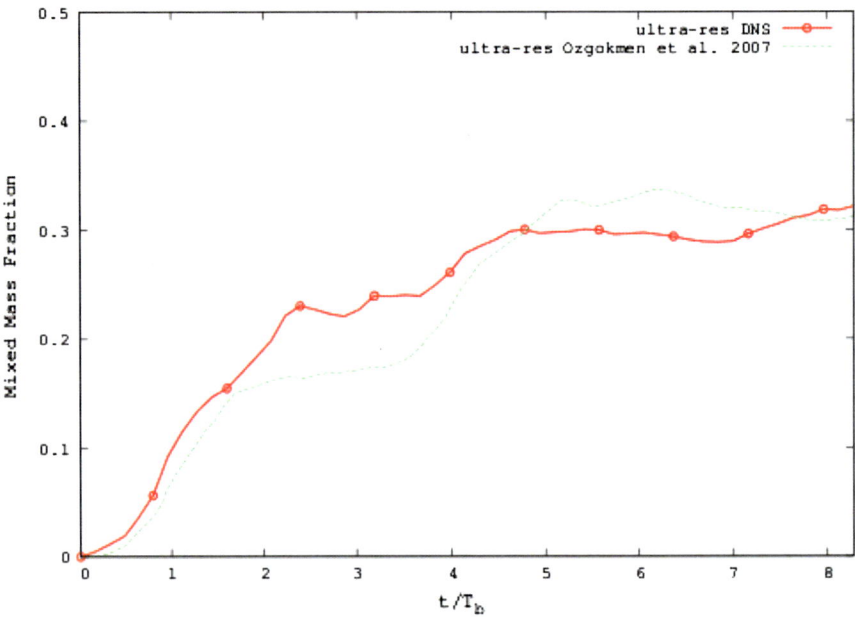

Figure 4
Time evolution of the mixed mass fraction. DNS results from the current study (*solid*) and from ÖZGÖKMEN *et al.* (2007) (*dashed*). Both simulations use the same mesh resolution

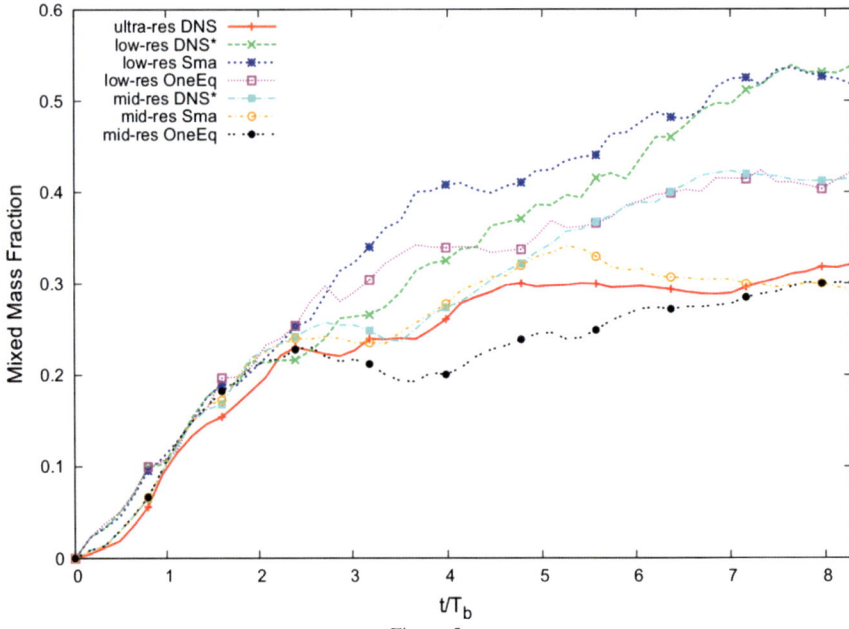

Figure 5
Time evolution of the mixed mass fraction with $\frac{1}{3} < \rho_s < \frac{2}{3}$ for the various low resolution LES models. The DNS results (*solid*) serve as a benchmark

$$\mathrm{PE}_0/g = 10, \qquad (8)$$

If we consider the initial distribution of fluid and particles in the stratified case, with $\rho_s(x, z, 0) = 0.01\,\rho_f(x, z, 0)$, and $\rho_{f,\text{homog.}} = 100$, we get

$$\rho_s(x, z) = \frac{1 - \eta\, z}{1 - \eta\,\frac{H}{2}}\,\mathcal{H}(-x), \qquad (9)$$

where $\mathcal{H}(x)$ is the Heaviside step function. Evaluating the same energies [as those in (8)] for the

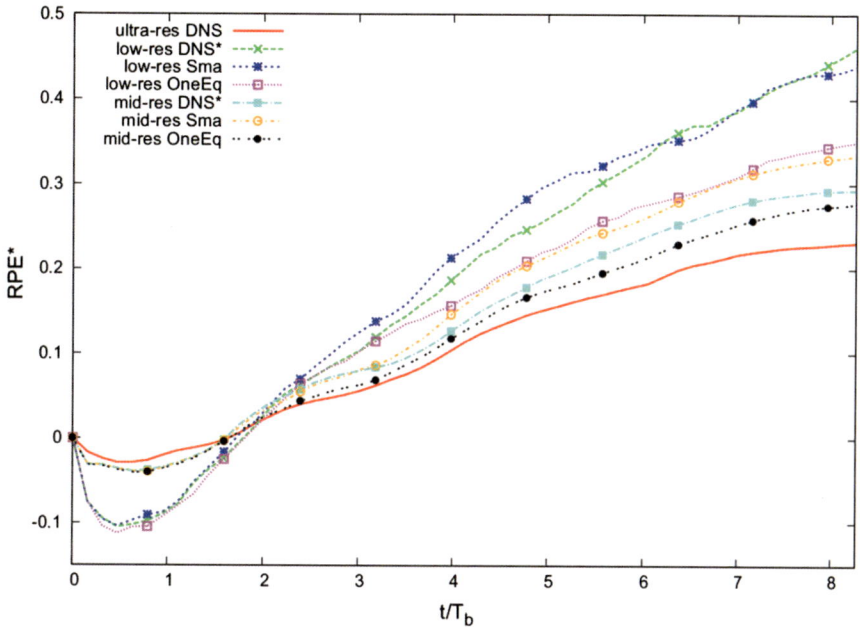

Figure 6

Time evolution of the non-dimensional background energy (RPE*), for the LES models at various resolutions. The DNS results (*solid*) serve as a benchmark. The time is normalized with T_b

stratified density distribution considered and using $H = 2$ and $L = 10$, we get

$$\mathrm{PE_{str.}}/g = \frac{10}{3}\frac{(3 - 4\eta)}{1 - \eta},\qquad (10)$$

and also

$$\mathrm{RPE^*_{str.}} = \frac{\mathrm{RPE_{str.}} - \mathrm{RPE_0}}{\mathrm{RPE_0}} = \frac{-\eta}{3(1 - \eta)} < 0.\quad (11)$$

These analytical computations show that the RPE of the stratified state is smaller than that of the homogeneous state. In the next section we will discuss this issue in more detail.

3.4. A Few Remarks on the Model Without the Barotropic Assumption

In this section, we compare the results of the previous sections with some low-res simulations obtained from the same test case, by using system (4), i.e., without the assumption of a barotropic fluid. The simulations with model (4) are more time-consuming, and so we performed them only at low-res [simulations with finer mesh resolution are in preparation and their results will appear in the forthcoming report (CERMINARA 2014)].

The barotropic assumption is based on the fact that the thermal and kinematic diffusion ($\nabla \cdot q$ and $\mathbb{T} : \nabla u$) in Eq. (4) are negligible, so that the entropy s of the system is constant along streamlines, i.e., $(\partial_t + u \cdot \nabla)s(x, z, t) = 0$ [cf. FEIREISL (2004) for the one-phase case and CERMINARA (2014) for the multiphase case]: This is a reversibility assumption. Indeed, the background energy can be considered as a sort of entropy, measuring the potential energy dispersed in the mixing (WINTERS *et al.* 1995). The fact that the transformation is reversible allows the background energy to decrease. On the contrary, if we remove this assumption, coming back to the full multiphase model (4) (including the energy equation), we find that the background energy becomes monotone, see Fig. 7. This figure suggests that the barotropic assumption may be not completely justified during the initial time-interval needed to adjust from the homogeneous to the stratified condition (probably this transformation can not be considered fully isoentropic). Nevertheless, the barotropic assumption seems justified after the time T_a.

Moreover, the stratified initial condition makes the simulation more stable and accurate, but also less

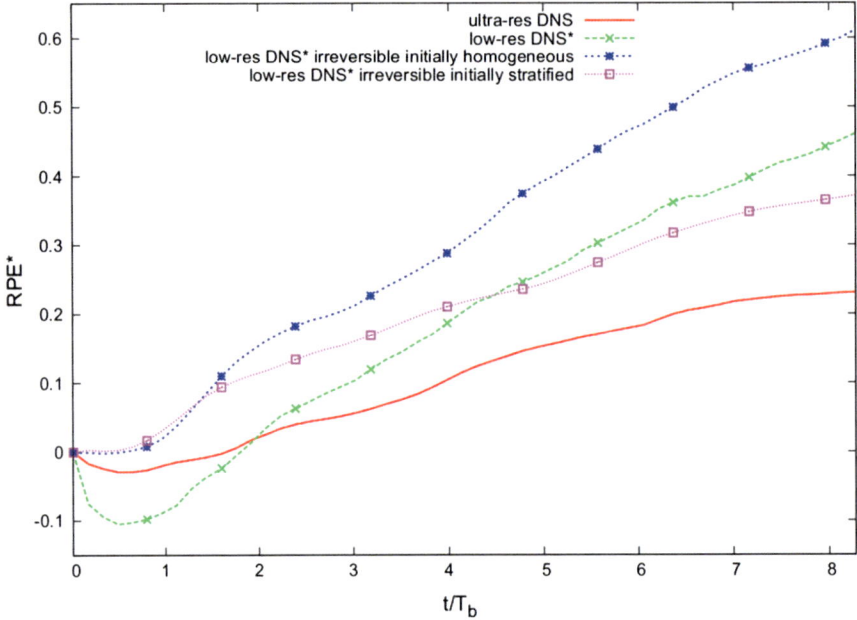

Figure 7

Plot of RPE* obtained by using model (4). Low res DNS* compared with the ultra-res DNS and the low-res DNS*. The line with "··□··" represents RPE* starting from the initial condition (9), while the line with "- -*- -" represents the same quantity starting from the homogeneous initial state. The *solid line* and the line with "- -×- -" are the RPE* obtained with the barotropic model (5) with homogeneous initial state, with the ultra-res DNS and the low-res DNS*, respectively.

diffusive, even at low-res. The RPE* is monotonically increasing when using model (4) (low-res irreversible), and, starting with the stratified initial condition, decreases the mixing and brings it closer to that of the DNS.

Note that the low-res DNS* irreversible with homogeneous initial data and the ultra-res DNS start from the same datum. Even if the low-res DNS* is under-resolved, the behavior of the RPE* is correct and it is monotonically increasing. The behavior, at the beginning of the evolution, is closer to the DNS than the behavior of the LES described in Fig. 6, obtained from the barotropic model (5). On the other hand, after this transient time the behavior becomes comparable with that of the previous low-res barotropic simulation (low-res DNS* vs. low-res DNS* irreversible and homogeneous). The comparison of the results obtained at various resolutions and with different LES models for the barotropic and non-barotropic equations deserves further investigation, and we plan to perform it in the near future.

4. Conclusions

We examined a two-dimensional dam-break problem where the instability is due to the presence of a dilute suspension of particles in half of the domain. The Reynolds number based on the typical gravity wave velocity and on the semi-height of the domain is 4, 300, the Froude number is $2^{-\frac{1}{2}}$, the Mach number is 10^{-2}, and the Prandtl number is 1. The particle concentration is 10^{-3}, and the Stokes number is smaller than 10^{-3} (fine particles). The importance of stratification, measured as the density gradient times the domain height ($-\partial_y\rho_f/\rho_f H$), is about a few percent ($\sim 5\%$). Even if the problem is quasi-incompressible and quasi-isothermal, we used a full compressible code, with a barotropic constitutive law. We employed a homogeneous and orthogonal mesh with three different grid refinements ranging from 10^4 to 10^6 cells. A posteriori tests confirm that the finer grid can resolve all the scales of the problem. The code that we used was derived from the Open-FOAM® C++ libraries.

We compared our quasi-isothermal two-phase simulations with the analogous mono-phase problem, where the mixing occurs between the same fluid at two different temperatures, as reported in ÖZGÖKMEN et al. (2007). As we showed in Sect. 2, this is possible since the two physical problems become mathematically equivalent in the regimes under study. As expected, we found a good agreement between the two sets of numerical results. We reported the evolution of the background (or reference) potential energy (RPE), a scalar quantity measuring the mixing between the two fluids. The main contributions of this report are the following: We implemented a multiphase Eulerian model (that can be used in more complex physical situations, with more than two phases, and also involving chemical reactions between species, as in volcanic eruptions). We also showed the effectiveness of the numerical results obtained programming with an open-source code. More importantly, we discovered that peculiar effects due to compressibility influence the mixing. In the literature, we found that the mono-phase, incompressible Boussinesq test case has a monotonically increasing RPE. On the other hand, in our numerical experiments with slightly compressible two-phase flow, we found that the RPE initially decreases because of the stratification instability, and then it increases monotonically because of the mixing between the particles and the surrounding fluid. Indeed, even if the flow is quasi-incompressible ($Ma = 0.01$), it turns out that stratification effects are not negligible. We reported the preliminary results in the two-dimensional case. We plan to perform three-dimensional numerical simulations of the same problem in a future study.

REFERENCES

S. BALACHANDAR and J.K. EATON. Turbulent dispersed multi-phase flow. Annu. Rev. Fluid Mech., vol. *42*, pp. 399–434. Annual Reviews, Palo Alto, CA, 2010.

L.C. BERSELLI, P. FISCHER, T. ILIESCU, and T. ÖZGÖKMEN. *Horizontal large eddy simulation of stratified mixing in a lock-exchange system.* J. Sci. Comput., *49*:3–20, 2011.

R. E. BRITTER and J.E. SIMPSON. *Experiments on the dynamics of a gravity current head.* J. Fluid Mech., *88*:223–240, 1978.

G.F. CARRIER. *Shock waves in a dusty gas.* J. Fluid Mech, *4*:376–382, 1958.

M. CERMINARA. Multiphase flows in volcanology. PhD thesis, Scuola Normale Superiore, 2014. To appear.

M. CERMINARA, L.C. BERSELLI, T. ESPOSTI ONGARO, and M.V. SALVETTI. Direct numerical simulation of a compressible multiphase flow through the eulerian approach. In Direct and Large-Eddy Simulation IX, vol. 12 of ERCOFTAC Series. Springer, 2013. At press.

T. CHACÓN REBOLLO and R. LEWANDOWSKI. Mathematical and numerical foundations of turbulence models and applications. Birkhäuser, Boston, 2014.

B. CUSHMAN-ROISIN and J.-M. BECKERS. Introduction to Geophysical Fluid Dynamics. Academic Press, 2nd edition, 2011. ISBN: 978-0-12-088759-0.

Ä. DÖRNBRACK. *Turbulent mixing by breaking gravity waves.* J. Fluid Mech. *375*:113–141, 1998.

T. ESPOSTI ONGARO, C. CAVAZZONI, G. ERBACCI, A. NERI, and M.V. SALVETTI. *A parallel multiphase flow code for the 3d simulation of explosive volcanic eruptions.* Parallel Comput., *33*(7–8):541–560, 2007.

E. FEIREISL. Dynamics of viscous compressible fluids, Oxford University Press, Oxford, 2004.

H.J.S. FERNANDO. Aspects of stratified turbulence. In: Kerr, R.M., Kimura, Y. (Eds.), Developments in Geophysical Turbulence, pp. 81–92, 2000.

J.H. FERZIGER and M. PERIĆ. Computational methods for fluid dynamics, revised ed., Springer Verlag, Berlin, 1999.

U. FRISCH. Turbulence, The Legacy of A.N. Kolmogorov. Cambridge University Press, Cambridge, 1995.

C. FUREBY. *On subgrid scale modeling in large eddy simulations of compressible fluid flow.* Phys. Fluids, *8*(5):1301–1311, 1996.

J. HACKER, P. F. LINDEN, and S. B. DALZIEL. *Mixing in lock-release gravity currents.* Dyn. Atmos. Oceans, *24*(1–4):183–195, 1996.

M.A. HALLWORTH, H.E. HUPPERT, J.C. PHILLIPS, and R.S.J. SPARKS. *Entrainment into two-dimensional and axisymmetric turbulent gravity currents.* J. Fluid Mech., *308*:289–311, 1996.

M.A. HALLWORTH, J.C. PHILLIPS, H.E. HUPPERT, and R.S.J. SPARKS. *Entrainment in turbulent gravity currents.* Nature, *362*:829–831, 1993.

R.I. ISSA. *Solution of the implicitly discretised fluid flow equations by operator-splitting.* J. Comput. Phys., *62*(1):40–65, 1986.

H. JASAK. Error Analysis and Estimation for the Finite Volume Method with Applications to Fluid Flows. PhD thesis, Imperial College, London, 1996.

L.H. KANTHA and C.A. CLAYSON. Small Scale Processes in Geophysical Fluid Flows, vol. 67 of Int. Geophysics Series. Academic Press, 2000.

D.A. KAY, P.M. GRESHO, D.F. GRIFFITHS, and D.J. SILVESTER. *Adaptive time-stepping for incompressible flow. II. Navier-Stokes equations.* SIAM J. Sci. Comput., *32*(1):111–128, 2010.

F. MARBLE. Dynamics of dusty gases. Annu. Rev. Fluid Mech., vol. 3, pp. 397–446. Annual Reviews, Palo Alto, CA, 1970.

B. R. MORTON, G. TAYLOR, and J.S. TURNER. *Turbulent gravitational convection from maintained and instantaneous sources.* Proc. R. Soc. Lond. A, *234*, 1–23 1956.

T. ÖZGÖKMEN, T. ILIESCU, and P. FISCHER. *Large eddy simulation of stratified mixing in a three-dimensional lock-exchange system.* Ocean Modelling, *26*:134–155, 2009a.

T. ÖZGÖKMEN, T. ILIESCU, and P. FISCHER. *Reynolds number dependence of mixing in a lock-exchange system from direct numerical and large eddy simulations.* Ocean Modelling, *30*(2):190–206, 2009b.

T. ÖZGÖKMEN, T. ILIESCU, P. FISCHER, A. SRINIVASAN, and J. DUAN. *Large eddy simulation of stratified mixing in two-dimensional dam-break problem in a rectangular enclosed domain.* Ocean Modelling, *16*:106–140, 2007.

J.J. RILEY and M.-P. LELONG. Fluid motions in presence of strong stable stratification. In Annu. Rev. Fluid Mech., vol. 32, pp. 613–657. Annual Reviews, Palo Alto, CA, 2000.

D.A. SIEGEL and J.A. DOMARADZKI. *Large-eddy simulation of decaying stably stratified turbulence.* J. Phys. Oceanogr., *24*:2353–2386, 1994.

Y.J. SUZUKI, T. KOYAGUCHI, M. OGAWA, and I. HACHISU. *A numerical study of turbulent mixing in eruption clouds using a 3D fluid dynamics model.* J. Geophys. Res.: Solid Earth, *110*(B8):B08201, 2005.

S.A. VALADE, A.J.L. HARRIS and M. CERMINARA. *Plume Ascent Tracker: Interactive Matlab software for analysis of ascending plumes in image data.* Comput. & Geosci., *66*(0):132–144, 2014.

K.B. WINTERS, P.N. LOMBARD, J.J. RILEY, and E.A. D'ASARO. *Available potential energy and mixing in density-stratified fluids.* J. Fluid Mech., *289*:115–128, 4 1995.

(Received March 21, 2014, accepted June 26, 2014, Published online July 27, 2014)